Virtue Epistemology Naturalized

SYNTHESE LIBRARY

STUDIES IN EPISTEMOLOGY, LOGIC, METHODOLOGY, AND PHILOSOPHY OF SCIENCE

VOLUME 366

For further volumes:
http://www.springer.com/series/6607

Abrol Fairweather
Editor

Virtue Epistemology Naturalized

Bridges Between Virtue Epistemology and Philosophy of Science

Springer

Editor
Abrol Fairweather
Department of Philosophy
San Francisco State University
San Francisco, CA, USA

ISBN 978-3-319-04671-6 ISBN 978-3-319-04672-3 (eBook)
DOI 10.1007/978-3-319-04672-3
Springer Cham Heidelberg New York Dordrecht London

Library of Congress Control Number: 2014939307

Printed on acid-free paper

Springer is part of Springer Science+Business Media (www.springer.com)

Contents

Bridges Between Virtue Epistemology and Philosophy of Science 1
Abrol Fairweather

Part I Epistemic Virtue, Cognitive Science and Situationism

The Function of Perception 13
Peter J. Graham

Metacognition and Intellectual Virtue 33
Christopher Lepock

Daring to Believe: Metacognition, Epistemic Agency and Reflective Knowledge 49
Fernando Broncano

Success, Minimal Agency and Epistemic Virtue 67
Carlos Montemayor

Towards a Eudaimonistic Virtue Epistemology 83
Berit Brogaard

Expanding the Situationist Challenge to Reliabilism About Inference 103
Mark Alfano

Inferential Abilities and Common Epistemic Goods 123
Abrol Fairweather and Carlos Montemayor

Part II Epistemic Virtue and Formal Epistemology

Curiosity, Belief and Acquaintance 143
Ilhan Inan

Epistemic Values and Disinformation 159
Don Fallis

Defeasibility Without Inductivism 181
Juan Comesaña

Part III Virtues of Theories and Virtues of Theorists

Acting to Know: A Virtue of Experimentation 195
Adam Morton

Is There a Place for Epistemic Virtues in Theory Choice? 207
Milena Ivanova

Bridging a Fault Line: On Underdetermination and the *Ampliative Adequacy* of Competing Theories 227
Guy Axtell

Epistemic Virtues and the Success of Science 247
D. Tulodziecki

Experimental Virtue: Perceptual Responsiveness and the Praxis of Scientific Observation 269
Shannon Vallor

A Matter of Phronesis: Experiment and Virtue in Physics, A Case Study 291
Marilena Di Bucchianico

Part IV Understanding, Explanation and Epistemic Virtue

Knowledge and Understanding 315
Duncan Pritchard

Understanding as Knowledge of Causes 329
Stephen R. Grimm

Knowledge, Understanding and Virtue 347
Christoph Kelp

Bridges Between Virtue Epistemology and Philosophy of Science

Abrol Fairweather

The essays collected here seek to establish bridges between virtue epistemology and philosophy of science (broadly construed, including the history of science, the use of specific scientific results to construct naturalistic philosophical theories, formal epistemology, modeling, theory choice, etc.). Since Ernest Sosa's ground breaking essay "The Raft and the Pyramid" (1980) and Linda Zagzebski's *Virtues of The Mind* (1996), epistemologists have become increasingly interested in the normative aspects of knowledge, justification, understanding and other epistemic states. Virtue epistemologists ground our evaluation of human cognition in a general commitment to aretaic (or virtue theoretic), rather than deontological or consequentialist norms.[1] Two broad defining features of virtue epistemology are often understood through the following principles: (a) Knowledge and other important epistemic concepts are essentially normative and (b) epistemically valuable states of agents confer epistemically valuable properties on their beliefs, not the other way around.[2] Virtue epistemology thus borrows liberally from the rich tradition in virtue ethics for a range of normative resources that have proven quite useful for epistemologists interested in addressing traditional problems regarding epistemic luck and epistemic value. While much more will be said about virtue epistemology below, and there are indeed many species of virtue epistemology on offer in contemporary literature, what unifies this movement can fruitfully be seen through the unique way virtue

[1] This is not to suggest that overtly normative epistemology was not happening prior to Sosa and Zagzebski's work, as Roderick Firth (1978) and Roderick Chisholm had nicely articulated to rule-consequentialist structure of reliabilist theories and the deontological structure of internalist theories respectively.

[2] The second commitment is typically described as 'reversing the direction of analysis' for terms of epistemic appraisal.

A. Fairweather (✉)
Department of Philosophy, San Francisco State University, San Francisco, CA, USA
e-mail: afairweather@gmail.com

A. Fairweather (ed.), *Virtue Epistemology Naturalized: Bridges Between Virtue Epistemology and Philosophy of Science*, Synthese Library 366, DOI 10.1007/978-3-319-04672-3_1,

epistemology foregrounds the normativity of knowledge and places the agent at the center of the analysis.[3]

It will not be my aim to provide a thorough overview of the subject here,[4] but rather to look in new directions. The success of early virtue epistemology lead to a broader "value turn" in the last 15 years of literature in epistemology (Riggs 2008). Value driven (or "axiological") epistemic inquiry become quite complex in the large literature on the *value problem* (and the related *Meno Problem*), which examines whether the value of knowledge can be reduced to the value of any proper subset of its parts (Kvanvig 1992; Zagzebski 1996; Pritchard 2012). As noted, this ambitious value driven approach has also been quite successful in meeting more traditional problems in epistemology, such as Gettier Problems (Zagzebski 1996; Turri 2011) and problems of epistemic luck more generally, as well as the structure of knowledge (as etiological rather than foundational or coherentist), and Chisholm's problem of the criterion (Riggs 2009).[5]

The virtue turn in epistemology that started with the early work of Sosa (1980) and Zagzebski (1996) has now produced a large and mature literature in normative epistemology. However, there are more than corners and pockets left to investigate, as fundamental issues still call to be examined, in particular the empirical adequacy of all this normative epistemology. Over the span of 34 years since the publication of "The Raft and The Pyramid" and the ensuing rise of virtue epistemology, there has been an equally impressive increase in empirical work on the nature of personality traits in psychology[6] and the metaphysics of dispositions (see Byrd, Mumford, and recently Greco on 'powers'). While both developments might appear to be yet *more* good news for virtue epistemology, this has to be shown. In particular, to properly heed Anscombe's (1958) admonition to always consider the psychological plausibility of a moral theory one wants to endorse, virtue epistemologists must also show that the ontological framework (dispositions, skills, habits,) and forms of explanation (e.g., when a success is sufficiently 'due to' virtue) are at least consistent with work in the relevant sciences and developments in disposition theory. The constitutive commitments of virtue epistemology above (the normativity of knowledge and the

[3] Although there are virtue epistemologists like Jason Baehr (2011) and Roberts and Wood (2007) who overtly reject the traditional project of providing an analysis of knowledge. Greco, Pritchard and Sosa clearly show interest in using virtue epistemology to pursue traditional epistemic projects such as answering the skeptic, providing an analysis of epistemic terms and properly handling 'cases'.

[4] See an excellent overview from Heather Batally and a recent reader on virtue epistemology from MIT Press (Greco and Turri).

[5] Additional topics salient in the virtue epistemology literature include: epistemic agency (Sosa, Zagzebski, Greco), the role of motivations and emotions in epistemology (Hookway, Zagzebski, Fairweather) the nature of abilities (Greco, Millar, Pritchard), skills (Bloomfield, Greco), and competences (Sosa), the value understanding (Kvanvig, Grimm, Riggs), wisdom (Riggs, Zagsebski), curiosity (Whitcomb, Inan) and even education policy and practice (Baehr).

[6] See Alfano and Fairweather (2013) for an overview of situationism and virtue theory.

redirection of analysis toward the agent) can be sustained in light of the best available and relevant science, where this might include work on traits in social psychology, research on meta-cognition, bounded rationality, evolutionary theory, and even the history and practice of science itself. This is to request a empirical justification of virtue theoretic approaches to knowledge and simultaneously points the way to overtly naturalistic forms of virtue epistemology.

While the essays collected here cover much more ground than justifying the epistemic psychology presupposed by virtue epistemology, as there are essays on the history of science, formal epistemology and scientific practice as well, they all speak to a general interest in connecting a very successful movement in normative epistemology to the sciences. Exploring and creating bridges between virtue epistemology and the sciences is promising not only because this is an underexplored area in the field, but the continued success of virtue epistemology will require showing that, despite being essentially normative, it can nonetheless meet the empirical constraints that Anscombe and other naturalistically inclined philosophers have pressed on other normative perspectives, for example ethics and aesthetics. If virtue epistemologists succeed here, there will be a clear path to developing naturalistic forms of virtue epistemology, and this would be an important broadening of the field. The essays collected here explore a number of connections between the flourishing work in virtue epistemology and the sciences broadly construed, and thus we begin our search for bridges (and precipices) between scientific knowledge and epistemic virtue.

1 Bridge 1: Empirically Informed Theories of Epistemic Virtue

One clear way of bridging work on epistemic virtue and work in the relevant sciences is to grounds their epistemic psychology (person level cognitive dispositions and the processes that count as their manifestation) in current work in evolutionary biology and cognitive science. In "The Function of Perception", **Peter Graham** argues that human perceptual systems have reliably producing accurate perceptual representations as a biological function, and defends this thesis against Tyler Burge's recent criticisms in defense of reliabilist virtue epistemology for perception. **Chris Lepock**'s essay "Metacognition and Intellectual Virtue" examines the impact of recent research in cognitive science on meta-cognition (roughly, cognitive processes whose function is to "think about thinking") on theories of epistemic virtue, and argues that *monitoring* and *control* emerge as the two primary functions of the epistemic virtues. Lepock argues that meta-cognition does not require meta-representation, just an adequate modelling of first order cognition. **Fernando Broncano** furthers the connection between meta-cognition research and virtue epistemology in "Daring To Believe". He argues that a mere juxtaposition of performance skills and epistemic competence is insufficient to characterize an epistemic agent, an integration of faculties at a personal level that is sufficient to evaluate the agent's epistemic risk in particular situations is necessary as well. While meta-cognition research might appear promising for an empirically informed account of epistemic agency, **Carlos Montemayor** draws upon divided agency research "divided agency" to enter a cautionary word about using robust agential requirements for knowledge in his essay "Success, Minimal Agency and Epistemic Virtue". Montemayor argues that virtue epistemologists must be careful to recognize importantly different forms of agency and the trade-offs involved in employing either when constructing a theory of epistemic virtue. Stronger challenges from empirical research come from **Berit Brogaard** and **Mark Alfano.** In "Towards A Eudaimonistic Virtue Epistemology", Brogaard uses current research in social psychology to argue against the attributability of robust cross situationally stable character traits favored by responsibilist virtue epistemologists and the inability of virtue reliabilism to meaningfully distinguish itself from virtue responsibilism. She defends a novel 'eudaimonistic' account of epistemic virtue which takes intellectual flourishing as the fundamental epistemic good. In "The Situationist Challenge to Reliabilism About Inference", **Alfano** discusses research on inductive reasoning by Kahneman and Tversky that raises a challenge to the reliability of human inductive reasoning, which would be an essential commitment of any reliabilist virtue epistemology. Alfano argues that virtue reliabilism for inference is untenable in the face of situationist challenges. In "Inferential Virtues and Common Epistemic Goods", **Abrol Fairweather and Carlos Montemayor** utilize "bounded rationality" research to mount a response to Alfano's situationist challenge by examining a range of basic inferential abilities involved in knowledge of syntax, assertion, communication and action. They argue that some form of virtue reliabilism can be supported by these more encouraging empirical results which speak in favor of the reliability of some very important human inferential abilities.

We see diverse forms of engagement on this first bridge, some friendlier than others. Graham and Lepock present compelling accounts of how philosophical theories of epistemic virtue can be grounded in research in biology or psychology. Both appear promising for virtue theoretic approaches to knowledge. Tempering any undue enthusiasm, Montemayor's arguments suggest important constraints and distinctions for virtue epistemologists regarding agency in epistemology when developing. Alfano and Brogaard go further and argue that virtue epistemology cannot be empirically grounded, at least not without serious revision.

In Part II of this volume, fruitful resources for virtue epistemologists of a different kind are found between formal epistemology and virtue epistemology. **Don Fallis'** game-theoretic inquiry into "deception" and "misinformation" provides useful formal results for the ethics of belief debate and accounts of epistemic responsibility. **Ilhan Inan** develops a Russelian semantics for *curiousity*, defends a taxonomy of importantly distinct forms of curiousity and important correlations between curiosity and other mental states such as ignorance, certainty, knowledge, acquaintance, and imagination. **Juan Comesana** argues that four plausible principles relating to doxastic justification give rise to a contradiction, and ultimately require abounding inductivism. He argues the rejection of inductivism is, however, compatible with defensibility and endorsing the virtue of open-mindedness

While this first bridge shows both prospects and problems for virtue epistemology on the empirical front, there might be a general concern that, even if the more optimistic lines win out, empirical work in the relevant sciences will replace authentic epistemic theorizing, and will essentially result in the end of virtue epistemology once the relevant scientific accounts have been fully worked out. This is the familiar worry about 'imperialist' naturalism, or replacement naturalism.[7] This takes us to broader questions about the nature of naturalism, which are very important but beyond the scope of this Introduction. However, the second and third bridge discussed below suggest interesting ways in which the imperialist worry is actually less threatening for virtue epistemology because epistemic virtues are essential elements in the success of the sciences. We see this in recent work on virtue theoretic solutions to underdetermination, work on theory virtues and in specific discoveries in the history of science itself. This suggests a fruitful partnership, but now with virtue epistemology informing the epistemology of science, rather than scientific results informing virtue epistemology. We explore these connections below.

2 Bridge 2: Virtue Theoretic Solutions to Underdetermination

The problem of empirical underdetermination of theory choice by empirical evidence[8] challenges our intuition that the best scientific theory will enjoy a unique epistemic standing not shared by other theories of its kind. Duhem's famous argument against the Baconian idea of a "crucial experiment" appears to show that theory confirmation is holistic, and thus there will always be multiple revision strategies available to scientists confronting what appear to be falsifying evidence against a given hypothesis. Duhem uses his theory of 'good sense', a cluster of moral and intellectual virtues, to provide a virtue-theoretic solution to the vexing problem of underdetermination (see Stump 2007). The virtue-theoretic solution is roughly this: amongst two or more competing theories which are not distinguished by evidential support or support from theory virtues, the choiceworthy theory will be the

[7] See Flanagan (2006) for an interesting discussion of the varieties of naturalism, including *imperialist naturalism*, which is strongly reductive.

[8] See Duhem's (1954) classic argument from confirmation holism and of course much of Quine's philosophy, although Duhem was far more modest than Quine in the conclusions he drew.

one that scientists employing intellectual virtues would actually choose. This solution clearly exhibits the direction of analysis characteristic of virtue epistemology, which confers good making properties on beliefs (in this case scientific theories) by virtue of their connection to the good making features of believers (in this case scientists exercising intellectual and moral virtues). However, this has been a controversial point. In "Is There a Place for Epistemic Virtues in Theory Choice", **Milena Ivanova** argues against the virtue theoretic readings of Duhem's 'good sense' defended by Stump (2007) and Fairweather (2011). She claims that Duhem's position is that observational evidence ultimately decides which theory most closely approximates a 'natural classification', and that both theory choice and 'good sense' provide pragmatic but not properly 'epistemic' standing to a theory. In "Bridging a Fault Line: On Underdetermination and the Ampliative Adequacy of Competing Theories" **Guy Axtell** takes issue with Ivanova's account of the epistemic significance of theory virtues and uses Ernst McMullen's work on theory choice and argues for a substantive connection between theory virtues and intellectual virtues. In "Epistemic Virtues and the Success of Science", **D. Tulodziecki** uses a case study regarding the transmissibility of puerperal fever to argue that whether theory virtues are epistemically potent or not is an empirical issue, and this in itself undermines much of the anti-realists argument.

3 Bridge 3: Epistemic Virtues in the History of Science

Another source of illumination is the way in which the epistemic virtues of scientists, or the scientific community, have been essential in specific episodes in the history of science and in aspects of current scientific practice. In these ways, the virtues of individual scientists and the broader scientific community will have important roles in the epistemology and success of science. **Adam Morton** opens this section with "Acting to Know", where he argues that scientific experiments are epistemic actions that have a lot in common with a variety of everyday activities, such as asking for the time or wiping your glasses. The important feature in both cases is that the act succeeds only if knowledge results, and the capacities involved in doing this well are thus both epistemic and practical virtues. Morton explores one central virtue which he calls 'experiment-shopping', the virtue of knowing if an experiment is worth performing. In "Experimental Virtue: Perceptual Responsiveness and the Praxis of Scientific Observation", **Shannon Vallor** draws upon the work of Husserl to argue that the virtue of *perceptual responsiveness* is essential to successful experimental science. In "*A Matter of Phronesis: Experiment and Virtue in Physics, a Case Study*", **Marilena DiBucchianico** presents the story of the balkanization of the scientific community in Condensed Matter Physics (CMP) and superconductivity research to show that a certain kind of phronesis is necessary to generate progress on this vexing issue.

4 Bridge 4: The Value of Understanding

Virtue epistemology and science both aim at understanding, one of the topics virtue epistemology has opened up in contemporary literature. **Duncan Pritchard** in "Knowledge and Understanding", and **Stephen Grimm** in "Understanding as Knowledge of Causes" both examine the view, especially common in philosophy of science, that understanding is a species of knowledge, in particular 'knowledge of causes'. On this view, which goes back to Aristotle and has recently been defended by Lipton, understanding is not some sort of 'super knowledge', but rather a very specific kind of knowledge. We go from knowing that p to understanding why p when we know the cause of p. Pritchard argues that this plausible sounding view, defended recently by Grimm, is flawed. Grimm answers the objections to the 'knowledge of causes' account of understanding by defending a more nuanced account of what it is to have knowledge of causes. Grimm argues that, properly understood, the knowledge of causes account can be sustained. In "Knowledge, Understanding and Virtue", **Christoph Kelp** examines the powerful combination of the knowledge of causes account of understanding and a virtue theoretic account of knowledge. If both can be sustained, Kelp argues that we have a powerful, unified account of two of the most important epistemic standings that can be achieved in our cognitive lives. Virtue epistemology can explain the nature and value of both knowledge and understanding. However, this attractive package will have to sustain objections to both elements. Kelp argues that these objections can be handled, and the powerful UK+VK package can be sustained.

5 Going Natural

The essays collected here make a wide range of connections between virtue epistemology and recent results in cognitive science, social psychology, evolutionary theory, decision theory, and the history and practice of science. These inquiries should allow the reader a glimpse into the rich fabric that will inform any fully developed naturalized virtue epistemology. This volume is intended to constitute part of that fabric and to contribute to the development of naturalistic perspectives in virtue epistemology. Virtue theory in ethics has seen well-developed empirically focused accounts in the last few years, and a number of well-funded research projects involving the virtues are underway as this volume goes to press. The current volume aims to create bridges and partnerships between Virtue Epistemology and the sciences that run in both directions, striking a balance between papers that ground an account of epistemic virtues in the sciences (or argue that this project faces challenges) and those that show the need for epistemic virtue in explaining the success of the sciences themselves (virtue theoretic solutions to underdetermination, epistemic virtues in experimental practice and in the history of scientific discovery).

Some of the essays included here may be challenging and unsettling to virtue epistemology, but it is my firm conviction that working through these difficult issues will bear fruit.

References

Alfano, M. 2012. Expanding the situationist challenge to responsibilist virtue epistemology. *The Philosophical Quarterly* 62(247): 223–249.

Alfano, M. 2013. *Character as moral fiction*. New York: Cambridge University Press.

Anscombe, G.E. 1958. Modern moral philosophy. *Philosophy* 33(124): 1–19.

Axtell, G. 2010. Agency ascriptions in ethics and epistemology: Or, navigating intersections, narrow and broad. *Metaphilosophy* 41(1–2): 73–94.

Baehr, Jason S. 2011. *The inquiring mind: On intellectual virtues and virtue epistemology*. New York: Oxford University Press.

Bloomfield, P. 2000. Virtue epistemology and the epistemology of virtue. *Philosophical and Phenomenological Research* 60(1): 23–43.

Doris, John M. 2002. *Lack of character: Personality and moral behavior*. Cambridge: Cambridge University Press.

Duhem, Pierre Maurice Marie. 1954. *The aim and structure of physical theory*. Princeton: Princeton University Press.

Fairweather, A. 2011. Epistemic motivation. In *Virtue epistemology: Essays on epistemic virtue and responsibility*, 63–81. New York.

Flanagan, O. 1991. *Varieties of moral personality: Ethics and psychological realism*. Cambridge: Harvard University Press.

Flanagan, O. 2006. Varieties of naturalism. In *The Oxford handbook of religion and science*, ed. Clayton Philip and Simpson Zachory, 430–452. Oxford/New York: Oxford University Press.

Flanagan, O. 2009. Moral science? Still metaphysical after all these years. In *Personality, identity, and character: Explorations in moral psychology*, ed. Darcia Narvaez and Daniel Lapsley, 54–65. Cambridge: Cambridge University Press.

Goldman, A.I. 1994. Naturalistic epistemology and reliabilism. *Midwest Studies in Philosophy* 19(1): 301–320.

Greco, J. 1993. Virtues and vices of virtue epistemology. *Canadian Journal of Philosophy* 23(3): 413–432.

Greco, J. 1999. Agent reliabilism. *Noûs* 33(s13): 273–296.

Greco, J. 2010. *Achieving knowledge: A virtue-theoretic account*. New York: Cambridge University Press.

Grimm, S.R. 2006. Is understanding a species of knowledge? *The British Journal for the Philosophy of Science* 57(3): 515–535.

Harman, Gilbert. 1999. Moral philosophy meets social psychology: Virtue ethics and the fundamental attribution error. *Proceedings of the Aristotelian Society* 99: 315–331.

Hookway, C. 2003. Affective states and epistemic immediacy. *Metaphilosophy* 34(1–2): 78–96.

Kim, J. 1988. What is "naturalized epistemology?". *Philosophical Perspectives* 2: 381–405.

Korsgaard, Christine M. 2008. Aristotle's function argument. In *The Constitution of Agency*, 129–150. Oxford: Oxford University Press.

Kvanvig, Jonathan L. 1992. *The intellectual virtues and the life of the mind: On the place of the virtues in contemporary epistemology*. Savage: Rowman and Littlefield.

Kvanvig, J. L. 2003. Zagzebski, L. 2001. Must knowers be agents. *Virtue epistemology: Essays on epistemic virtue and responsibility*, 142–157.). *The value of knowledge and the pursuit of understanding*. Cambridge, UK: Cambridge University Press.

Merritt, M. 2000. Virtue ethics and situationist personality psychology. *Ethical Theory and Moral Practice* 3(4): 365–383.

Millar, A. 2008. Perceptual-recognitional abilities and perceptual knowledge. In *Disjunctivism: Perception, action, knowledge*, 330–347. New York.
Miller, C. 2013. *Moral character: An empirical theory*. Oxford: Oxford University Press, forthcoming.
Moore, G.E., and T. Baldwin. 1993. *Principia ethica*. Cambridge: Cambridge University Press.
Pritchard, D. 2012. Anti-luck virtue epistemology. *Journal of Philosophy* 109(3): 247.
Quine, W.V., and W.V.O. Quine. 1969. *Ontological relativity and other essays*, vol. 1. New York: Columbia University Press.
Riggs, Wayne. 2008. The value turn in epistemology. In *New waves in epistemology*, 300–323. Palgrave: Macmillan.
Riggs, Wayne D. 2009. Understanding, knowledge, and the meno requirement. In *Epistemic value*. Oxford: Oxford University Press.
Roberts, R.C., and W.J. Wood. (2007). *Intellectual virtues: An essay in regulative epistemology*. Oxford University Press.
Roderick, Firth. 1978. Are epistemic concepts reducible to ethical concepts? In *Values and morals: Essays in honor of William Frankena, Charles Stevenson, and Richard Brandt*, ed. Goldman Alvin and Kim Jaegwon, 215–229. Dordrecht: Kluwer.
Russell, D.C. 2009. *Practical intelligence and the virtues*. Oxford: Oxford University Press.
Slingerland, Edward. 2011. The situationist critique and early Confucian virtue ethics. *Ethics* 121: 390–419.
Snow, N.E. 2010. *Virtue as social intelligence: An empirically grounded theory*. New York: Taylor & Francis.
Sosa, E. 2007. *A virtue epistemology: Apt belief and reflective knowledge, volume I*. Oxford: Oxford University Press.
Sosa, E. 1980. The raft and the pyramid. In *Studies in epistemology*, ed. P.A. French, T.E. Uehling Jr., and H.K. Wettstein. Minneapolis: University of Minnesota Press.
Stump, David J. 2007. Pierre Duhem's virtue epistemology. *Studies in History and Philosophy of Science* 18(1): 149–159.
Turri, John. 2011. Manifest failure: The Gettier problem solved. *Philosophers' Imprint* 11(8): 1–11.
Zagzebski, L. 1996. *Virtues of the mind: An inquiry into the nature of virtue and the ethical foundations of knowledge*. New York: Cambridge University Press.

Part I
Epistemic Virtue, Cognitive Science and Situationism

The Function of Perception

Peter J. Graham

1 Fitness, Functions, and Functional Analysis

I believe human perceptual systems—especially visual systems—have producing reliably accurate perceptual representations as a biological function (Graham 2010, 2012, 2014). I defend this against an argument from Tyler Burge. Burge argues that perceptual states cannot have representing accurately as a biological function, for there is a "root mismatch" between representational success and failure, on the one hand, and biological success and failure, on the other. Truth and accuracy are semantical, not practical, matters, and biology only cares about practical matters. Representational success and failure thus cannot be biological functions of any psychological state or system. Burge is not alone, as many have argued that truth and accuracy cannot be biological functions.[1]

I shall argue this isn't necessarily so. In the first Sect. 1 say a few words about biological functions before saying why, in the second, I think human perception has accurately representing the environment as a biological function. In the third and fourth I state Burge's case for thinking this isn't so. In the fifth I explain why Burge's grounds do not make his case and then in the sixth I critically examine an example Burge offers to buttress his case. In the seventh I say why the issue matters to Burge and why, even though I reject his argument, we are not at cross-purposes.[2]

There is an everyday sense of 'fitness' and a technical sense. I trust we would all like to stay fit—to stay in shape. And so many of us go to fitness centers to exercise and work out. That's the everyday notion of fitness. But it's not the sense in biology.

[1] E.g. Churchland 1987, Cruz and Pollock 2004, Plantinga 1993, Stich 1990.

[2] And since we are not at cross-purposes, there is always a chance I've misinterpreted his argument. This is especially true when interpreting a philosopher as subtle and sophisticated as Burge, who is often fighting on many fronts. And so I shall quote as extensively as the occasion demands.

P.J. Graham (✉)
University of California, Riverside, USA
e-mail: peter.graham@ucr.edu

A. Fairweather (ed.), *Virtue Epistemology Naturalized: Bridges Between Virtue Epistemology and Philosophy of Science*, Synthese Library 366, DOI 10.1007/978-3-319-04672-3_2,

In biology, fitness is all about survival and reproduction; it's all about getting your genes in the next generation. Being "fit" in the first sense may contribute to being "fit" in the second, but not always. You may be in great shape without having many—if any—children, and you may be in terrible shape in the everyday sense but have more than your fair share of offspring.

The two main theories of biological functions connect functions with survival and reproduction. On the first, the function of a trait supervenes on its *propensity* to contribute to the fitness of its bearer. This theory "looks forward" to future fitness-enhancing effects. On the second, the function of a trait supervenes on its *past* contributions to fitness in ancestors undergoing natural selection, contributions that then partly explain via heredity why the trait exists in current organisms. This theory "looks backward."[3] Though I prefer the second, which one is correct does not matter for present purposes, as we'll see.[4]

To assign functions, biologists engage in what Robert Cummins calls "functional analysis" (1975). According to Cummins, a functional analysis explains how any system is able to produce an effect by *analyzing* the system. Suppose we want to know how a factory produces cars along an assembly line. An analysis breaks the faculty down into its parts and how they interact.

> Production is broken down into a number of distinct tasks. Each point on the line is responsible for a certain task, and it is the function of the workers/machines at that point to complete that task. If the line has the capacity to produce the product, it has it in virtue of the fact that the workers/machines have the capacities to perform their designated tasks, and in virtue of the fact that when these tasks are performed in a certain organized way—according to a certain program—the finished product results. Here we can explain the line's capacity to produce the product…by appeal to certain capacities the workers/machines and their organization into an assembly line. (Cummins 1975: 74)

Applied to biology, functional analysis is "essentially similar" (Cummins 1975: 74). Living organisms survive and reproduce. A functional analysis explains why. Start with the whole organism and then break it down into its major systems: digestive, circulatory, respiratory, reproductive, immune, nervous, and so on. They then break those into their components. The digestive system, for example, breaks down into the mouth, esophagus, stomach, liver, pancreas, intestines, and colon. Then break those down. The mouth, for example, includes saliva glands, teeth and tongue. The tongue in turn involves muscles, sensory receptors, and so on. Then explain how all the parts interact so as to contribute to fitness. The cells, by making

---

[3] The first, propensity theory is associated with Bigelow and Pargetter (1987) where functions are adaptive effects. The second, etiological theory is associated with Wright (1973), Millikan (1984), Neander (1991), Godfrey-Smith (1993) and others. These papers are anthologized in Buller (1999). For discussion of both theories and important elaboration of the etiological theory, see McLaughlin (2001). For more recent discussion, see Lewens (2004). For my preferred statement of the etiological account, see my 'Functions, Warrant, History' (2014).

[4] Burge agrees: "There are many explications of the notion of biological function. But the differences are not important for present purposes" (2010: 299). However, he seems to prefer the etiological account. See page 320, note 44.

up the muscles of the tongue, make it possible for the tongue to move food around our mouths, so that our teeth may masticate the food. The muscles in turn also assist in swallowing. Once broken down and swallowed, food passes through the esophagus to the stomach, where the stomach in turn processes the food. Each part has various capacities that contribute, through their role in the system, and the system's role in the whole organism as it interacts with other systems, to the ability of the whole organism to survive and reproduce in its natural habitat. Biological functions are then the capacities of the parts that contribute to fitness, the capacities that explain how the system is able to survive and reproduce in its natural habitat. As a result, biological traits often have more than one function. For they often contribute to survival and reproduction in many ways. The tongue helps us eat nutritious food. But it also helps us talk. Our hands also help us eat. But they also help us find food in the first place.

On the propensity theory, biological functions are the capacities of a trait that enter into a functional analysis of propensities to survive and reproduce (Lewens 2004). On the etiological theory, biological functions are the capacities of a trait that enter into a functional analysis of ancestor's survival and reproduction, and so enter into evolutionary explanations of the trait (Griffiths 1993). And so on either theory, when looking for functions look for the fitness enhancing capacities of the trait in a functional analysis of survival and reproduction.

2 The Biological Utility of Vision

What, then, is the biological function of vision? In the first chapter of his textbook *Vision Science*, Stephen Palmer asks what human vision is for.

> [We] should ask what [visual perception] is *for*. Given it biological importance to a wide variety of animals, the answer must be that *vision evolved to aid in the survival and successful reproduction of organisms*. (1999: 5)

How does perception contribute to survival and reproduction? What role does perception—especially visual perception—play in a functional analysis of our ability to survive and reproduce? Palmer continues:

> Desirable objects and situations—such as nourishing food, protective shelter, and desirable mates—must be sought out and approached. Dangerous objects and situations—such as precipitous drops, falling objects, and hungry or angry predators—must be avoided or fled from. Thus, to behave in an evolutionarily adaptive manner, we must somehow get information about what objects are present in the world around us, where they are located, and what opportunities they afford us. All of the senses—seeing, hearing, touching, tasting, and smelling—participate in this endeavor.
>
> There are some creatures for which nonvisual senses play the dominant role—such as hearing in the navigation of bats—but for *homo sapiens*, as well as for many other species, vision is preeminent. The reason is that vision provides spatially accurate information from a distance…It gives a perceiver highly reliable information about the locations and properties of environmental objects while they are safely distant. (1999: 6)

Vision helps by accurately representing objects, properties and relations in the environment. Vision benefits humans because it produces accurate representations:

> Evolutionarily speaking, visual perception is useful only if it is reasonably accurate. … Indeed, vision is useful precisely because it is so accurate. By and large, *what you see is what you get*. When this is true, we have what is called veridical perception…This is almost always the case with vision. (1999: 6)

Palmer concludes:

> [The] evolutionary role of visual perception is to provide an organism with accurate information about its environment. (1999: 15)

Palmer is far from alone. Witness Andrew Parker:

> Today, vision is the most universally powerful sense in its impact on animal interactions and behavior. With the evolution of the first eye, the size, shape, color, and behavior of animals were revealed for the first time—the position and movement of animals could be accurately tracked. Hence, the introduction of vision can be considered to be the launch of the most powerful weapon on Earth…Since [the] first eye, vision has remained on Earth. Although only 6 of the approximately 37 animal phyla possess eyes, more than 95% of all species belong to these. Vision has been a powerful weapon and a successful innovation in the animal kingdom. (2010: 441)

Witness too Ludwig Huber and Anna Wilkinson of the University of Vienna:

> Perception is a universal phenomenon. It functions primarily as a means of allowing an organism to process changes in its external environment. Thus, perception has substantial survival value and can be observed in all living species. (2010: 401)[5]

So just as the heart contributes to survival and reproduction by pumping blood, and just as the lungs contribute to survival and reproduction by taking in oxygen and removing carbon dioxide, human perceptual systems contribute by

[5]Huber and Wilkinson continue: "The primary function of the brain is to compute dynamic, predictive models of the environment. Across the animal kingdom, organisms are able to rapidly evaluate their current situation and respond appropriately to it. This suggests that the perceptual constructions of the external world provide meaning or functional significance to object and situations. As humans, we perceive an object as having a particular shape or color and we perceive it as a dog, or tree (or whatever it is). Being able to identify objects as members of known categories allows the organism to respond to them in appropriate ways." (2010: 404) Hugh Foley and Margaret Matlin say in their textbook on sensation and perception that "Our senses evolved over time to enable us to succeed in responding to the environment…our senses are functional. We live in a physical world and our well-being is very much dependent on our ability to safely negotiate that world…For example, each sensory system serves to detect change in the world…As you can surely imagine, it is often vital to notice changes in the world ("that car is heading toward me")… Most of the time, our perceptions are sufficiently accurate to enable us to interact successfully with the world" (2010: 9–10). And John Frisby and James Stone write in their textbook on vision that by seeing "…we know what objects we are looking at…we are able to describe their various features—shape, texture, movement, size—or their spatial relationships one to another. Such abilities are basic to seeing—they are what we have a visual system for, so that sight can guide our actions and thoughts" (2010: 11).

reliably representing objects, properties and relations. Just as our teeth break down food for further processing, vision helps us identify food for consumption in the first place.[6]

3 The Root Mismatch

Burge argues this isn't so. Burge argues that perceptual systems do not, for they cannot, have accurate representation as a *biological* function.

Burge grants that perceptual systems and "some of their states" have biological functions. He even holds that "biological function is relevant to understanding both the content of perceptual states and their relation to actions that serve biological needs" (2010: 229).[7] But he denies that perceptual systems have producing accurate perceptual states as a biological function. For there is "a *root* mismatch" between *representational* success and failure and *biological* success and failure:

> Biological functions are functions that have ultimately to do with contributing to fitness for evolutionary success. Fitness is very clearly a practical value. It is a state that is ultimately grounded in benefit of its effects for survival and reproduction. Explanations that appeal to biological function are explanations of the practical (fitness) value of a trait or system. But accuracy is not *in itself* a practical value. (2010: 301)

Consider an accurate perceptual representation. Accuracy is a *semantic* relationship between representation and represented object. *As such, in itself*, accuracy contributes no good or benefit to the perceiving organism; this *semantic* fact is not a *practical* fact. And so *in itself, as such*, accuracy is not a biological good or benefit.

---

[6] Some people think the fallibility of perception—the possibility of perceptual illusion—undermines this conclusion. This, Palmer says, would be a mistake: "It is easy to get so carried away by illusions that one starts to think of visual perception as grossly inaccurate and unreliable. This is a mistake. As we said earlier, vision is useful to the extent that it is accurate—or, rather, as accurate as it needs to be. Even illusory perceptions are quite accurate in most respects. For instance, there really are two short horizontal lines and two long oblique lines [in a horizontal line drawing]. The only aspect that is inaccurately perceived is the single illusory property—the relative lengths of the horizontal lines—and the discrepancy is quite modest. Moreover, illusions such as these are not terribly obvious to everyday life; they occur most frequently in books about perception. All things considered, then, it would be erroneous to believe that the relatively minor errors introduced by vision overshadow its evolutionary usefulness" (1999: 8). Most perceptual errors, Palmer thinks, occur when the perceptual system is outside of normal conditions: "[P]erceptual errors produced by these illusions may actually be relatively harmless side effects of the same processes that produce veridical perception under ordinary circumstances" (1999: 9). So that under "most everyday circumstances…normal visual perception is highly veridical" (1999: 23–4).

[7] "An individual's perceptual capacities are individuated partly through causal and practical relations that the perceiver's perceptual system bears (normally in its evolutionary history) to elements in the environment" (Burge 2010: 256). "I believe that biological basic actions—eating, navigating, mating—along with whole animal biological needs figure epistemically and constitutively in background conditions for perception, representation, and empirical objectivity" (Burge 2010: 292). See also pp. 24, 69–71, 94, 211–15, 275–6, 319–20, 320–1, 324, 330–1, 345, 373.

> It is repeatedly said that the biological function of a sensory state [or perceptual representation] is to 'detect' [or accurately represent] the presence of some distal condition (perhaps a predator). Given this claim, any failure of correlation with the distal condition is in itself a biological failure at some level of explanation. But *in itself* detection [or accurate representation] does literally *nothing* to contribute to fitness...Being present when a certain condition obtains cannot *in itself* be a contribution to biological success...One cannot assimilate issues of accuracy and inaccuracy to issues of practical use. Functioning to be accurate is not *in itself* a biological function, at any level. Biological functioning is not a semantical matter. It is a practical matter, a matter of fitness for reproduction. (2010: 301, n. 17, n. 18)

Since accuracy *in itself* does literally *nothing* to further fitness, being accurate cannot be a biological function of a perceptual state. Burge thinks it is a just a mistake to attribute accuracy as a biological function to perceptual states.

What then is the connection between accuracy, on the one hand, and biological function, on the other? Burge says it lies in the *further effects* of the sensory or perceptual state. Concerning detection, Burge says:

> I do not doubt that biological functions can involve detection relations to distal conditions. I do doubt that biological functions, as ordinarily understood, ever reside strictly in detection by itself, or in mere correlation with distal conditions. A biologically more accurate description would be that the function is to initiate some sequence of states that ultimately issues in some response to the distal condition. Sensory states that are predator detectors, for example, have the biological function of initiating a chain of avoidance behavior, given further states and conditions, with respect to the predator. It is this initiation, not the detection per se, that contributes to biological success. (2010: 301)

Predator "detectors" do not have the function of detecting predators, but rather the function of initiating predator-avoidance behavior. For "detection" *in itself* has no practical significance, whereas avoiding a predator clearly does.

Concerning perception, Burge says:

> Although accuracy in perception[s]....usually contribute[s] to fitness, [accurate perceptions] are not in themselves contributions to fitness. When they do contribute, it is not the accuracy per se that makes the contribution. The tendencies of the state to produce efficient response to *need* or, more precisely, tendencies to produce evolutionary fitness—not the veridical aspects of the state—make the contribution. (2010: 302)

And so it's not accuracy per se or the "veridical aspects of the state" that helps the organism survive when it perceives its environment. Rather it is the further effects on behavior in the organism's environment that helps the organism survive.

Burge concludes:

> There is no question that biological structures that underlie perceptual and cognitive systems evolved and were selected for. These structures were selected for not because they are or underlie representational systems per se—systems for accurately representing the world (to within some degree of accuracy). They were selected for because they yielded results that were good enough to further fitness. Evolution does not care about veridicality. It does not select for verdicality per se. (2010: 302–3)

Palmer—and countless others in perceptual psychology and evolutionary science—has made a subtle error; confusing the biological utility of vision—a further practical effect of a perceptual state—with its representational accuracy—a semantical, non-practical relation between mind and world. The biological function of perception lies in its further practical effects; it cannot reside in its representational power.

4 The Argument

I grant that evolution does not care about veridicality per se, that nature does not select for truth and accuracy *as such*. I grant that semantical relations to the environment do not, in themselves, further fitness. Even so, I think perceptual states contribute to fitness by accurately representing the environment, and so have accurately representing the environment as a function; semantical matters are also sometimes practical matters. And so I think the argument I've just attributed to Burge makes a mistake.

To find the mistake, it will prove helpful to make the reasoning behind the argument explicit and fully general. Here's my interpretation:

1. Nature does not care about capacity F of trait T *as such*; F does not further fitness *in itself*.
2. F is a biological function of trait T only if nature cares about F *as such*, only if F furthers fitness *in itself*.
3. So F cannot be a biological function of trait T.

Of course trait T may have been selected for, or may contribute to fitness, and so may have other capacities as biological functions.

4. So if T has a biological function, it must reside in further effects or capacities of T, in the organism/natural habitat.
5. But to satisfy (2), those further capacities or effects of T must be ones that nature cares about *as such*; they must further fitness *in themselves*.

And so it's natural ask what capacities or effects of biological traits nature cares about *as such*. What capacities or effects of biological traits further fitness *in themselves*? What capacities or effects of biological traits further survival and reproduction *as such*?

Nature certainly cares about the capacity to survive and reproduce *as such*, and the capacity to survive and reproduce certainly furthers surviving and reproducing *in itself*. But this is trivial and non-explanatory. Functions, recall, are capacities that enter into a functional analysis of the organism's capacity to survive and reproduce, where the functional analysis *explains* how the organism is able to survive and reproduce in terms of the capacities of the parts and how they interact, given the organism's habitat. Nature may care about survival and reproduction *as such*, and so the capacity to survive and reproduce may meet the condition premise (2) lays down on biological functions, but since functions are *explanatory*, survival and reproduction are not the capacities we're looking for.

At this point the four Fs come to mind: Feeding, fleeing, fighting, and reproducing. For it seems empirically true that an organism can only survive if it eats nutritious food, flees from dangerous predators, successfully fights off real competitors, and finds fertile and cooperative mates. For these and related capacities seem empirically necessary for an organism to survive and reproduce. All known organisms need food. All known organisms need to avoid being eaten. All known organisms with competitors need to fight from time to time. And all known organisms that sexually reproduce need to find cooperative and fertile mates to reproduce their kind. Living organisms survive and reproduce by having their biological needs met, by feeding, fleeing, fighting and reproducing. These are all clearly practical matters, matters of great importance to survival and reproduction. And so it seems we have discovered four capacities that nature cares about *as such*, capacities that further fitness *in themselves*.

6. The explanatory capacities nature cares about *as such* are the four Fs (or other capacities at the very same level of explanation, capacities that nature clearly seems to care about *as such, in themselves*).
7. Given (2) and (6), capacity F of trait T is a biological function of T only if F is one of the four Fs.

In other words, the only biological functions of traits are feeding, fleeing, fighting and reproducing. Biological functions consist in the capacities or effects that most obviously serve survival and reproduction: finding *nutritious* food, finding *fertile* and *cooperative* mates, *successfully* fleeing from *dangerous* predators, fighting *effectively* with *competitors* for mates, food, shelter and so on. Biological functions consist in meeting or fulfilling biological *needs*. For these are all obviously—if not analytically—*practical* goods, goods that clearly contribute to, if not comprise, survival and reproduction.[8]

Applied to our question the consequence is clear: since nature does not care about representational accuracy *as such* (accurately representing the environment is not *the same as* [or at the same level as] eating nutritious food, fleeing from danger, fighting off a rival or predator), accurately representing the environment cannot be a biological function of perceptual states. Their functions lie in their further effects, in their further contributions to practical needs. And so the biological functions of perceptual states are to contribute to finding food, fleeing from predators, fighting off rivals and predators, and finding cooperative and fertile mates, and so on. Our perceptual systems were selected for because they were good enough to further fitness, not because they accurately represent the environment. Accuracy is not, for it cannot be, a biological function.

[8]There is some suggestive but inconclusive evidence that Burge identifies functions with needs, or the fulfilling of needs. On page 371 he says individuals fulfill "basic whole-animal functions" and on page 292 he calls eating, navigating and mating "biologically basic actions…along with whole-animal biological needs." And on page 94 he describes processes that "are ecologically relevant to the individual's basic functions—functions such as eating, navigating, and fleeing danger." Combined, passages such as these at least suggest a tendency to identify biological functions with capacities that obviously, if not constitutively, contribute to survival and reproduction.

5 The Mistaken Premise

Given the interpretation, it's pretty clear how the first premise supports the conclusion. But it's also pretty clear that the argument doesn't go anywhere without the second. I grant the first; I reject the second. I reject the claim that a capacity F of a trait T can be a biological function only if capacity F *in itself* furthers fitness. True, biological functions are contributions to fitness. True, the four Fs are empirically necessary contributions to fitness. Even so, they are not the only biological functions of biological traits. For the biological functions of traits are the capacities that enter into a functional analysis of how the trait contributes to feeding, fleeing, fighting and reproducing. The capacity of the trait that contributes to the four Fs is the function, even if the trait, *as such* and *on its own*, does nothing to further fitness. Biological functions of traits are capacities of the trait that *explains* how it contributes to fitness, where most traits only contribute to fitness as a contingent, empirically determined matter of fact, given their capacities, their role in the organism, and the broader environment, even though by themselves, all on their own, taken in isolation, they contribute literally nothing to fitness. Whether they contribute to fitness *as such* and *in themselves* does not matter. What matters is whether, as a matter of fact, they contribute to fitness given their role in the system and the system's role in the broader environment. Biological functions are explanatorily relevant capacities of traits that contribute to fitness, whether contingently or necessarily so; it does not matter whether they are means towards that end *as such, in themselves*; they need only be means to that end. The second premise is false.

Consider surface coloration. An organism's surface color *as such* and *in itself* clearly does not further fitness. Being red, white or blue isn't the same as eating nutritious food, or fleeing from a dangerous predator. Even so, surface coloration often makes a huge difference to fitness. Take a polar bear's white fur. Its function is to camouflage the bear as it stalks its prey. How does it do that? By matching the background snow. And so matching the background is functional for the bear. A brown polar bear would fail to stay hidden for very long; mismatching the background is obviously dysfunctional. Matching the environment then enters into a functional analysis of the bear's ability to survive and reproduce; matching the environment explain how it provides camouflage, which then in turns explains how it successfully stalks its prey, which then it turn explains how it gets enough food to eat. And so matching the environment is a biological function of the bear's fur, for matching the environment enters into a functional analysis of its ability to survive and reproduce.

Is matching the background environment a contribution to fitness *in itself*? No. Not at all. Camouflage is not the same thing as eating nutritious food, finding a mate, avoiding a predator, and so on. Camouflage *as such* is not a fitness enhancing effect; coloration *as such* is not a practical good. Camouflage *as such* is an *aesthetic*, not a practical, matter. Even so, the polar bear's white fur is supposed to camouflage the bear. Camouflage is often also a practical matter.

Surface coloration often makes a huge contribution to fitness in countless species, as a contingent matter of fact, even though it doesn't make a difference *as such*.[9] Some species identify mates by skin color. Some species hide from predators by skin color. Some species avoid detection as they stalk their prey by color. Surface color partly explains why these species survive and reproduce. The contingent, not-necessarily functional capacities of surface colors often enter functional analyses of how organisms survive and reproduce.

We can make the same points about accurate representations. Accuracy *as such* does not contribute to fitness. But does it follow that representational accuracy can never, in any circumstance, contribute to fitness? Does it follow that representational accuracy cannot enter into a functional analysis of an organism's ability to survive and reproduce? True, accuracy *as such* is not a practical good. Representational accuracy is not eating nutritious food, finding a mate, avoiding a predator, and so on. Even so, can representational accuracy make a contingent contribution to finding food, finding mates, avoiding predators, and so on?

Yes. Representational accuracy often makes a huge contribution to fitness in countless species. We all know this. Some species rely on perception to identify mates. Misrepresenting a predator as a mate can bring your life to an early end. Some species identify and flee or hide from predators by first accurately representing them as predators or as danger. Misrepresenting predator as prey can be just as bad or worse as misrepresenting a predator as a mate. Organisms rely on their perceptions to navigate their environments. Accurate representations are better guides. Just as white fur helps the bear because white matches its environment, accurate perceptions help countless creatures because accurate perceptions match their environments. The accuracy of perceptual representations—especially visual representations in humans—plays a role in the functional analysis of how organisms with perceptual systems are able to survive and reproduce. Getting it right often contributes to fitness, as a contingent, empirically determined matter of fact, in countless creatures with perceptual systems. Just take away accuracy but leave everything else intact and see what happens. Would you rather walk towards a cliff with accurate, or inaccurate, representations as your guide? If you find yourself at all puzzled by this, re-read the second section, including the notes.

Burge says "there is no question that biological structures that underlie perceptual" systems underwent natural selection. He says they were not selected because they "underlie representational systems per se" but rather they were selected because they further fitness. "Evolution," he says "does not care about veridicality" per se. But evolution also does not care about coloration per se; it does not care about pumping blood per se; it does not care about sharp teeth or long legs per se; it

[9]We can imagine cases where coloration is completely irrelevant to survival and reproduction. Think of animals in lightless caves. These animals do not use vision to identify anything, and so they do not use color to identify food, mates, predators, etc. Nor do they use skin color to avoid predators or to avoid detection by their prey. Their color makes no difference whatsoever to their chances for survival and reproduction, both in their current environment and in their evolutionary history. In such a case, color makes no contribution whatsoever to fitness. *A fortiori* it makes no contribution whatsoever to fitness *as such*.

does not care about oxygen diffusion or photosynthesis per se. Evolution only cares, per se, about contributions to fitness and reproduction. It does not follow from any of this that evolution did and does not care, *as a matter of fact*, about coloration, pumping blood, oxygen diffusion, sharp teeth and long legs. It does not follow from any of this that coloration, pumping blood, oxygen diffusion, sharp teeth and long legs cannot enter into the functional analysis of an organism's ability to survive and reproduce. And so it does not follow from the fact that evolution does not care about veridicality per se that it does not care about veridicality as a contingent, empirically well-established matter of fact. All the point shows is that if accurate representations did not contribute to fitness, nature would not have cared about them. But since they do, nature cares.[10]

I think the tendency to infer from the fact that evolution does not care about a capacity *as such* to the conclusion that the capacity cannot be a biological function results from not thinking through the functional analysis of the trait in the overall economy of the organism and the its natural habitat. If biological functions are the capacities and effects of traits that contribute to meeting needs, albeit contingently given their role in the organism in its natural habitat, then many capacities are biological functions even if they don't contribute to fitness *as such, in themselves*. To suppose otherwise is to *identify* biological functions with needs, and thereby *exclude* the capacities or effects of the traits of the organism that, often as a contingent matter of fact, *explain* how those needs are met. Then only the four Fs would fall under the category of function. The function of the heart would not be to pump blood, but only to assist in fleeing, feeding, fighting. The function of the kidneys would not be to remove wastes. The function of the eyes would not be to see. Though biological functions are necessarily *associated* with survival and reproduction—with meeting practical needs—it does not follow the biological functions are restricted to those capacities that contribute to need *as such*.[11]

[10] Burge says predator detectors have the function of "initiating a chain of avoidance behavior with respect to the predator." But the organisms *succeeds* at avoiding the predator by first detecting it; it relies on detection of the predator to avoid the predator. No detection, no initiation of avoidance behavior. And so detection enters into the functional analysis of how the organism avoids predators. Detection is not epiphenomenal when explaining fitness.

Burge says accurate perceptions are not *in themselves* contributions to fitness. Burge says the "tendencies of the state to produce" evolutionary fitness and "not the veridical aspects" of the perception "make the contribution" to fitness. True, the tendencies of the state to produce fitness make the contribution to fitness; that's trivial, but for the same reason non-explanatory; we want explanations of how traits contribute to fitness. And it's pretty clearly true that "veridical aspects" often partly explain why perceptual states contribute to fitness. It is true that without the further effect on behavior the perceptual state would not contribute to fitness. The perceptual state does not contribute to fitness all on its own, *by itself* or *as such*; further behavior is required too. But it is also true that that the behavior contributes to fitness partly because guided by an accurate perceptual state. The accuracy of perceptual states is not epiphenomenal in the explanation of how perceptual systems contribute to fitness. Getting it right often matters.

[11] It may be helpful to sketch out the following in the margins, just to the right. At the top of the paragraph write 'survive and reproduce'. Then write down 'the heart' at the bottom. Then write the 'four Fs' just under 'survive and reproduce'. Now think about how the heart contributes to the four Fs, and so to survival and reproduction, and write them in as well. You will end up writing down

I agree that nature per se only cares about fitness-enhancing traits. But nature cares, as a contingent matter of fact, about countless capacities of traits for their contingent, matter of fact contributions to fitness. It is because they contribute to fitness that nature cares about them. So from the fact that nature does not care about a particular capacity or effect per se shows nothing about whether nature, as a contingent, empirical matter of fact, cares about that capacity or effect. Nature may not care about veridicality per se, but for all that it may care about veridicality a great deal. Veridicality can be, and surely is, a *biological* function of many of our perceptual states and perceptual systems. The second premise is just false.

6 Burge's Example

This concludes my discussion of Burge's argument. Burge also provides an example to make his point. And since examples are sometimes more compelling than arguments from general principles, it would be wrong not to discuss his case. Burge's example purports to show that misrepresentation does not entail failure of biological function, so that correct representation is not a biological function.

Burge imagines a creature like a rabbit that relies on a detection mechanism to avoid predators. Such mechanisms are often unreliable, for false positives ("danger is present" when there is nothing to fear) outnumber true positives ("danger is present" when it's time to run). Burge further imagines that every triggering increases strength and agility: being frightened spurs the exercise required to stay in shape (like having a workout buddy that drags you to the gym everyday, or an alarm clock that reminds you it's time to go to the gym). And so every triggering

the steps in a functional analysis of the heart's contribution to survival and reproduction. In my sketch I wrote 'pump blood' just above 'the heart'. Moving up, I wrote 'moving blood and other stuff through the organism'. I then wrote 'and so assist the organism in fighting diseases, providing energy to the muscles, removing dangerous wastes, etc.' I then wrote 'and so contribute to digestion, locomotion, cognition, etc.' And by doing all of that it contributes to fleeing, feeding, fighting and reproducing. Voila, a crude functional analysis of the heart's role in the organism's ability to survive and reproduce.

I call all of the contributions at all of the different levels "vertical" functions of the trait. And so there are biological functions at the highest level and the lowest and all the levels in between. It's a biological function of the heart to contribute to survival and reproduction, to contribute to fleeing and feeding, to bring oxygen to the brain and take wastes to the kidneys, ... and to pump blood by beating regularly. It contributes to survival and reproduction by contributing to meeting needs, and it contributes to meeting needs by driving circulation of blood and oxygen, among other things, through the body, and it does all that by pumping blood. Though pumping blood—or pumping anything at all—is not *as such* a practical good (taken in isolation pumping fluid is but a mechanical property) it's pretty clearly, as a contingent matter of fact, a practical good. If it followed that F is not a function of a trait because nature does not care about F per se, then pumping blood is not a function of the heart. That cannot be right (Graham 2011, 2012; Fodor 1998).

contributes to fitness, for it keeps the animal in tip-top shape, so the animal is more likely to avoid predators when really present. So when the mechanism fires and there's no predator to avoid—in a case of representational error—there would be "no biological sense in which the mechanism failed to fulfill a biological function…The biological function is to contribute to a fit response to the predator—which entails contributing to avoiding predators" which is exactly what this inaccurate perception does (2010: 302).

Though Burge does not do so, we can put the example in terms of a functional analysis of survival and reproduction. How does an inaccurate perception contribute to fitness? The inaccurate perception spurs exercise, which contributes to strength and agility, which contributes to its capacity to evade predators, which contributes to survival and so to fitness. Spurring exercise figures in a functional analysis of the creature's capacity to survive and reproduce, and so is a function of the detection mechanism. And in the very case Burge imagines, the inaccurate perception fulfilled that function. Burge claims the case shows that perceptual states do not have the biological function of representing accurately, for even though the creature misrepresented its environment there is no biological sense in which the mechanism failed to fulfill a biological function. Representational error without biological error entails that representational success and error is not a species of biological success and error.

I do not think the example works. Burge has overlooked the possibility that the mechanism has more than one function. Spurring exercise may be one function, representing danger another. Many traits have more than one function: think of how your tongue helps you eat as well as talk, the way your hands help you communicate, eat, fight, climb, and so on. From the fact that a trait fulfills one function nothing automatically follows about whether it succeeded or failed in fulfilling other functions. The creature's mechanism may have the function of spurring exercise (so as to run quickly from predators) as well as accurately representing the presence and location of predators (so as to run at the right time in the right way from predators). As long as accuracy plays a role in the functional analysis of the danger-detection mechanism's contribution to fitness, the danger-detection mechanism would have accurately representing the presence of danger as a biological function.

And surely accuracy plays a role. Imagine that the detector failed to represent the presence of danger when danger was present. Then the animal would be in big trouble indeed. Or imagine that though it correctly represented danger, it represented it in the wrong location. The animal might then run into the open arms of its predator. Being full of strength and agility wouldn't help at all.

Danger detectors in many animals have, I believe, the biological function of detecting danger, for detecting danger—even if they are not very reliable at it—plays a role in the functional analysis of how the detector contributes to the capacity of the animal to survive and reproduce. From the fact that they sometimes or even usually misrepresent, or from the fact that there are cases where misrepresentation has very little costs, or from the fact that the device might contribute to fitness in other ways and so have more than one biological function, it does not follow that when they misrepresent with no obvious immediate costs, that there is "no biological sense in which the mechanism failed to fulfill a biological function." If representing accurately is a biological function of the detection device, then every representational error is also a biological error, even if there are no obvious or immediate biological costs.

Burge seems to be reasoning as follows. Suppose the exercise of an avoidance mechanism, in each and every case, increases strength and agility, and so increases the effectiveness of predator avoidance behavior. Suppose it fires on an occasion when danger is not present. Then, Burge concludes, because the device contributed to fitness, there was no biological sense in which it failed to fulfill a biological function. To see that something has to be wrong with Burge's example, consider an analogous case. Suppose the exercise of a sperm producing device increases, in each and every case, its own health and vitality, and so on average it would produce more sperm over time, and so it would fertilize more eggs. Suppose it fires on an occasion and fertilizes no eggs on that occasion. Then, by parallel reasoning, we should be entitled to conclude that there was no biological sense in which the device—and the sperm it produced—failed to fulfill a biological function. But that, of course, is absurd. From the fact that a device may fulfill one biological function on an occasion, nothing follows about whether it succeeded or failed in fulfilling its other biological functions. Burge has not imagined a case where representational accuracy plays no role in contributing to fitness, and so he has not imagined a case where representational accuracy is not, or cannot be, a biological function of a representational state.[12]

Burge's example exploits a fact about functions that is worth making explicit: the biological functions of traits are not always reliable capacities or effects. The trait may have a function that it only fulfills once in a blue moon. Ruth Millikan uses the example of sperm to make this point (1984). The biological function of sperm is to fertilize eggs. However, the vast majority of sperm never come close to an egg. The biological function of an item is what it does *often enough* to contribute to fitness, even if it hardly ever does. Nature settled on a mechanism for reproduction—sperm and egg—where countless sperm are produced for every egg. As a result, though each and every sperm is supposed to fertilize an egg—that is its function—nature is perfectly okay with nearly every sperm failing to fulfill its function; a male can reproduce if only one of millions of its sperm fulfills its function. For sperm, success

[12] The function of triggering exercise is, I think, a decoy. It's there to get us to agree that the device contributed to fitness, despite the error. But imagine a case where the animal doesn't need to exercise to stay in shape. Or imagine a case where the animal doesn't get any "fitter" in the colloquial sense from exercise. Or just imagine that the detector mechanism isn't there in the animal because it helps the animal stay in shape. In all of these cases, there would be no fitness-enhancing benefit to running away when there is no danger to runaway from. And so imagine cases where triggering exercise doesn't enter into the functional analysis of the animal's ability to survive and reproduce, and so isn't a biological function of the device. And so when the animal misrepresented the presence of danger and sprinted away, there would be a clear sense in which the device failed to fulfill a biological function.

once in a blue moon is success often enough.[13] Though *effective*—sperm do indeed fertilize eggs—they are not very reliable.

Many predator detectors work like this, where the representation of danger is not very reliable; it often represents the presence of danger when there is nothing to fear. Though *effective*—they "fire" almost every time danger is present and so keep the organism safe from harm, or at least give the animal a fighting chance—they frequently fire when danger isn't present, and so are not very reliable. Nature has settled on such a way of avoiding predators because false negatives ("there is no danger present; I'm safe" when danger is lurking) are so much worse than false positives ("danger is present, run!" when there's nothing to fear). If the animal overestimates the chances of danger and runs away at the slightest sign, it will effectively avoid predators when they are present, even if it frequently runs away when, in fact, it is perfectly safe. That's why the detector is *effective* (when danger is present it usually says it is) but *unreliable* (most of the time it's mistaken and there is nothing to fear). Burge's example exploits the fact that false positives are often pretty cheap. But the low cost of false positives does not diminish the high cost of false negatives. And it is the cost of false negatives, as well as the low cost of false positives, that explains why nature settled on an unreliable, but nevertheless effective, danger detection device. Accurate detection obviously matters—it explains why the device is effective—even if the device isn't very reliable. Nature settled on an unreliable but effective device, effective because accurate often enough. Most of our perceptual states and systems, however, are not like this. Most are reliable, and contribute to fitness by being reliable. Unreliable danger detectors are the exception that, so to speak, proves the rule.

7 At Cross Purposes?

I've critically discussed Burge's argument and his supporting example at length because the issue matters to me. But why does the issue matter so much to Burge?

The answer involves one of the main themes of his book, the distinction between sensation and sensory systems, on the one hand, and genuinely perceptual systems, on the other. Burge notes the popularity of "Sensation and Perception" as a textbook title, but laments the lack of a good account of the difference. And so in his book he sets out to provide one. But his account, as we will see, comes under threat from "deflationary" accounts of perceptual representation, accounts that effectively reduce perception to sensation where sensation in turn reduces to biological function. Burge then uses the

[13] Not so for the human heart. Not only must the heart pump blood once in a while to contribute to fitness, it must pump blood all the time. Unlike sperm, "often enough" for the heart is all the time. When I've made this point before, I've said that the heart, unlike sperm, not only has a certain effect as its function—pumping blood—the heart has producing that effect *reliably* as its function; the heart isn't just supposed to pump blood, its supposed to pump blood reliably (Graham 2012). Most organisms with hearts can survive and reproduce only if their hearts pump blood all the time. I believe most of our perceptual systems have reliably representing as a function.

premise that nature does not care about accuracy as such to block the reduction. That's (at least one reason) why Burge cares about the issue.

Here's Burge's account of sensation. According to Burge, non-perceptual sensory discrimination—sensation—involves functional information carrying. Information carrying is a broadly law-like correlation between a property of the signal and property of the source. For example, the rings of a tree (the signal) carry information about the age of the tree (the source), because of a law-like correlation between the number of rings in a tree and the age of a tree. Or, to take another example, a ringing doorbell (the signal) carries the information that someone is at the door (the source), because of a law-like correlation between ringing doorbells and visitors at the door. It's been widely recognized that our sensory systems carry information in this sense. Just as iron responds to the presence of oxygen, so too our skin responds to the presence of hot and cold temperatures, to light and dark illumination, and so on. When we touch something with our hands our sensory transducers respond to the change in shape and texture. When light enters our retinas the pattern of light absorbed causes regular changes in our visual system. Changes in the world cause changes in our retina, which in turn cause changes in our visual systems, all in a law-like way, such that changes in us (the signals) carry information about changes in our environment (the source). Our sensory organs—eyes, skin, ears, nose, tongue—not only carry information about our surrounding environment, they are *supposed* to, unlike iron and the rings of a tree. Carrying information, Burge holds, is one of the biological functions of our sensory organs: sensory states involve sensitivities to the environment that are "biologically functional for the individual" (2010: 293); the sensory systems of organisms are systems that are supposed to carry information (2010: 317).

Perception, however, differs. For functional information carrying is "pre-perceptual" and does not require that animal be able to perceptually represent its environment. Plants are sensitive to changes in their environment and respond in functionally useful ways. Likewise bacteria are sensitive to light, oxygen, and magnetic fields, and respond appropriately in turn. But plants and bacteria do not perceive. They do not genuinely represent distal objects, properties and relations.

"Deflationary" accounts of perceptual representation would fail to mark this difference between sensation and perception. Deflationary accounts seek to reduce perceptual representational content to the information that a state or system is supposed to carry. If a state is supposed to carry the information that the source is F, then the state, on these accounts, represents the source as F. If the source is F, then the state has fulfilled its biological function. If the source is not F, then the state has failed to fulfill its biological function. Deflationary accounts of representation then identify perceptual accuracy with fulfillment of biological function to carry information and identify perceptual error with failure to fulfill this function. On these accounts representational success is necessarily and essentially biological function fulfillment, and representational failure is necessarily and essentially failure to fulfill a biological function. Such accounts are associated with the work of Fred Dretske, among others.

Burge accepts such accounts for sensory registration. He rejects them for perception, as accounts of genuine objective sensory perceptual representation as of particulars in the distal environment. These accounts are deflationary, according to Burge, for they do not involve any genuinely psychological terms—iron carries information and nature is full of functions in non-psychological beings—and the accounts apply equally to the non-perceptual sensory systems of mollusks, paramecia and worms. They wrongly assimilate sensory *perception* to mere sensory *registration*. And so when Burge argues that nature does not care *as such* about accuracy, he's largely out to undermine deflationary views of representation that assimilate representation to biological function.

It is not my goal to defend these views, or to advance the general philosophical outlook from which they arise. It is not my goal to "naturalize" perception or epistemology in the sense of 'naturalize' Burge intends (2010: 296–8). It is not my goal to dispute Burge's distinction between sensation and perception or to dispute his account of perception. Nor is it my goal to replace the explanatory enterprise of perceptual psychology with the explanatory enterprise of evolutionary biology. My purposes are not at odds with Burge's.

Burge thinks that perceptual psychology and evolutionary biology ask different questions. I agree. Perceptual psychology asks how veridical and illusory perceptual representations of a distal environment are formed from proximal sensory registration on sensory transducers. Evolutionary biology asks about origins and fitness enhancing effects of perceptual systems and perceptual states. Psychology and biology ask different questions and offer different explanations about overlapping subject matters. Compare physiology and evolutionary biology. Physiology asks about the biochemistry and functional role of organs within an organism and evolutionary biology asks about origins and fitness enhancing effects of those organs within the organism and its natural habitat. They ask different questions and offer different explanations about overlapping subject matters, without the former reducing to the latter. My goal is to defend a claim about the fitness enhancing effects of perceptual states, not to reduce the very nature of perceptual states and perceptual representations to fitness enhancing effects.

Even so, for the sake of argument I can accept that representational content does not reduce to biological function. I can accept that "perceptual accuracy does not necessarily and constitutively contribute to biological" success. I can accept Burge's claim that "functioning in interacting successfully with respect to a beneficial or detrimental distal condition is not the same thing as accurately detecting the condition" (Burge 2010: 302).

Though I critically discussed Burge's example, we can imagine another Burge-inspired case where a perceptual state misrepresents without failure of biological function. Imagine an animal with veridical perceptions but the animal does not use them to control behavior in any way. Perhaps the creature over evolutionary time has become immobilized, has no predators, receives nutrition like a plant, and reproduces asexually. In such a case, we can imagine that the perceptions play no role in the functional analysis of the organism's ability to survive and reproduce, either in its current propensity to survive and reproduce or in its evolutionarily recent past. Perhaps its perceptual system is a vestigial, non-functional trait. And so on both accounts of functions, its perceptions would lack a biological function. *A fortiori* its representational successes and failures are not "fulfillments or frustrations of biological functions" (Burge 2010: 308). Unlike our perceptual states that contribute to fitness by accurately representing our environment, its perceptual states make no contribution at all.

And so I am not at odds with Burge's opposition to deflationary accounts of perceptual representation. Nor I am at odds with his main premise that nature does not care about truth and accuracy *as such*. Even so, our perceptual systems have the contingent, empirically established biological function of producing reliably accurate perceptual systems. Or so I have argued.[14]

References

Bigelow, J., and R. Pargetter. 1987. Functions. *Journal of Philosophy* 84: 181–196.
Buller, David. 1999. *Functions*. Albany: SUNY Press.
Burge, Tyler. 2010. *Origins of objectivity*. New York: Oxford University Press.
Churchland, Patricia. 1987. Epistemology in the age of neuroscience. *Journal of Philosophy* 84: 544–553.
Cruz, Joseph, and John Pollock. 2004. The chimerical appeal of epistemic externalism. In *The externalist challenge*, ed. R. Schantz. Berlin: de Gruyter.
Cummins, Robert. 1975. Functional analysis. *Journal of Philosophy* 72: 741–764.
Fodor, Jerry. 1998. Is science biologically possible. In *Critical condition*. Cambridge: The MIT Press.
Foley, Hugh, and Margaret Matlin. 2010. *Sensation and perception*. Boston, MA: Allyn & Bacon.
Frisby, John, and James Stone. 2010. *Seeing: The computational approach to biological vision*. Cambridge, MA: The MIT Press.
Godfrey-Smith, Peter. 1993. Functions: Consensus without unity. *Pacific Philosophical Quarterly* 74: 196–208.
Graham, Peter J. 2010. Testimonial entitlement and the function of comprehension. In *Social epistemology*, ed. A. Haddock, A. Millar, and D. Pritchard. Oxford: Oxford University Press.
Graham, Peter J. 2011. Does justification aim at truth? *Canadian Journal of Philosophy* 41: 51–72.
Graham, Peter J. 2012. Epistemic entitlement. *Nous* 46: 449–482.
Graham, Peter J. 2014. Functions, warrant, history. In *Naturalizing epistemic virtue*, ed. A. Fairweather. New York, NY: Cambridge University Press.
Griffiths, Paul. 1993. Functional analysis and proper function. *British Journal for the Philosophy of Science* 44: 409–422.
Huber, Ludwig, and Anna Wilkinson. 2010. Evolutionary approach. In *Encyclopedia of perception*, ed. E. Goldstein. Thousand Oaks, CA: Sage Publications.
Lewens, Tim. 2004. *Organisms and artifacts: Design in nature and elsewhere*. Cambridge, MA: The MIT Press.
McLaughlin, Peter. 2001. *What functions explain*. Cambridge: Cambridge University Press.
Millikan, Ruth Garrett. 1984. *Language, thought, and other biological categories*. Cambridge, MA: The MIT Press.
Neander, Karen. 1991. The teleological notion of 'function. *Australasian Journal of Philosophy* 69: 454–468.

[14] I benefited from comments from Zach Bachman and Colleen Macnamara on previous drafts. I presented an earlier version at the Academica Sinica in Taipei, Taiwan. I am grateful for comments on that occasion. And I am deeply grateful to Tyler Burge for getting me interested in these topics and for his continued encouragement and support.

Palmer, Stephen. 1999. *Vision science: From photons to phenomenology*. Cambridge: The MIT Press.
Parker, Andrew. 2010. *Encyclopedia of perception*, ed. E. Bruce Goldstein. Thousand Oaks, CA: Sage Publications.
Plantinga, Alvin. 1993. *Warrant and proper function*. New York, NY: Oxford University Press.
Stich, Stephen. 1990. *The fragmentation of reason*. Cambridge, MA: The MIT Press.
Wright, Larry. 1973. Functions. *The Philosophical Review* 44: 409–422.

Metacognition and Intellectual Virtue

Christopher Lepock

1 Reliability and Virtue

One chief motivation for virtue epistemology is to explain why not all reliable processes yield knowledge. BonJour's (1980) unwitting clairvoyants do not intuitively know the deliverances of their reliable powers, nor does the victim of Plantinga's (1993) brain tumour that causes its host to believe he will soon die and then kills him. The moral seems to be that knowledge only arises from a proper subset of reliable processes.

"Intellectual virtue" was the standard term for these knowledge-generating processes. One advantage of this usage is that it highlights parallels between moral and intellectual virtues. Another is that virtues have turned out to be useful for understanding a wide range of other intellectual activities and evaluations besides knowledge.[1]

The problem that arises, however, is that there are significant differences between two types of intellectual virtues. On the one hand we have knowledge-generating processes like perception, memory, and deduction; on the other, traits of intellectual character like conscientiousness, humility, originality. Baehr (2006b) calls these 'faculty' and 'character' virtues, respectively; in Lepock (2011) I use the blander but more accurate terms 'low-level' and 'high-level'. Corresponding to the two types are two branches of virtue epistemology: virtue reliabilists are primarily concerned with the first type of virtue, while virtue responsibilists focus on the latter.

There has been great difficulty in giving a unified account of both types of virtues. Space will not permit an exhaustive survey of the differences between the

[1] Axtell (2008) and Axtell & Carter (2008) survey some of the diversity of recent work along these lines.

C. Lepock (✉)
Athabasca University, Edmonton, Canada
e-mail: clepock@gmail.com

A. Fairweather (ed.), *Virtue Epistemology Naturalized: Bridges Between Virtue Epistemology and Philosophy of Science*, Synthese Library 366, DOI 10.1007/978-3-319-04672-3_3,

levels. The most important for my purposes is that it is highly problematic to see high-level virtues as *sources* of beliefs in the way that perception and memory are. Rather, they appear to describe ways that an agent treats or uses their sources of information. For instance, one may form a belief courageously in the face of unwarranted opposition from others. But it is not virtuous to form beliefs just from the opposition of others. Hopefully, the courageous belief arose from good evidence or some other reliable source.[2]

It seems that because of this difficulty in unifying these two conceptions of virtue, more recent work on the low-level virtues refers to them as 'abilities' or 'competences' rather than 'virtues' (e.g., Greco 2010; Sosa 2007). The advantage of this usage is that it emphasizes parallels between knowledge and success through ability in other contexts. Since it appears that this is only a new name for the same concept, in this paper I will use 'ability' and 'virtue' more or less interchangeably.

Perhaps the most crucial aspect of intellectual abilities or virtues is that they belong to agents rather than being mere mechanisms or properties of small parts of cognitive systems. We say that *I* can ride a bike or play a competent game of Scrabble, rather than attributing these capacities to narrow parts of me. The deliverances of an agent's abilities are attributable to her, and she can receive credit for their successes or discredit for their failures. They are not "alien" to the cognitive agent (Hookway 2003: 184), unlike, to take an extreme example, the effects of malignant carcinoma.

Thus having a mechanism that reliably outputs true beliefs does not immediately entail having an ability to acquire true beliefs. However, the difference between reliable processes and intellectual abilities has not been specified very clearly.

According to Greco (2003), virtuehood is a matter of "cognitive integration…a function of cooperation and interaction, or cooperative interaction, with other aspects of the cognitive system" (2003: 474). This makes it at least a matter of the range of beliefs generated by the disposition, the extent to which outputs of the disposition are related to other beliefs instead of peripheral to the belief-set, and the sensitivity of the disposition to defeating evidence. Greco acknowledges, however, that this is by no means sufficient detail.

Sosa proposes that virtues arise from an agent's "inner nature", which he fleshes out as "a *total* relevant epistemic state, including certain stable states of her brain and body" (1991: 285, e.i.o.). It seems he thinks that our intrinsic nature gives us a stock of fundamental virtues, and any new cognitive dispositions derived from those basic virtues are also virtues (see 1991: 278). It is not clear, however, just which

[2] Baehr (2006b) notes that we can attribute a belief to a high-level virtue if it was the most salient causal factor in the belief's production. If you would have missed a certain implication of your data were you not so conscientious, then it would be right to say that you acquired the belief because you were conscientious. Nonetheless, you cannot employ your high-level virtues without a source of belief (in this case, reasoning from evidence) to apply them to. It seems the converse does not hold: you can employ perception, say, without manifesting any high-level virtues.

stable states of brain and body are part of our nature, or what constraints there are on the derivation of new virtues from fundamental ones.

Finally, Zagzebski (1996) espouses a broadly Aristotelian view of both moral and intellectual virtues, and argues there is no fundamental difference between the two types. See Baehr (2006a) and Lepock (2011) for why this won't work as an account of knowledge-generating processes.

I will argue here that the difference between a mere reliable process and an intellectual virtue is effective metacognitive control. A controlled process is one that is integrated with the cognitive system in such a way that the system can make use of the process in its endeavours to acquire true beliefs and avoid false ones. Effective regulation is thus a plausible foundation for intellectual ability.

2 Basic Principles of Cognitive Management

I will begin with a general account of metacognition. Nelson and Narens (1990) propose three principles that characterize metacognition and that seem to be broadly accepted by researchers in the field. First, cognitive processes are split into an object level and a metalevel. Second, the metalevel is a model of the object level, but not the other way round. Third, the two levels are connected by relations of monitoring and control. In monitoring, information flows from the object level to the metalevel and informs the latter's activity. In control, the metalevel exerts causal relations on the object level that regulate its behaviour by initiating, sustaining, or terminating activity at the object level. This can include preventing object-level activity from eventuating in actions or beliefs, or permitting it to do so.[3]

The next question, then, is what functions agents would need metaprocessing for. I'll briefly describe three tasks that require some degree of central control. The need for performing these tasks is nearly ubiquitous in actual human cognition, outside of fanciful counterexamples. Successful cognizers whose processes are like ours need capacities for cognitive management.

2.1 Selective Application

The reliability of human cognitive processes varies tremendously across different environments. Vision is not reliable in a funhouse or at twilight in a landscape littered with barn façades; hearing is not reliable under water or on the moon; memory and deduction are only reliable when their inputs are reliably formed. One's

[3] "Process" here refers to something narrower than a belief-forming process or disposition, since belief-formation normally involves both object-level and metalevel processes. It is rather closer to how we speak of sources of belief in ordinary language.

processes also vary in reliability depending on the content of the information. Your sense of smell will not inform you of the presence of carbon monoxide; medical science was greatly advanced when it was realized that something can be dirty even though it looks spotlessly clean; and no number of observations of grue emeralds justifies the conclusion that all emeralds are grue. To acquire beliefs reliably, agents must have the capacity to *selectively apply* their cognitive processes: to be able, most of the time, to use cognitive processes to form beliefs only in environments in which and for contents for which the process yields true beliefs (Lepock 2006).

Henderson and Horgan (2007) call the capacity for selective application a requirement of "modulational control". Sosa (2007) proposes that agents with the sort of knowledge that is characteristically human have an implicit perspective on their faculties that manifests not in explicit or reflectively accessible beliefs, but a capacity to selectively apply those faculties.[4]

Another aspect of the capacity to selectively apply processes is to be able to initiate them when they are likely to form (desired) true beliefs. When apportioning study time to material, one crucial consideration is to spend enough time reviewing difficult material to be able to recall it accurately. This requires exposing oneself to the material to the extent necessary for understanding and retention (see Nelson and Narens 1990). (The other part of the problem is apportioning limited study time so that one can cover all the necessary material, a type of problem discussed below.)

2.2 *Conflict Resolution*

Inconsistent beliefs are unfortunately commonplace. To exist in the same subject, however, inconsistent beliefs have to be insulated from each other in various ways. One might fail to draw out the implications of one's beliefs to the extent necessary to discover the inconsistency. Or inconsistent beliefs might become occurrent at different times or with different prompts. I might believe that I will be home by eight and believe that I will be much later than that, provided that the latter occurs when I think of how much work I have to do and the former occurs when I talk to my partner. There are also more complicated strategies for maintaining inconsistent beliefs. A person might be able to honestly assert that not-p but act as if she believed

[4]This is perhaps clearest in Sosa (2009a). It may be worth noting that if I am right that effective regulation is constitutive of virtue, then Sosa doesn't need separate accounts of epistemic competences and the perspective an agent has on them. When a subject has a suitable implicit perspective on a process, that process will thereby be a competence. Such a move would remove so-called "animal" or "servomechanical" knowledge from Sosa's taxonomy of epistemic statuses. There are those of us who would find that intuitively plausible, but it might force revisions to Sosa's (2007, 2009b) responses to skepticism. A detailed discussion of this issues would take us too far afield, however.

that p by inventing rationalizations for her actions that indicate why they stem from beliefs other than a belief that p.

Nonetheless, *some* inconsistencies among beliefs are not possible, particularly among occurrent beliefs. The problem is that belief-states are tied in myriad ways to action, inference, assertion, and so on; incompatible states can inhibit the connections to other psychological states that each one has, so that neither state can qualify as a belief. Suppose, for instance, that the belief that p plus the rest of the agent's beliefs would lead her to do A, and the belief that not-p plus the rest of her beliefs would lead her not to do A. (Suppose, for instance, she wishes to honestly assert whether she believes p.) If both belief-states occurred at once, she would presumably have to both do A and not do A; but outside of Zen koans, that seems unlikely. She might be *inclined* to believe in incompatible directions, but this is not the same as actually having conflicting beliefs. If two inconsistent beliefs sever enough such connections, then neither can properly be said to be a belief-state.[5]

Cognitive processes generating belief-states conflict with each other when they provide the sort of incompatible data that cannot all be believed at once. A certain amount of central control is necessary to adjudicate between the conflicting processes and actually generate a belief. Conflict resolution procedures can be so simple as to hardly deserve to be regarded as metacognitive. Hesitation, or simply not forming a belief either way, is one possibility; another is to choose randomly between conflicting processes (see Carruthers 2008). Successful epistemic subjects, however, need conflict resolution capacities that are reliable, that permit the subject to form meaningful doxastic states that are likely to be true. That seems to require capacities to distinguish trustworthy from untrustworthy data, and thus to require metacognition.

[5] Delusional persons are sometimes described as having inconsistent occurrent beliefs—for instance, believing at the same time that the same person is both dead and in a room down the hall. Many such manifestations of beliefs are not inconsistent in the way I describe above. It is perfectly possible to be simultaneously disposed to honestly answer "Yes" when asked whether p and when asked whether not-p. It's just not possible to be disposed to answer both "Yes" and "No" when asked whether p. Delusions, however, often do not cause actions in the ways that ordinary beliefs would. (For instance, victims of Capgras delusions, who believe that a loved one has been replaced with a look-alike imposter, generally do not report the imposters to the police. See Davies & Coltheart 2000.) One might tentatively suppose that many of the connections that these beliefs would involve have been suppressed by inconsistencies with other psychological states.

For my purposes here, it seems best to say that delusions are belief-like but should not count as beliefs. With all the different connections a given state might or might not have, one imagines that there are many, many different kinds of belief-like states; but folk psychology offers us shockingly few terms with which to describe them. We are concerned here with states that are amenable to epistemic appraisal as "known" or "unknown", "reliable" or not, etc. We don't apply such appraisals to delusions and the like: they're *beyond* unjustified. It thus seems legitimate to rule them out of consideration here.

2.3 Resource Management

Real agents have limited cognitive resources—e.g., working memory and attentional capacities—and limited time in which to form significant beliefs. Thus, it is important for them to have the ability to apply their processes *efficiently*: to be able to select strategies and initiate processes that will lead to true beliefs while using minimal cognitive and temporal resources. Sometimes this is just a matter of maximizing the power of our cognitive capacities while minimizing effort. But without careful management, we sometimes cannot form the beliefs we need at all. Creatures who distinguish red lights from green by explicit logical deduction from sense data have a pathetic but praiseworthy tendency to die without reproducing (as Quine put it in another context).

Resource management overlaps with the two functions already described. Effective management requires being able to determine not just whether a faculty can be trusted, but whether it will produce a trustworthy answer at all, in order to avoid the pitfall of initiating processes that will not terminate with a suitable output. Suppose, for instance, that you are trying to remember whether one heard a particular sentence on the news. There are two available strategies for solving the problem. You can determine whether the sentence's content is plausible given what one remembers of the content of the news story, which is a moderately reliable strategy. Alternatively, you can try to remember as much of the story as possible and see if you retrieve the sentence in question. This strategy is quite reliable immediately after watching the news, and drops as a function of elapsed time afterward. Good resource management involves determining which strategy is more likely to yield a correct answer, in order to avoid wasting time estimating the plausibility of something that could be retrieved, or trying to retrieve something no longer available.[6] But determining this is of course also part of selectively applying one's processes.

3 Varieties of Metacognition

Talk of understanding virtue in terms of metacognition is liable to sound like internalist wine in naturalist bottles. It can suggest a hierarchical structure in which executive faculties govern lower-level object faculties. If metacognition is thought to consist of conscious processes regulating unconscious ones, we get a sort of Cartesian picture on which conscious mind reflects upon and judges the deliverances of implicit processing.[7] If virtue requires metacognition, our account would seem to entail that conscious deliberation is necessary for knowledge.

[6] See Cary & Reder (2002) for discussion of the experimental paradigm.

[7] Lepock (2006) defends this Cartesian conception of metacognition, but we all do foolish things when we are young.

This is, of course, just the view that virtue epistemology was developed to provide an alternative to.

In this section I will try to assuage such worries. Effective regulation just is successfully accomplishing the functions I described above, and there is a tremendous range of possible ways of doing that. What is essential is that processes be regulated by the rest of the cognitive system. In particular, there are three potential misunderstandings regarding metacognition that I hope to correct: that it must be conscious, that it involves the construction of explicit models of object-level processing, and that it involves dedicated metaprocesses. These are all possible ways that regulation might be accomplished, but none is a necessary feature of the picture.

First, "metacognition" is sometimes used to mean specifically conscious regulation of implicit processing (e.g., in Koriat 1994). It is clear, however, that monitoring and regulation can occur implicitly in humans (Cary and Reder 2002; Kentridge and Heywood 2000). There are perhaps good reasons for researchers to focus on conscious metacognition: it is easier to study, since one can make use of subjects' own introspective reports; it is more easily distinguished from distributed cognition; and for good or ill, the Cartesian picture has its attractions. We should not suppose, however, that conscious regulation is in any way obligatory.

Reflection and explicit reasoning play important roles in regulating belief-formation (see Thompson 2009), and conscious experience carries valuable information in the form of sensory evidence, "feelings of knowing" (see Koriat 1994), and the like. But this just tells us that conscious reflection is of great instrumental value to cognitive management; it does not tell us that there is any *a priori* reason why an agent couldn't make do without it.

It is important to remember that one can exert effective control over one's processes without causally interfering with their operations. Consider a visual belief formed from clear, crisp data, and plausible in light of the subject's background beliefs. Such a belief might be formed spontaneously without any metacognitive input, since it is formed in circumstances in which vision is highly trustworthy and does not conflict with any other processes. But the processes could still be under control if it is the case that were the belief not plausible, or not formed in auspicious circumstances, the agent would not have automatically formed the belief. Control can be manifested just as much in what one could have done, but didn't have to, as in what one did do.

The second potential misunderstanding arises from Nelson and Narens' second principle, according to which the metalevel functions as a model of the object level; i.e., that there is a mapping from events at the object level to responses at the metalevel.[8] Consider the behaviour of the governor on a steam engine. Assuming that the governor is optimal—responds immediately to changes in the engine's

[8] The principle is based on Conant & Ashby's (1970) theorem. Let R be the simplest optimal regulator of a system S. Let $\sigma(i)$ be S's response to input i and $\rho(i)$ be R's response to input i. Then there is a mapping $h: S \rightarrow R$ such that $\forall i,\ \rho(i) = h[\sigma(i)]$.

speed, etc.—there will be a mapping from the speed of the engine to the rate of intake the governor allows. Thus the behaviour of the governor functions as a model of the engine's speed. An imperfect governor—one that is insensitive to certain changes in speed, or slow to respond—imperfectly models the engine's speed. We can expect, of course, that human metacognition is generally imperfect. But we should nonetheless expect metalevel events to model object level events approximately.

This principle does not mean that the metalevel must *construct* a model of the object level. A perfect governor is a model of its engine's speed, but it nowhere represents the engine's behaviour. Constructing models is difficult and resource-intensive. It is necessary for some cognitive tasks (Clark and Toribio 1994; Grush 2003), but when cognitive processes are directly coupled with their objects, feedback loops can be more adaptive and efficient (see Clark 1997). There is a wide range of latitude in what metaprocesses can monitor and what sorts of models they need to generate in order to perform their functions.

More generally, metacognitive monitoring should be judged only by its contribution to control capacities. There is thus no fear of doxastic ascent with regard to monitoring. Agents need not have justified beliefs about object-level processes, since they need have no beliefs at all; they are only required to represent whatever is necessary to effectively control their own cognitive activity.

Even if we grant that metaprocesses can be unconscious and need not construct detailed models of the object level, one may still worry that requiring effective regulation amounts to imposing an architectural constraint on cognitive systems. It would seem that to have intellectual abilities or virtues, a system would have to have a hierarchical structure, with object-level processes at the bottom and metaprocesses at the top. But whether an agent has an ability of a certain sort should presumably depend on what the agent can *do*, rather than how the agent goes about accomplishing it.

This concern arises, I think, from supposing that metacognition requires dedicated metaprocesses, processes whose only function is to regulate. But the distinction between metalevel and object-level can occur on a case-by-case basis. Processes can be metalevel with respect to some processes but object-level with respect to others.

For instance, processes responsible for making inductive inferences are probably involved in evaluating the chances of success for alternative strategies for solving a problem.[9] But these processes would themselves be object-level with respect to metaprocesses that watch for statistical blunders or the projection of unprojectible predicates.

Likewise, some of conflict resolution (and intertwined aspects of selective application) can be resolved by the simple expedient of having processes feed their outputs into a network with simple rules for maintaining coherence among beliefs.

[9] This would explain the correlation between success in strategy selection and aptitude at inductive reasoning (see Schunn & Reder 1998).

In such a structure, there is no dedicated metalevel. Rather, each process is regulated by other object-level ones.

In fact, this latter example in some ways provides a better model of effective regulation than the Cartesian, hierarchical picture. Metacognition can be *holistic* rather than hierarchical. What matters in whether one has an intellectual ability is whether object-level processes are integrated with the rest of the system in such a way that they can be made to contribute to the agent's overall end of acquiring true beliefs. The metaprocesses are just those parts of the system that are responsible for ensuring this integration. They may be central processes, dedicated specifically to this integrative function, but there is no reason why they must be.

4 Metacognition and Knowledge

The difficulty of performing metacognitive functions—of selectively applying processes, managing resources, and avoiding conflicts—varies depending on the nature of the object-level processes and other aspects of the problem. For instance, resource management is reasonably easy for vision, since it is fast and competes minimally with other processes for resources. Selective application is greatly assisted by the fact that we usually experience degraded images when vision is unreliable. On the other hand, consciously weighing the evidence for a proposition is slow and attention-intensive. It is subject to many biases, some of which are easier to correct for than others. Many appear to derive just from reasoning too quickly, and can be corrected just by giving subjects an incentive to reason carefully. Other biases—such as ignoring base rates and other probabilistic fallacies—are made *worse* by providing such incentives (Lerner and Tetlock 1999).

Despite this variability, the usual sorts of human processes are ones that require some metacognitive control in order to be reliable. That is, in our usual situation when a subject S has a reliably formed belief, she also has a stable capacity to exert control over her processes in a way that allows her to form true beliefs and avoid forming false ones. Thus in paradigm cases of knowledge, when we appraise reliability we also appraise the agent's capacities to regulate the processes underlying her belief-formation.

It appears that the counterexamples to reliabilism are cases where subjects are able to form beliefs reliably despite lacking effective metacognitive control. They involve processes for which selective application is trivially easy; that is, processes that can be blindly trusted and still yield a great enough proportion of true beliefs in relevant environments for those beliefs to be reliably formed. Call these "easy management" cases. If selective application is trivial, then conflict resolution can also be easy. For instance, information arising from the easily managed process can just trump all other considerations.

The author of a guide to the *I Ching*, the ancient Chinese method of divination, wrote

> Someone once asked me if I did not worry about being too dependent on the I Ching. On consulting it, it replied a shame not to make use of that friend. [sic] (Anthony 1988)

Now suppose that consulting the *I Ching* is in fact a reliable method of divination. Even if it is, we would not want to say that your true beliefs arising from it are knowledge if the *I Ching*'s deliverances are the only reason you have for trusting it. It should be easy to see that the problem is a lack of metacognitive control. If the *I Ching* were unreliable, the author would still use it to acquire beliefs because of the unreliably generated belief, acquired from the *I Ching*, that the *I Ching* is to be trusted. The agent lacks the sort of control that we exhibit with our usual ways of thinking.

Examining reliable brain tumours and other strange processes[10] yields similar conclusions. A fatal brain tumour might very well cause only the belief that the victim is going to die, but it might just as easily have caused the belief that the victim's stock holdings were worthless or that his mother did not really love him. The subject's lack of control is evinced by his inability to avoid believing falsely in these nearby possibilities.

For a third example, consider Goldman's (1986: 51–2) Humperdink, who is tutored in mathematics by a man that he has been warned is a fraud with no academic credentials. Humperdink learns various problem-solving algorithms, only one of which is actually correct. He forms a belief from applying the one correct algorithm to an appropriate problem. While Humperdink does happen to use a reliable process, if it were not reliable in the circumstances, he would have no capacity to refrain from using it. Given how he learned the algorithm and his current inability to detect that most of what he was taught was wrong, he might just as easily have applied an unreliable algorithm to the problem.

5 High-Level Virtues

Appealing to metacognition allows us to explain why certain possible reliable processes do not yield knowledge. Our metacognitive capacities do more than just help to avoid forming false beliefs, however. Capacities for conflict resolution and resource management are crucial to the task of making a bundle of disconnected

[10] BonJour's (1980) reliable but unwitting clairvoyants present a more complicated case, since (as Sosa 1991 and Greco 2003 argue) it is possible for clairvoyance to be a virtue even if the clairvoyant is unaware of his power's reliability. But the intuitive appeal of denying knowledge in BonJour's case arises, I think, from the feeling that blindly trusting a sudden urge to believe that the President is in New York City evinces a lack of metacognitive control. It's plausible that Norman could not avoid forming false beliefs in environments in which his clairvoyance was unreliable, or if his clairvoyant powers were to degrade over time. BonJour's account also doesn't address whether Norman would be able to handle informational conflicts. If Norman seemed to be shaking the President's hand at the White House at the same time his clairvoyance told him the President was in New York City, it's not clear what he would believe.

processes into a well-integrated agent able to achieve her cognitive goals in her environment. They provide the ability to turn potentially inconsistent information into beliefs about the world, and to manage resources so that we can find trustworthy answers to the questions we wish to answer. It thus seems plausible that metacognitive capacities play a crucial role in defining one's cognitive character.

Axtell (2008) argues that there are two branches of virtue responsibilism with distinct ways of understanding the high-level intellectual virtues. The *phronomic* conception (most clearly articulated by Zagzebski 1996), draws heavily from Aristotle. Intellectual virtues are aspects of a unified cognitive character exemplified by ideal virtuous agents, and driven by the motivation to believe the truth. On the *zetetic* conception (defended by Axtell, Hookway 2006 and Morton 2004) virtues are dispositions to conduct inquiry well: to employ effective strategies for inquiry and to exhibit diachronic rationality even when one lacks conscious access to the grounds of one's beliefs or the truth-conduciveness of one's strategies.

On the zetetic conception, it is easy to see high-level virtues as metacognitive capacities. For instance, Roberts and Wood (2003) argue that intellectual humility is a dispositional lack of concern with the status that goes with intellectual achievements or with dominating the thinking of others, and a disposition not to claim unwarranted entitlements on the basis of one's intellectual excellence. This is readily seen as a stable capacity to control belief-formation to prevent it from being biased or misled by desires for status and dominance. Similarly, open-mindedness is (approximately) a capacity to control belief-formation so as to assign due weight to alternative positions or beliefs, or unexpected problems or questions. Perseverance, which Hookway calls an "ability to acknowledge the consequences of [one's] views without wavering" (2003: 187), seems to be a capacity to successfully resolve conflicts between the deliverances of reasoning and one's prior expectations or preferences about what to believe.[11]

6 A Theory of Intellectual Virtue

The above considerations suggest the following general account of intellectual virtue:

> S has an intellectual virtue iff she has a stable capacity to exert metacognitive control over her processes in a way that allows her to attain her cognitive goals.

This definition is meant to be broad enough that we can say that our capacities to regulate vision, memory, and the like so that they are reliable are virtues; and we can also say that a trait like humility, which controls for a particular sort of bias, is also a virtue. This broadness arises from identifying virtues with capacities to control

[11] In Lepock (2011), I identify a number of process desiderata (reliability, power, portability, and significance-conduciveness) and use them to try to explain why high-level virtues like those mentioned above are so valuable.

underlying processes rather than with belief-forming processes or dispositions. Thus (on the zetetic conception) a trait like conscientiousness is immediately a virtue. The difference between merely reliable processes and knowledge-generating intellectual abilities is that the latter are regulated by virtues (which can be narrow, faculty-specific, or subpersonal regulatory mechanisms rather than broad traits like conscientiousness).

In keeping with this, a necessary condition for knowledge should be that S believes out of intellectual virtue, i.e.:

S believes B out of intellectual virtue iff:

(a) B is reliably formed, and
(b) B's being reliably formed is in part due to S's having some virtue that controls the processes involved in generating B.

It may be helpful to remind the reader again that control does not require causal interference, but only that metacognitive capacities are capable of intervening when necessary.

7 Control

It will no doubt be objected that the notion of 'metacognitive control' I appealed to above is no more precise than 'cognitive integration' or 'inner nature'. We can, however, make this notion more precise by examining the sort of control that we have over our cognitive processes, and particularly by comparing our capacities to those of subjects with reliable processes that do not yield knowledge. In doing so, we can make use of empirical investigations of metacognition. Space will only permit a brief sketch of how this will work.

We have the capacity to avoid trusting our ordinary processes in circumstances in which they would be misleading. Our usual methods of belief-formation would not be reliable otherwise. In the counterexamples to bare reliabilism, however, these circumstances are either too distant from normal or too rare to prevent the processes from being reliable.[12] Effective control requires such an ability, however. Suppose that aluminum-foil hats scramble clairvoyant signals, leading to deranged intuitions; but Norman the reliable clairvoyant, being very stylish, would never wear such a thing. Then one sign that Norman can effectively control his clairvoyance is that he would avoid believing the results of the scrambled signals even in the highly unusual or rare circumstances in which he might end up wearing an aluminum-foil hat.

Another important feature of control is the ability to learn about when one's processes are reliable or unreliable—to identify signs of reliability, but also to be able to recognize when those signs are no longer veridical. For instance, cognitive

[12] I am using a propensity-type measure of reliability here, according to which reliability is a matter of the proportion of true beliefs generated over a range of relevant or normal situations. But my discussion should be applicable to truth-tracking and safety accounts, *mutatis mutandis*.

fluency, or the ease of cognitive processing, is an important metacognitive signal (see Benjamin and Bjork 1996). Fluent processing is interpreted as more reliable, to the point that statements written in a clear font are more likely to be judged true than those that are more difficult to read. However, relying on fluency appears to be learned; at the very least, subjects placed in a situation where fluency judgements are regularly wrong will learn not to rely on them (Unkelbach 2006).

When we measure a process's reliability, we hold the process type constant. Ordinary humans, however, are able to handle minor perturbations or gradual degradations in our object-level processes without being too likely to be misled. Thus it makes sense that control arises in part from an ability to cope with variations in process type in nearby worlds, such as might arise if Humperdink had been taught a slightly different and unreliable algorithm, or if Norman's clairvoyance became myopic over time.

One part of selective application is the ability to initiate processes when they would be likely to yield valuable true beliefs. Norman the reliable clairvoyant has a passive capacity to be reliably right as to the location of the President. But if what he *really* wants to know is whether his financial advisor is in a country with no extradition treaty, he's out of luck. It also seems to be important for control to be able to resolve conflicts between processes in a way that does not conflict with selectively applying them. I noted above that in "easy management" cases, conflicts can be resolved through very simple rules, such as allowing information from one's clairvoyance or reliable tumour to trump the deliverances of all other faculties. But this sort of simple rule has the consequence that those other faculties are not being permitted to inform belief-formation in circumstances in which they might be reliable. This is, intuitively, irresponsible behaviour. Suppose Norman was in Washington shaking the hand of a man who has been introduced as the President, looks exactly like him, is guarded by a Secret Service detail; at the same time, his clairvoyance tells him that the President is in New York. Effective control requires taking the deliverances of his senses and background knowledge seriously, rather than trusting only the promptings of his clairvoyance.[13]

Various sorts of capacities for rendering one's belief-set coherent will also be relevant to whether one has control over one's processes. Given that ordinary non-reflective subjects appear not to have particularly coherent beliefs, it seems that the

[13] In effect, this is to say that control requires a sensitivity to disconfirmatory evidence. It is more fruitful, I think, to put the requirement in terms of selective application and conflict resolution. Not just any sensitivity to disconfirmation is relevant for control. Suppose that Humperdink would refrain from believing the result of his algorithm if subjected to a mathematical intervention in which friends, family, and experts in mathematics would testify to the error of his ways and how it makes them feel. That is not sufficient to change our appraisal of Humperdink's situation, since it is simply too unlikely to happen. One must be sensitive to disconfimatory evidence that is sufficiently likely to arise in situations in which one uses one's processes. Given that it is the (consciously accessible) deliverances of our processes and our accessible background knowledge that constitute the evidence we have, by considering the needs of conflict resolution and selective application we should be able to determine what "sufficiently likely" is going to mean.

ability to fit the deliverances of a process into a coherent belief-set will be sufficient for control,[14] but not necessary.

Let me also mention a factor over which virtuous agents are apparently not expected to have control. A number of studies have found that subjects are reasonably good at monitoring the accuracy of their general knowledge. Koriat (2008), however, found that subjects' confidence in their answers to general-information questions was correlated with the chance that the majority of participants in the study would give the same answer rather than with the truth of the answer. It is generally reliable to evaluate one's general-information beliefs by considering whether others would agree, since most of the time, if other persons believe that p, then p is true. But it suggests that for a broad range of beliefs, we may have difficulty controlling for the possibility that the bulk of our epistemic community would be wrong. That makes it plausible that virtue does not require us to be able to cope with the possibility of widespread error regarding p in the community (as long as the subject's belief is reliably formed given the actual situation).[15] Similar considerations show that we do not need to control against the possibility of evil demons, that we are dreaming, and other extreme skeptical scenarios. We only take such possibilities seriously when attempting to refute them; this suggests that virtuous belief-formation does not require controlling against those possibilities.

I have argued that we should understand intellectual virtues as capacities for metacognitive control. The difference between reliable processes that yield knowledge and those that do not appears to be one of effective control. Moreover, the "high-level" virtues are easily seen as capacities for controlling or regulating inquiry and belief-formation. Finally, we can use empirical studies of metacognition to inform our understanding of what is necessary for virtue. This methodology will allow us to make the notion of intellectual virtue more rigorous, and should open up a wide range of avenues for future research in virtue epistemology.[16]

---

[14] Assuming coherence is defined in such a way that one cannot preserve coherence by merely discounting or ignoring disconfirmatory or problematic evidence; otherwise, coherence-maximizing subjects may lack control in the way described in the previous paragraph.

[15] In some cases, we will want to deny knowledge to agents upon learning that their belief-formation is only likely to be correct if the rest of the community is likely to be right. But it seems (at least according to my own intuitions) that when we do that, it is not because of worries about lack of control, but worries about lack of reliability. For instance, we would not say that S knows that members of his ethnic group have not perpetrated atrocities against members of another ethnic group if his only reason is that most members of his group believe they have not. This seems to be not because of any lack of control, but because community beliefs of this sort are often self-serving.

[16] Versions of this paper were presented at the Canadian Philosophical Association, the University of Calgary 2007 Philosophy Graduate Conference, and the 2008 Bled Philosophy Conference. At these venues I received invaluable criticism and suggestions from audience members too numerous to list. I am also indebted to Adam Morton, Jennifer Nagel, Eric Dayton, Bruce Hunter, David Henderson, and Vladan Djordjevic. This research was supported by the Social Sciences and Humanities Research Council of Canada.

References

Anthony, C.K. 1988. *A guide to the I Ching*. Stow: Anthony Publishing.
Axtell, G. 2008. Expanding epistemology: A responsibilist approach. *Philosophical Papers* 37(1): 51–87.
Axtell, G., and J.A. Carter. 2008. Just the right thickness: A defense of second-wave virtue epistemology. *Philosophical Papers* 37(3): 413–434.
Baehr, J.S. 2006a. Character in epistemology. *Philosophical Studies* 128: 479–514.
Baehr, J.S. 2006b. Character, reliability, and virtue epistemology. *Philosophical Quarterly* 56: 193–212.
Benjamin, A., and R. Bjork. 1996. Retrieval fluency as a metacognitive index. In *Implicit memory and metacognition*, ed. L.M. Reder, 309–338. Mahwah: Erlbaum.
BonJour, L. 1980. Externalist theories of empirical knowledge. *Midwest Studies in Philosophy* 5: 53–73.
Carruthers, P. 2008. Meta-cognition in animals: A skeptical look. *Mind and Language* 23(1): 58–89.
Cary, M., and L.M. Reder. 2002. Metacognition in strategy selection. In *Metacognition: Process, function, and use*, ed. P. Chambres, M. Izaute, and P.-J. Marescaux, 63–77. Dordrecht: Kluwer.
Clark, A. 1997. *Being there: Putting brain, body, world together again*. Cambridge, MA: MIT Press.
Clark, A., and J. Toribio. 1994. Doing without representing? *Synthese* 101(3): 401–431.
Conant, R.C., and W.R. Ashby. 1970. Every good regulator of a system must be a model of that system. *International Journal of Systems Science* 1(2): 89–97.
Davies, M., and M. Coltheart. 2000. Introduction: Pathologies of belief. *Mind and Language* 15(1): 1–46.
Goldman, A. 1986. *Epistemology and cognition*. Cambridge, MA: Harvard University Press.
Greco, J. 2003. Further thoughts on agent reliabilism. *Philosophy and Phenomenological Research* 64(2): 466–488.
Greco, J. 2010. *Achieving knowledge: A virtue-theoretic account of epistemic normativity*. Cambridge, UK: Cambridge University Press.
Grush, R. 2003. In defense of some 'Cartesian' assumptions concerning the brain and its operation. *Biology and Philosophy* 18(1): 53–93.
Henderson, D., and T. Horgan. 2007. Some ins and outs of transglobal reliabilism. In *Internalism and externalism in semantics and epistemology*, ed. S. Goldberg. Oxford: Oxford University Press.
Hookway, C. 2003. How to be a virtue epistemologist. In *Intellectual virtue: Perspectives from ethics and epistemology*, ed. M. DePaul and L. Zagzebski, 183–202. Oxford: Oxford University Press.
Hookway, C. 2006. Epistemology and inquiry: The primacy of practice. In *Epistemology futures*, ed. S. Heatherington, 95–110. Clarendon, Oxford: Oxford University Press.
Kentridge, R.W., and C.A. Heywood. 2000. Metacognition and awareness. *Consciousness and Cognition* 9: 308–312.
Koriat, A. 1994. Memory's knowledge of its own knowledge: The accessibility account of the feeling of knowing. In *Metacognition: Knowing about knowing*, ed. J. Metcalfe and A.P. Shimamura, 116–135. Cambridge, MA: MIT Press.
Koriat, A. 2008. Subjective confidence in one's answers: The consensuality principle. *Journal of Experimental Psychology: Learning, Memory, and Cognition* 34(4): 945–959.
Lepock, C. 2006. Adaptability and perspective. *Philosophical Studies* 129(2): 377–391.
Lepock, C. 2011. Unifying the intellectual virtues. *Philosophy and Phenomenological Research* 83(1): 106–128.
Lerner, J.S., and P.E. Tetlock. 1999. Accounting for the effects of accountability. *Psychological Bulletin* 125: 255–275.
Morton, A. 2004. Epistemic virtues, metavirtues, and computational complexity. *Noûs* 38(3): 481–502.

Nelson, T.O., and L. Narens. 1990. Metamemory: A theoretical framework and new findings. *Psychology of Learning and Motivation* 26: 125–173.

Plantinga, A. 1993. *Warrant and proper function*. New York: Oxford University Press.

Roberts, R.C., and W.J. Wood. 2003. Humility and epistemic goods. In *Intellectual virtue: Perspectives from ethics and epistemology*, ed. M. DePaul and L. Zagzebski, 257–279. Oxford: Oxford University Press.

Schunn, C.D., and L.M. Reder. 1998. Strategy adaptivity and individual differences. *Psychology of Learning and Motivation* 38: 115–154.

Sosa, E. 1991. *Knowledge in perspective*. Cambridge, UK: Cambridge University Press.

Sosa, E. 2007. *A virtue epistemology: Apt belief and reflective knowledge*, vol. 1. Oxford, UK: Oxford University Press.

Sosa, E. 2009a. Replies to Brown, Pritchard, and Conee. *Philosophical Studies* 143: 427–440.

Sosa, E. 2009b. *Reflective knowledge: Apt belief and reflective knowledge*, vol. II. Oxford, UK: Oxford University Press.

Thompson, V. 2009. Dual-process theories: A metacognitive perspective. In *In two minds: Dual processes and beyond*, ed. J. Evans and K. Frankish, 171–196. Oxford: Oxford University Press.

Unkelbach, C. 2006. The learned interpretation of cognitive fluency. *Psychological Science* 17: 339–345.

Zagzebski, L. 1996. *Virtues of the mind*. Cambridge, UK: Cambridge University Press.

Daring to Believe: Metacognition, Epistemic Agency and Reflective Knowledge

Fernando Broncano

1 Introduction

In this paper, I will be arguing for a view of knowledge as a true belief that manifests a competent (epistemic) agency. Beyond a mere juxtaposition of performing skills, epistemic competent agency requires an integration of faculties at a personal level that is sufficient to evaluate the agent's epistemic risk in particular situations. I will propose that, in order to meet this requirement, agency must scale to a personal level where the agent's engagement in epistemic situations manifests a competent endorsement of her beliefs. This view can deal with the predicaments of Virtue Epistemology in a naturalistic atmosphere by changing the emphasis from representation to agency, and by considering knowledge as an expression of achievement. This interpretation faces two related problems: first, the issue of self-knowledge in agency, and second, the problem of the integration of competencies from the personal standpoint of a unified agent. In this paper, I will only be dealing with the second problem.

I will begin by bearing in mind certain criticisms to the notion of reflective knowledge (as proposed by Sosa's version of virtue epistemology). From a naturalistic perspective, the empirical evidence about the sub-personal nature of many mechanisms involved in knowing, puts virtue epistemologists under the pressure to give an account for the integration problem. By itself, reflective knowledge would not be well-integrated in the epistemic character of a knower along these naturalistic lines, thus becoming an unnecessary or insufficient requirement for knowledge. Answering this problem leads us to question the perspective that the agent takes in her reflection: it is either a theoretical, third-person perspective, or, by contrast, a deliberative, first-person perspective. The very notion of agency would support the latter; but, then, how does the first-person perspective work through the

F. Broncano (✉)
Department of Humanities, Universidad Carlos III de Madrid, Madrid, Spain
e-mail: Fernando.Broncano@uc3m.es

A. Fairweather (ed.), *Virtue Epistemology Naturalized: Bridges Between Virtue Epistemology and Philosophy of Science*, Synthese Library 366, DOI 10.1007/978-3-319-04672-3_4,

process of forming a practical intention or an epistemic judgment? Though a distant, theoretical perspective could be taken into account forensically with the deliverances of the subject's multiple cognitive devices, the problem that such an approach poses is that it precludes the agential perspective.

I will then consider as a possible answer the case of *metacognition* as a naturalistic candidate for the seat of reflective knowledge. I will discuss the features of this mechanism, in particular the function of refraining from action when the system's performance is jeopardized. Animals can possess certain forms of metacognition, but it is far from clear that they possess agency. Though metacognition provides the agent with skills of monitoring and control, it does not yet completely solve the problem of reconciling an objective evaluation of the agent's competencies with an agential first-person perspective on epistemic judgments. What we need, I will suggest, is a scale of progressive stages in order to achieve a full agential perspective. I will be thus proposing that metacognition is relevant to agency only if it delivers meta-coherence as a way of calibrating epistemic risks. However, meta-coherence is a property that is still in need of being integrated into a unified agential structure. I will then end by adding that the first-person experience of knowing is a necessary component of epistemic agency; and, finally, that evaluating epistemic risks is a task that is involved in reflective knowledge, analogous to the process of forming a reason for an action. This implies integrating the evaluation of personal faculties with the demands of the situation and the actual ability to respond to these demands.

2 Reflection and the Agential Turn in Virtue Epistemology

Recent trends in Virtue Epistemology are prone to consider knowledge from an agential point of view rather than an intellectualist one. In this way, some leading authors regard certain expressions of agency to achieve knowledge as essential, such as, *motivation* (Zagzebski 1996), *success from ability* (Greco 2010), or *performance from competence* (Sosa 2011). This sort of "agential turn"[1] makes epistemic agency a central topic of epistemology, and renews its attention to related issues, such as authority, control, or autonomy. As a result, these recent trends tend to displace the traditional central role of justification to different epistemological interests (Axtell and Carter 2008), and thus shed new light upon some traditional discussions in analytical epistemology. One of these is the discussion about the naturalization of epistemology. If knowledge becomes a manifestation of agency, such naturalization should also involve a further naturalization of essential aspects of human agency itself. Thus, instead of the "externalist/internalist" debate, the split between "personal/sub-personal" levels of agency appears in this naturalization program as a more relevant topic than it is

[1] Although the term "Value Turn" is much more extended in characterizing these new approaches, I consider that other pragmatic interests are also involved in this "Second Wave" of Virtue Epistemology; and hence, "agential" could be a term with a broader scope that better describes this change.

commonly accepted. To put it crudely, the question that naturalism now poses to Virtue Epistemology is the problem of how empirical findings about the sub-personal devices involved in the buildup of decisions can match normative claims epistemologically for the agent considered as a whole at a personal level. Presumably, such empirical results seem to draw an image of the epistemic agent that undermines the epistemic intuitions guiding the conceptual approaches of virtue epistemologists. Although this seems to affect very different approaches in Virtue Epistemology (Weinberg et al. 2008; Kornblith 2002), I will be focusing on the sort of reflective knowledge that Sosa identifies as "full knowledge". Sosa, along with action theorists (Frankfurt 1988), considers a second-order stance as a sign of agency. However, what happens with the basic working components underneath the reflective level? A particularly vivid scenario for the tension between the philosophical and the allegedly scientific images of the epistemic agent can be easily drawn from this suspicion.

Sosa contends in the spirit of this pragmatic turn that "belief is a kind of performance" (Sosa 2011, p. 1). Consequently, he defines knowledge as a kind of success of this performance, also referred to as an apt belief: at the animal level, an apt belief, as other kinds of practical achievements, spring from the organism's competencies. However, practical achievements themselves admit different degrees of agential success. For instance, although running can be considered a success for someone recovering from an injury, to refrain from running in order to help somebody in danger, in spite of some understandable panic, can still be a greater agential achievement. In cases such as these, the subject will need to reflect about her own position in order to make a final judgment or decision. By taking such a reflective stance, the agent obtains a reason to make up her own mind and, thus, she reaches a result in a fully intentional way. Suppose, though, that this runner is asked: "Would you dare to run?",or "Can you help me?" If the questions are accepted, the addressee will elicit a conscious answer, yes or no. Such a choice demands from her a careful evaluation of her own cognitive (practical) position, or her ability to answer to the request. This is what Sosa considers "reflective knowledge".

Although the overall reflective process presents different facets, a minimal condition is that some sort of self-knowledge must be available to the agent. However, this putative self-knowledge raises two controversial questions that are highly debated in the current literature. The first one is: what degree of transparency should the mental states involved in bringing about the requested answer have (and, correspondingly, what degree of self-transparency is necessary for the agent to be a reflective creature is necessary)? The second related question is: what degree of integration of the competencies in charge of such reflective process is necessary for the epistemic agent to act consciously?

Even though the first transparency issue has an epistemological importance as it is connected with the well-known sceptic requirement of knowing that one knows, it also has a cognitive dimension that has been regarded as troublesome for conceptual analyses of knowledge based on epistemic intuitions. Some philosophers have thus emphasized the devastating epistemological consequences a possible lack of transparency to introspection could have due to the sub-personal activity of biases and mental heuristics in reasoning (Kornblith 2002, chap. 4). If introspection results to be an unreliable source of information about the agent's own states and procedures, a second level of reflective knowledge cannot guarantee or imply an improvement in her success (Kornblith 2010).

The second doubt concerning the integration of our cognitive faculties was already raised by Montaigne. In his *Essays* he writes: "We are all patchwork, and so shapeless and diverse in composition that each bit, each moment, plays its own game. And there is as much difference between us and ourselves as between us and others" (Montaigne 1958, II, 1, p. 244). His categorical contention does not lack severe consequences for agency:

> A man who does not have a picture of the whole in his head cannot possibly arrange the pieces. What good does it do a man to lay in a supply of paints if he does not know what he is to paint? No one makes a definite plan of his life; we think about it only piecemeal. The archer must first know what he is aiming at, and then set his hand, his bow, his string, his arrow, and his movements for that goal. Our plans go astray because they have no direction and no aim. No wind works for the man who has no port of destination. (Montaigne 1958, II, 1, p. 243)

Montaigne entertained the sceptical suspicion under which each part of the organism could have its own agenda that need not be dependent on the agent's higher aims. Montaigne alleged multiple experiences of involuntary actions of some organs, and from here, he extended his observation to a more general lesson concerning the nature of agency. That is, that the lack of integration is independent of the proper functioning of each organ because the issue lies on the need for the integration of all the parts in a common goal. Montaigne's suspicion can be further extended to all kinds of cognitive aspects involved, like those considered by the recent psychology of reasoning.[2] Accordingly, it would be plausible to imagine that a non-well-integrated mind could arrive to a right belief; and nevertheless, from a normative point of view, such belief could still fail to attain the success that knowledge is supposed to represent. Thus, the problem of integration poses a question of epistemological force. For instance, let us imagine an inebriated driver that considers it risky to drive her car. Let us consider, for the sake of argument, that in spite of having taken two glasses of wine, she is actually able to drive safely because her course of travel does not offer any risk. Her moral convictions, her abilities, and her decisions would be badly integrated, although her actual decision of driving, in those circumstances, was safe enough.

Let us notice that, although transparency and integration could be stated as a priori requirements, from a naturalistic point of view, we should also take into account the numerous empirical findings concerning sub-personal and automatic mental mechanisms and cognitive biases, as when having risky beliefs or taking

[2] In spite of the great variety of mechanisms and biases in psychological literature, I would quote here (because of its close relationship with the problem currently addressed) some biases such as the illusion of control and the optimistic bias, that usually act in most risky behaviours and decisions. Some authors have considered moral luck as relevant to the topic: Enoch (2010), Royzman and Kumar (2004), Domsky (2004). Some of the empirical findings about risky behaviours, such as in health perception, are exposed in Klein and Helweg-Larsen (2002), Harris and Middleton (1994).

decisions like the one considered in the example above. Transparency and integration cannot, therefore, be reasonably taken for granted. At the most, both properties could be considered as modest symptoms of a putatively "healthy" degree of agential status. That is, they could be taken as empirical signs of the adequacy of one's perspective vis-à-vis a challenging task, or more precisely, as signs of epistemic or agential "quality". In these terms, reflection would be a stance that the agent takes in order to assess the quality of her position. The term seems to entail that the agent "reflects" about her own forces before accepting some belief, or before acting.

Nevertheless, the term "reflection" can be confusing depending on how we consider the nature of the process in which the agent is involved. For example, we could take "to reflect" to mean an introspective process of inspecting one's own mental states. Apparently, this conception exploits a "visual" metaphor, as the very etymology of "introspection" suggests "looking within". This is the notion of reflection the Cartesian tradition seems to picture. "Reflecting" would, then, amount to taking some evidence from one's mental states concerning their epistemic or agential quality. But such "evidentialist" notion of reflection raises a new sequence of questions, since the "visual" metaphor appears to be as intuitively weird as metaphysically unjustifiable: Who looks? Where is she looking? What is seen as evidence?, etcetera. What I want to consider is how to preserve a reflective stance without falling in a Cartesian image that presupposes a homuncular self looking transparently inside herself.

To begin with, there is a certain feeling of paradox that emerges from the notion of reflection taken under Cartesian lights. Let us recall as an analogy Vermeer's well-known canvas *The Art of Painting*. Beyond a curtain, there is a painter, putatively Vermeer himself, with his back visible to the viewer of the painting while he portrays a maiden symbolizing the Muse of History. When we ask what is represented on the picture, this famous canvas presents us with an intriguing paradoxical appearance: how is it possible to portray one's own back while one is painting on a canvas? Certainly, one could regard the picture as if it represented the percept held by some indeterminate observer. But how is it possible to depict the perceptions of other minds? Clearly, between the two impossibilities, one of them understands the canvas as a representation of an imaginative simulation of a mind. That is, the canvas portrays a meta-representation. But this would lead us to inquire about the agent holding this meta-representation. Perhaps it was Vermeer portraying himself? Or was he picturing another person looking at him while he was painting? Or, finally, was he depicting a general scene of somebody looking at someone else painting? The picture is the same, but the represented subjects are very different in the three cases.

3 Sosa's Reflective Knowledge

Let us now consider virtue epistemology, as exposed by Ernest Sosa. Curiously, the jacket book of Sosa 2007, a main exposition of his views, shows Vermeer's painting. It is not surprising because E. Sosa addresses the question of how could reflective knowledge avoid the threats of a sceptic who would claim that, in order to know that *p*, one must also know that one is not dreaming in a similar scenario. Sosa's answer is relevant because this requirement is deeply entrenched in the very folk notion of knowledge. Sosa distinguishes between "to know that one knows that *p*" and to have a reflective competence for assessing the risk one takes when accepting the deliverances of one's own cognitive faculties. In order to attain a "virtuous" evaluation of her epistemic status in a certain circumstance, the agent is gifted with meta-competencies that eventually will produce reflective knowledge. Such reflective knowledge is the response to philosophical scepticism according to E. Sosa's virtue epistemology. It is relevant to notice that such a faculty must be a kind of *meta-competence* that reaches further than the mere coherence of beliefs ("coherence might conceivably be detached from the environing world of the thinker, so as to deprive him of reliable access to truth") (Sosa 2007, p. 190). Fleeing from the threats of circle or regress, Sosa stipulates that this meta-competence is endowed with the function of examining the quality of the agent's epistemic position. Reflective knowledge, then, is postulated in order to exclude luck from the epistemically apt formation of true belief. This level would provide a higher quality for the knowledge attained:

> Reflective knowledge goes beyond animal knowledge, and requires also an apt apprehension that the object-level perceptual belief is apt. What competence might a believer exercise in gaining such meta-apprehension? It would have to be a competence enabling him to size up the appropriateness of the conditions. (Sosa 2007, p. 108)

Thus, reflective knowledge becomes the achievement of an epistemic meta-competence; that is, of a faculty or disposition to evaluate aptly the agent's epistemic position and her circumstances of knowing.

But the question would be whether the evaluation could discriminate among evaluative results, as the three possible interpretations of the canvas exemplify, depending on how this competence is conceived. In other words, what is the process that the meta-competences evaluates. On the one hand, it could be assessed as "a portrayed subject" from an alien third-person point of view, external to the first-person point of view; or, on the contrary, it could be assessed as a phenomenal picture of the painter's own situation. The point is that in each case the subject is involved in very different ways. Therefore, the questions I will be addressing are: first, to what degree the subject must be involved in reflective knowledge, and second, in which ways could this task be accomplished.

Sosa contends that reflective knowledge seems to confer a better epistemic status and value to the overall process of knowing. In his view, reflective knowledge adds justification to the first-order aptness as it strengthens cognitive success by reducing the role luck might have in cognitive achievements. As the subject is able to form the judgment that she justifiably knows that *p* properly, she meets a "principle of criterion": "In order to know fully well that *p*, one must be justified in believing (at least implicitly or dispositionally, if not consciously) that one's belief that *p* is formed in a way that is at least minimally reliable, that it has at least minimally a reliable source (if the proposition that one's source is thus reliable is within one's grasp)" (Sosa 2007, p. 122). Thus, such reflective knowledge endows the agent with a reflective justification. Certainly, a kind of unreflective justification can be conferred by her first-order aptness, but this second-order meta-aptness also provides a rational justification: "*Reflective* rational justification, by contrast, is acquired at least in part through rational endorsement: *either* through endorsement of the *specific reliability* or one's basis (or at least the *safety* of one's basis, of the fact that it would not lead one astray in delivering the deliverance that *p*) *or* through endorsement of the *generic reliability* of one's basis" (Sosa 2009b, p. 239). Rational endorsement is then the sign of having reached a reflective stage in knowing. Recently, by underscoring the agential status of knowledge, Sosa has expounded the analogy of knowing with the case of Diana the huntress when she manifests a first-order competence in shooting, as well as a second meta-competence when evaluating if she might take the risk of shooting under the actual circumstances. A shot is meta-apt if and only if it is well-selected (Sosa 2011, p. 8); shot selection involves here an endorsement, and apt endorsement means a better justification and, thus, a fuller agency.

It is worthwhile noticing that Sosa allows for rational endorsement of reliability to be produced in unconscious or implicit ways. His reason is that reflective knowledge springs out from a meta-competence, that is, a disposition to evaluate correctly the faculties' reliability, and this disposition could work in different ways. Reflective as well as unreflective knowledge can both produce apt true beliefs; and this production, Sosa argues, does not depend on the degree to which subjects are voluntarily engaged. The sole condition is that the subject be confident about her reflectively obtained belief, and this could be a result of her overwhelming disposition to believe confidently.

The archery analogy clearly shows that such meta-competence is a kind of control that is manifested even if the agent withholds her performance, for instance, when she chooses not take the risk of shooting. But this kind of control could be exerted even below the voluntary and explicitly conscious level. Performance manifests a meta-competence, at least to the degree of eventually forbearing to act if the monitoring and control of action are efficiently exerted. However, Montaigne's threat still hangs over this purported meta-level of reflective knowledge because it threatens that such control could be exerted in a non-integrative way. Empirical findings from the Cognitive Sciences show that this control can be a mere sub-personal managing mechanism of performances without necessarily amounting to a full stage of agency.

4 The Problem of Integration

Let us imagine William Tell doubting about his ability to safely hit the apple on his son's head at the very moment of shooting. Let us consider the content of the following propositions:

1. I am skilfully prepared to shoot.
2. William Tell is skilfully prepared to shoot.

The belief that he is prepared to shoot is the same, at least under a certain interpretation, but obviously, the possibilities are very different according to their consequences. William Tell is confronted with only three options open to him:

> (…) (a) "No, I don't know that," or (b) "Who knows whether I know it or not; maybe I do, maybe I don't," or (c) "Yes, that is something I do know." (Sosa 2007, p. 115)

Here is where the subject's integration problem appears. According to Sosa, "Answer (a), and even answer (b), would reveal a certain lack of integration in that stretch of consciousness; only answer (c) entirely avoids disharmony within that consciousness at that time" (Sosa 2007, p. 115). Suspension of judgement or forbearing shooting when one is competent shows a kind of failure. "The huntress who forbears taking a shot that she obviously should take fails in her performance of forbearing" (Sosa 2011, p. 7). Something occurs if one is objectively competent but lacks enough self-confidence for the performance or asserting that one knows. This could point to an underlying lack of integration between the possible state of knowing and its positive assertion, that is, between (1) and (2) as possible propositions ascribing knowledge to William Tell.

Why does this question of integration bring us back to the seemingly paradoxical scenario of Vermeer's painting? In order to answer this question we must observe that the meta-competence that reflective knowledge manifests has to evaluate acts of believing that are relative to a broad context in which a certain space of possibilities is involved. Some of these possibilities include stable dispositions to reach success, but others introduce luck and instability in the relation between the agent's performances and environmental circumstances. By focusing on William Tell's self-confidence when shooting, our example explicitly blends together cases (1) and (2). Tell's purpose is, thus, evaluated in this particular circumstance, but this evaluation is made against the background of a space of possibilities within which a counterfactually purported relationship between skill and success is established: "this successful shot is due to the archer's abilities".

The feeling of paradox would disappear if an external referee were to evaluate the archer's merits, but the problem arises because, in our example, the referee and the subject evaluated is the same person. The knowing subject is ascribing to himself a competence in those circumstances, but, is the content of his evaluation the belief (1) or is it, rather, the belief (2)? The difference between these two possible answers the agent can give himself shows the degree in which he is involved as an autonomous agent when delivering his judgement. In the first case ("I am skilfully prepared to shoot"), it is as if one could endow the agent with some kind of self-ownership. But, in the second case, it would seem that the agent, in spite of being skilfully prepared to shoot, did not identify himself with such ability.

There is a close analogy between what I am pointing out to and the well-known objection concerning strange and fleeting processes that has been addressed against reliabilism. Bernecker (2008), for instance, considers the case of Norman, an agent endowed with a stable and reliable clairvoyance. One could raise a relevant question for virtue epistemology if one were to evaluate, for example, whether ascribing knowledge to Norman when he believes that p out of his clairvoyant deliverances in spite of having evidence against *p*. Traditional internalist epistemologists require the agent to have subjective coherence in order to be justified in believing that *p*. However, in this case, what Norman suffers is not lack of coherence but lack of integration. The point in question is what relation is there between a personal level of reflection and the underlying mechanisms which produce the agent's deliverances.

According to Sosa, an epistemic perspective would be the way to bypass the dilemma between reliable dispositions and subjective justification:

> Knowing fully well thus requires some awareness of the status of one's belief, some ability to answer that one does know or that one is epistemically justified, and some ability to defend this through the reliability of one's relevant competence exercised in its appropriate conditions. (Sosa 2007, p. 132)

We should notice that Sosa takes "some awareness of the status of one's belief," or to have "some ability to answer that one knows", as well as to have "some ability to defend this" as sufficient symptoms for possessing reflective knowledge. On the other hand, he contends that "reflective knowledge manifests not just modular deliverances blindly accepted, but also the assignment of proper weights to conflicting deliverances, and the balance struck among them" (Sosa 2004, p. 291). John Greco has objected to a requirement on the grounds that it demands an implausible level of psychological complexity (Greco 2004, p. 97). Certain cases, as the William Tell case, can dramatically stage sophistication, but it would be quite implausible to require such complexity for children or even for most every-day assertions of knowledge in which a metarepresentation or second-order attitude is putatively not activated. For Greco, reliability, proper motivation (of reaching true beliefs), and a well-integrated character are sufficient to achieve knowledge, without the strong further requirement of a reflective stance (Breyer and Greco 2008). But the question is not one of psychological sophistication, but of the agent's integration of her personal and sub-personal levels. What is precisely at stake is the quality of epistemic agency that knowledge requires. On the one hand, character is something that needs time and learning. Children supposedly possess temperament, that is, innate dispositions, but not yet character, precisely the aim of education. On the other hand, though, to take a well-integrated character as a requirement could be insufficient unless we can specify who is the subject evaluating such integration. Let us suppose that William Tell is a reliable archer, that he is properly motivated to aim at the target, and that he has an integrated character, but that, faced with the dramatic decision of aiming at the apple on his son's head, he is overwhelmed by unconscious mechanisms that induce in him a lack of self-confidence. The question, thus, is what degree of personal involvement is needed in knowledge ascriptions.

Hilary Kornblith has nicely placed this question in what is the role that reflective endorsement plays for improving the reliability of a non-reflective, first-level knowledge:

> But there are mechanisms involved in reflection which sometimes act as sub-personal cognitive yes-men, endorsing whatever beliefs the first-order mechanisms produce. These mechanisms do not improve our reliability, but merely further entrench our first-order

> beliefs, however reliable or unreliable they may be. In addition, of course, there are mechanisms of reflection which interfere with the smooth working of reliable first-order processes of belief acquisition: they lower the reliability of the overall belief acquisition process. So the suggestion that first-order belief supplemented by reflection is more reliable than first-order belief alone is simply mistaken. (Kornblith 2010, p. 4)

Children and animals, for instance, can achieve success in their aims without reflection, and reflective agents can be opaquely get muddled up in their endorsements.

A possible answer to this sceptical scenario is to consider that reflection introduces a difference of degree rather than of class:

> (…) It does lead me to think of the difference between the two sorts of knowledge, the animal and the reflective, as difference of degree. The higher brutes may be credited, along with small children, with some minimal degree of perspectival, reflective knowledge, of the implicit, subconscious sort, which largely resides in hosted inference patterns. (Sosa 2003, p. 129)

Such degree of reflection could be connected with a parallel degree or quality of agency. Even if "in richness, explicitness, and explanatory power, that falls short of the reflective knowledge to which a human can aspire, especially someone philosophically inclined" (Sosa 2003, p. 129), still the significant point is that reflective knowledge can be attributed to creatures lacking higher degrees of deliberative consciousness. Surely, things turn out differently when someone asserts belief in a public context, for example, when giving forensic testimony in court. In these cases, the reflectively held belief is just a part of the act of assertion. But, according to Sosa, full consciousness would not be amongst the conditions that are required to achieve the status of meta-apt belief, as the traditional coherentist epistemologist surely claims. The point at issue is how agential quality and reflection can be related by virtue of plausible and naturalistic psychological conditions. By agential quality I here mean not only practical (or epistemic) success, but also the degree in which the agent is involved as an autonomous creature in such success. Is there some psychological condition that explains this connection?

5 Metacognition and Reflective Knowledge

I will first consider an empirical candidate to accomplish such a function; I will argue, second, that a malfunctioning of this cognitive mechanism can shed some light on the integration problem; then, from an example, I will conclude, third, that Sosa needs a kind of involvement of the first-person point of view that is not necessarily equivalent to assertion. I will be thus claiming that a reliable first-person perspective equals that of an integrated epistemic agent who is in charge of the task of knowing.

The cognitive function I will be referring to is that which psychologists and neurologists have named as *metacognition* (Koriat 2000; Proust 2007; Metcalfe 1994; Necka and Orzechowski 2005). It is often described as a cognitive device that aims to "know(ing) about knowing, that is, a cognitive function to distinguish what one knows about one's own cognitive abilities, states of knowledge, and actual performance from the cognitive abilities, states of knowledge and performance per se" (Koren et al. 2006, p. 313). William Tell's hesitation about his shooting is, in my example, a relevant case of metacognition: "metacognitive processes are required for decision making, troubleshooting, strategy selection and performance of non-routine actions" (Fernández-Duque et al. 2000, p. 289).

Metacognition is probably the best candidate to exemplify one of the functions of reflective knowledge in cognitive systems. Although reflective knowledge can be characterized as a higher intellectual and conceptual process, it is also surely based upon more basic cognitive mechanisms as metacognition would be. Moreover, the proper working of these devices can help us clarify the question of the integration requirement. As a functional system or ability, metacognition is probably already present in some animals other than human beings. Some studies with simians, dolphins, and even rats have shown that many animals can refrain from acting when the cognitive conditions of a formerly well-known task have turned harder (Smith 2005; Foote and Crystal 2007). These experiments do not allow us to clearly conclude that certain animals are gifted with self-consciousness, but they are expressive enough to endow them with some degree of metacognition. These are not big news for Sosa's concept of reflective knowledge, as for him, "a minimal degree of perspectival, reflective knowledge" can be possessed by animals (Sosa 2009a, b, p. 239). We should notice, however, that a refraining behaviour points to the existence of strong links between the evaluation of information and the control of acts. These links do not amount to collapsing acts and beliefs in animals (among other things because they lack intentions both in actions and in beliefs). The issue here is that control is based in an appreciative feeling of ignorance.

Another interesting trait of metacognition that apparently supports Sosa's views is that it does not necessarily involve metarepresentation (Proust 2007, 2009; Carruthers 2009). Metarepresentation is a higher-order process that presupposes certain steps in cognitive development (children reach it at the age of three-and-half, at least in a rough version of metarepresentation) as it requires open consciousness and deliberative stances. If metacognition were equivalent to metarepresentation, Sosa could legitimately argue that he is postulating a more basic metacompetence or skill that is not equivalent to such a high degree of cognitive processing (although a full reflective stage of knowledge would always be possible). Nevertheless, metacognition is perhaps a good candidate to embody the skill to evaluate one's own cognitive perspective without presupposing the status of assertion or of deliberation to act. Furthermore, metacognition draws along structurally important functions, such as "a theory of the mind" (or simulation ability), memory retrieval, transmission of learning, executive control, etc. It thus amounts to a structural component of any cognitive task in which epistemic quality is involved. In fact, the lack of metacognitive skills is a symptom of mental disorder. This metacognitive deficit is called *Anosognosia*, and it is observed in patients suffering schizophrenia, Alzheimer's and other kinds of dementia (Cosentino and Stern 2005). Anosognosia is a sort of unawareness of one's own cognitive and functional impairment. Patients with this symptom are invited to discover their deficit by indirect means, because they deny their deficit and exhibit a very poor insight regarding their cognitive status.

According to the hypothesis proposed by most authors on the subject, metacognition is a faculty composed by two more basic functions: *monitoring* the cognitive status, and *controlling* the ulterior processing of the given information. For example, refraining from action may be one of the possible outputs of the control mechanism (as well as refraining from believing). Monitoring and control do not constitute two separate functions aimed at belief and action, respectively, but rather two necessary aspects of any meta-competence, even though they operate at conscious levels that are not too explicitly represented. This double aspect could, therefore, be considered as establishing some symmetry between belief and action at the basic level of their production: reflective knowledge would require both monitoring and control.

Maybe one would argue against the strategy of resorting to empirical findings when the issue is of an exclusive conceptual nature. I will not dispute now about the required degree of naturalistic considerations in epistemology. My point is that even in spite of the fact that the most basic levels of cognitive faculties worked well, we would still be facing the problem of a potential lack of integration. In this way, metacognition furnishes us with a framework where the integration problem can be formulated without reaching such higher levels.

6 Degrees of Personal Integration

The integration problem arises when we focus our attention on the personal level. Breyer and Greco 2008, consider the integration of cognitive competencies in these terms: "…a cognitive disposition counts as part of S's cognitive character (…) only if it is well enough integrated with other of S's cognitive dispositions" (p 178). They also consider that integration of cognitive dispositions and character jointly pertains to "belief ownership": "ownership (…) is essential to subjective justification" (p 178). I contend that this triple conjunction of cognitive dispositions, integration, and ownership can only be possible at a personal level. By personal level I do not necessarily mean here scaling to the highest levels of deliberative stances, but rather looking for a more constitutive stage where the issue is the overall equilibrium of the system. The idea is that a person is healthily constituted when her mental faculties are reasonably functioning and acceptably coordinated (obviously, disabilities, local malfunctionings, etc. cannot be discarded). In this sense, metacognition is a mechanism that can accomplish its function only in a systemic way; that is, it works insofar as other cognitive mechanisms are coherently at work. The coherence this integration requires is unlikely to be mere logical or informational coherence. Actually, we would say that the system demands a sort of metacoherence. Metacoherence, as metacognition, does not imply necessarily second-order attitudes. Breyer and Greco suggested that "cognitive integration does not result from either reflective endorsement or coherence among belief contents, but from the cooperative causal interaction of relevant cognitive dispositions" (Breyer and Greco 2008, p. 183). Furthermore, in the task of knowing, working with a higher degree of harmoniously systemic coordination is a requisite in order to engage the overall cognitive system in such a demanding task. This requirement derives from the very nature of knowledge itself because, according to Greco, knowledge in virtue epistemology is a kind of achievement springing from the agent's character. However, it is far from clear that metacognition and metacoherence can be considered as components of the agent's character, for such properties are rather part of her cognitive endowment at a more constitutive level than character. Character consists of stable dispositions acquired through the agent's cognitive history, but metacognition and metacoherence between sub-systems are, in a certain sense, orthogonal to character. Although metacognition reaches the personal level, it also drills down to deeper layers of the agent's cognitive constitution.

The conceptual point here is that several functions must be well-integrated, but also that the first-person perspective must be involved. I would claim that systemic coherence, as described from a third-person perspective, will not be sufficient. Something in the working system is required to ensure that it is the agent, and not a mere part of him, who is engaged in the task. In this regard, metacognition, when involved in reflective knowledge, always entails the first-person perspective regarding the agent's own cognitive processing. Hence, metacoherence, integration, and first-person perspective either hold together or they collapse together. Certainly, it is not difficult to grant that conscious deliberation is a symptom of autonomy in normal cases of cognitively healthy human adults. However, the question is whether there can be autonomous systems in lower steps of cognitive development. Although metacognition is a part of an assembled bundle of faculties that integrates an agent, when metacognition works, it indicates that such compound and the subject are the same agent. In cases of knowledge, the knower is self-ascribing a competence to herself, though such self-ascription need not be processed at higher deliberative layers. As I previously said, metacognition has the functions of monitoring and controlling the cognitive status when the organism confronts a particular task, but such functions can be quite automatically performed. It is rather irrelevant whether this double function is performed by a single mechanism or by two different ones. By contrast, what is relevant is that metacognition performs the twofold function as part of a singular but compound cognitive task. We should note that even though monitoring can be considered as a more passive engagement of the organism, such is not the case for control, as control spontaneously and actively engages the overall organism in the production of a right outcome. The function of metacognition would then be the evaluation of the organism's ability to deal with difficult cognitive tasks. It does not matter if deliberation precedes this function or not. The system works properly as far as it is able to detect a state of dangerous ignorance, that is, when it detects that the knowledge available is not enough.

Let us now compare this function with the William Tell case at the very moment of aiming at the apple on his son's head (instead of aiming at his son). In (a), Tell thoroughly assesses his action and consciously decides that the shooting will be safe. In (b), Tell does not even consider the question and simply shoots. He is very confident in his skills to hit the target. In both scenarios, reflective knowledge (at least metacognition) is involved, but in (b) awareness of the situation does not necessarily imply an explicit discursive deliberation. Tell will be trustworthy insofar as his control system works and he will be able to refrain from shooting when his accuracy is jeopardized. Therefore, an acceptable connection between the quality of agency and autonomy could be established by the following requirement:

> *Autonomy*: Autonomy involves a sufficient function control in order to refrain from continuing troubling processes.

This requirement does not demand, although it does not exclude, conscious deliberation between alternative possibilities. A minimal frontier of agential quality, as compared for instance with compulsive behaviour, is claimed here in the general capacity of refraining to act. The second point is that only a first-person perspective guarantees this required agential engagement in cognitive tasks.

In the case of metacognition, normal subjects report having a "feeling of knowing" when the system is working well. Consider, for example, the tip-of-the-tongue cases when, after some efforts to recall a name, one is able to retrieve it. Then, a feeling of knowing overwhelms us, and this feeling signals that the task is accomplished. However, the philosophical meaning of this feeling is not easy to assess. On the one hand, it can be considered as a symptom of a proper working of one's metacognitive skills. She who feels that she knows apparently achieves a fuller cognitive state than when just a mere belief occurs to her. I am not making an epistemological claim that knowledge requires this feeling to account for reflective knowledge (perhaps it is a mere psychological indicator), but certainly its occurrence tells us that something is happening, namely, that the agent is experiencing the ownership of knowledge.

To take stock of our steps, the personal level can be involved differently in this increasingly significant list: "metacognition", "feeling of knowing", "first-person perspective", and "reflective stance". Each stage amounts to a higher degree in the system's integration regarding its agential status. It is likely that each of them occurs at different diverse moments in the cognitive process. But the important point is that epistemic agency is characterized by a certain form of integration of the cognitive and executive systems, i.e., it is an agential shaping of behaviour in such a way that the subject turns out to be the owner of her outcomes. Even if this task does not reach a full deliberative and conscious status, agency is still supported by the self-confidence agents have in their capacity to attain their goal. The personal level is openly expressed in intentional action, as assertive discourse is, but it is also required in several tasks which can be performed in more implicit ways. In the case of knowledge, the personal level is required for a knowing subject when she ascribes to herself an epistemic competence. It should be now noted that the implicit/explicit dichotomy does not equal the personal/sub-personal dichotomy. The personal level is required when meta-coherence is at stake. William Tell, facing a dreadful experience that demands his maximum self-trust, is a paramount example of this requirement of a system's full integration. Although a fully explicit reflective stance can be a further stage for an agent who deliberates in an open, as well as internal forum, we should distinguish between the coherentist claim that full consciousness is required for both a full justification and this view about the first person engagement as a necessary trait of an integrated agent. Meta-coherence, in our sense, is a structural property of the personal level, and it need not imply a full Kantian approach. As it is well-known, this approach demands to not simply obey a rule but to do so because of the concept it embodies. Less-than-Kantian approaches such as mine simply require a good integration for a sufficient agency to occur, and in this case, it is an epistemic one.

7 Epistemic Agency and Personal Engagement

The requirement of a personal level for integration has to do with the status of being an agent. A subject enjoys a well-integrated status when she behaves as an agent in the achievement of knowledge. But knowledge, as opposed to mere information processing, implies the presence of an agent in the scene of theoretical or practical processes. The reason is that both share the common nature of agency as the distinctive trait expressing human autonomy. It means a capacity for self- determination in a particular state (in the world, as well in the own subject) as a result of competencies the subject possesses *qua* subject.

Thus, knowledge implies the subject's expression of being able to determine a distinctive mental state: the state of believing as the product of her own competencies. It might be that this judgment takes the form of an internal assertion, as well as that of an open avowal. In such cases, the outcome has been preceded by an explicit judgment, but it need not be so for the agent's self-expression. It will suffice that a well-integrated person makes up her mind even in an implicit way. Certainly, beliefs are very often not the products of such self-determination processes, for they come from perceptual, emotional, or automatic cognitive mechanisms. It is indeed possible that these beliefs deliver information and that they can even attain the status of knowledge. But in these cases, the merit is attributed to those component faculties and not to the subject itself. As in knowledge, in the case of action, there are forms of behaviour that do not attain a full intentional status: they can be forms of behaviour resulting from skills or ways of knowing how that do not call for a full agential engagement. We take for granted that this automatic behaviour constitutes a usual way of acting in daily life. The agent need not always express herself as an agent unless relevant circumstances demand such a higher manifestation of agency. Personal engagement is only needed when the agent's own position is at stake. Sometimes, defining her position elicits an explicit judgment in the form of an assertion or decision, but what is relevant is that this need is only activated when the agent is forced to assess the possibilities she has to attain her goal. When such a degree of control appears, we say that she is responsible, and becomes praiseworthy.

From a normative perspective, it is essential to consider such state-determinations as attaining a certain normative level of success. A success is a property that depends on a broader context in a creature's life span. Dewey explained in *Art and Nature* that "ends", in this normative sense, are characteristic points of special relevance to life. For example, births or deaths are indifferent points in the course of physical chains of causes and effects, but it is quite apparent that, in the frame of the life of living beings, these points are of an exceptional relevance. Similarly, agential self-determinations of the agent's mental states are of normative relevance when they are also relevant points that can be both qualified and certified as successes.[3]

In order to elaborate my understanding of reflective knowledge, I would underline that knowledge as an outcome is more valuable than true belief, and that this higher value is significant for agency itself. To reach knowledge is something more than believing truly. Thus, in order for a belief to reach the status of knowledge, it must be an achievement owed to the agent's own competencies. Moreover, such a degree of accomplishment claims a subject that he is worthy of it. What is significant here is that such achievement does not merely amount to arriving at the proposed end, but to attaining these ends from the active engaging of a well-integrated subject in the task of knowing (Broncano and Vega 2011).

How this way of considering knowledge gives us one of the best justifications for Virtue Theory is telling, as it provides a causal background for an agential concept of subject. In other traditions, such as those that obey more intellectualist approaches, the notion of the subject becomes metaphysical in a very strange way: although a substance is considered to be able to exhibit spontaneity, this spontaneity seems to come from nowhere. In contrast with this, Virtue Theory proposes that reflective competencies are the way in which an agent becomes responsible for the quality of her epistemic views; she becomes, thus, a sort of causal singularity in the universe. She is able to attain such a status because she controls the risk of believing, that is to say, she dares to believe given her own epistemic position in the world. This capacity to take epistemic risks turns the agent into an autonomous believer, namely, into an epistemic agent.

Now, how would assuming epistemic risks provide a criterion for epistemic agency? I would answer that the decisive aim of agency is to grant that a particular move makes sense for the subject. David Velleman has convincingly argued that the human way of acting is ordered as a self-understanding of the course of action one is choosing. For example, let me point out the following explanation to how an agent chooses a reason for deciding to do something:

> I believe that the process of improvisational self-enactment constitutes practical reasoning, the process of choosing an action on the basis of reasons. Why do I think that the self-enactor chooses his action? Because it is his idea, which he puts into action in preference to other ideas that he might have enacted, if this one hadn't made more sense. Why do I think that he chooses for reasons? Because he chooses his action in light of a *rationale* for it, which consist in consideration in light of which the action makes sense. (Velleman (2009), p. 18)

---

[3] Dewey deeply observes: "It is not easy to distinguish between ends, as *de facto* endings, and ends as fulfilments, and at the same time to bear in mind the connection of the latter with the former. We respond so directly to some objects in experience with intent to preserve and perpetuate them that it is difficult to keep the conception of a thing as terminus free from the element of deliberate choice and endeavour; when we think of it or discourse about it, we introduce connection" Dewey (1929), p. 111.

In contrast with Velleman's interpretation, it could be imagined that taking risks is, from the agential point of view, like mechanically trespassing some threshold of probability. But it could be argued that any mechanical engine, still lacking the status of agency, could take this chance. A reflective agent, by contrast, is one for whom it makes sense to dare to have certain beliefs or to take a certain decision. A reflective stance is not, then, a sort of faculty added sequentially to previous springs of beliefs or decisions. My previous discussion on metacognition showed that this faculty could also be possessed by different species of animals. If agency is a distinctively human feature, it must consist of something else: my contention has been that it consists of the expression of knowledge and action by the subject in question. Such expression can only occur when a well-integrated agent takes a course of action that makes sense to her. What is the risk she takes? Maybe it is increasing the chance of not being able, and failing, to achieve the aim, given the agent's cognitive resources and competencies, and given the circumstances of the task undertaken. The risk of failure is, therefore, a risk the agent must calibrate given her self-confidence, as well as her objective capacity to undertake her task.

As William Tell faces his tragic performance, an epistemic agent must decide to accept a belief that can be of an indeterminate relevance for other aims, theoretical or practical, but that is of a constitutive centrality to the task of knowing. Recall dramatic scenarios, as when doctors answer the question of their fearful patients: "Do you know that it is cancer?" The degree of engagement and attention the epistemic agent devotes to the quality of her epistemic perspective can depend on the type of demands present in the question involved, but in any case, the agent must calibrate her own powers in order to make an assertion. The extent to which whether the answer constitutes an epistemic achievement is something that would depend, not only on the ethical or practical relevance of the question, but also on the virtuous character of an agent who dares to believe.

References

Axtell, G., and J.A. Carter. 2008. Just the right thickness: A defense of second-wave virtue epistemology. *Philosophical Papers* 37(3): 413–434.

Bernecker, S. 2008. Agent reliabilism and the problem of clairvoyance. *Philosophy and Phenomenological Research* 76(1): 164–172.

Breyer, D., and J. Greco. 2008. Cognitive integration and the ownership of belief: Response to Bernercker. *Philosophy and Phenomenological Research* LXXVI/1: 173–184.

Broncano, F., and J. Vega. 2011. Engaged epistemic agents. *Critica* 43(128): 55–79.

Carruthers, P. 2009. How we know or own minds. The relationship between mindreading and metacognition. *Behavioral and Brain Sciences* 32: 121–182.

Cosentino, J., and Y. Stern. 2005. Metacognitive theory and assessment in dementia. Do we recognize our areas of weakness? *Journal of the International Neuropsychological Society* 11: 910–919.

Dewey, J. 1929. Nature, ends and histories. In *Experience and nature*. New York: Dover. 1959.

Domsky, D. 2004. There is no door: Finally solving the problem of moral luck. *The Journal of Philosophy* 101(9): 445–464.

Enoch, D. 2010. Cognitive biases and moral luck. *Journal of Moral Philosophy* 7(3): 372–386.

Fernández-Duque, D., J.A. Baird, and M.A. Posner. 2000. Executive attention and metacognitive regulation. *Consciousness and Cognition* 9: 288–307, p. 289.
Foote, A.L., and J.D. Crystal. 2007. Metacognition in rats. *Current Biology* 17(6): 551–555.
Frankfurt, H. 1988. *The importance of what we care about*. Cambridge: Cambridge University Press.
Greco, J. 2004. How to preserve your virtue while losing your perspective. In *Ernst Sosa and his critics*, ed. J. Greco. Oxford: Blackwell.
Greco, J. 2010. *Achieving knowledge: A virtue-theoretic account of epistemic normativity*. Cambridge: Cambridge University Press.
Harris, P., and W. Middleton. 1994. The illusion of control and optimism about health: On being less at risk but no more in control than others. *British Journal of Social Psychology* 33: 369–386.
Klein, C.T.F., and M. Helweg-Larsen. 2002. Perceived control and the optimistic bias: A meta-analytic review. *Psychology and Health* 17(4): 437–446.
Koren, D., L.J. Seidman, M. Goldsmith, and P.H. Harvey. 2006. Real-world cognitive and meta-cognitive in schizophrenia, a new approach for measuring (and remediating) more "right-stuff". *Schizophrenia Bulletin* 32: 310–326.
Koriat, A. 2000. The feeling of knowing: Some metatheoretical implications for consciousness and control. *Consciousness and Cognition* 9: 149–171.
Kornblith, H. 2002. *Knowledge and its place in nature*. Oxford: Oxford University Press.
Kornblith, H. 2010. What reflective endorsement cannot do. *Philosophy and Phenomenological Research* 80(1): 1–19.
Metcalfe, J. (ed.). 1994. *Metacognition: Knowing about knowing*. Cambridge, MA: MIT Press.
Montaigne, M. 1958. *The Complete Essays of Montaigne*. Trans. Donald L. Frame. Stanford: Stanford University Press.
Necka, E., and J. Orzechowski. 2005. Higher-order cognition and intelligence. In *Cognition and intelligence*, ed. R. Stenberg and J. Pretz. Cambridge: Cambridge University Press.
Proust, J. 2007. Metacognition and metarepresentation: Is a self-directed theory of mind a precondition of metacognition? *Synthese* 159: 271–295.
Proust, J. 2009. The representational basis of brute metacognition: A proposal. In *The philosophy of animal minds*, ed. R.W. Lurz. Cambridge: Cambridge University Press.
Royzman, E., and R. Kumar. 2004. Is consequential luck morally inconsequential? Empirical psychology and the reassessment of moral luck. *Ratio* 17(3): 329–344.
Smith, D. 2005. Studies on uncertainty monitoring and metacognition in animals and human. In *The missing link in cognition. Origins of self-reflective consciousness*, ed. H.J. Terrace and J. Metcalfe. Oxford: Oxford University Press.
Sosa, E. 2003. Knowledge, animal and reflective: A reply to Michael Williams. *Proceedings of the Aristotelian Society* 77(Suppl): 113–30.
Sosa, E. 2004. Replies. In *Ernest Sosa and its critics*, ed. J. Greco. Oxford: Blackwell.
Sosa, E. 2007. *A virtue epistemology. Apt belief and reflective knowledge*, vol. I. Oxford: Oxford University Press.
Sosa, E. 2009a. Respuestas a mis comentadores. *Teorema* XXVIII/1: 112–124.
Sosa, E. 2009b. *Reflective knowledge. Apt belief and reflective knowledge*, vol. II. Oxford: Oxford University Press.
Sosa, E. 2011. *Knowing full well*. Princeton: Princeton University Press.
Velleman, D. 2009. *How we get along*. Cambridge: Cambridge University Press.
Weinberg, J.M., S. Nichols, and S.P. Stich. 2008. Normativity and epistemic intuitions. In *Experimental philosophy*, vol. 29, ed. Joshua Knobe and Shaun Nichols, 429–460. Oxford: Oxford University Press.
Zagzebski, L.T. 1996. *Virtues of the mind: An inquiry into the nature of virtue and the ethical foundations of knowledge*. Cambridge: Cambridge University Press.

Success, Minimal Agency and Epistemic Virtue

Carlos Montemayor

1 Introduction

The consensus in contemporary epistemology seems to be that agents should play a much more important role in accounts of knowledge and knowledge attributions. Virtue epistemology is committed to this principle, by making epistemic agents the main target of epistemic evaluations. Epistemic agency and virtue epistemology are opening new possibilities for unified treatments of topics that seem hopelessly unconnected. For instance, some authors think that virtue epistemology is the only way to provide a theory of knowledge that is capable of solving the value problem. (See for instance Greco 2010) Moreover, reliabilist and internalist accounts of knowledge and justification seem to be incomplete without the explicit incorporation of epistemic agency.

It has also been argued that the two master intuitions in epistemology, driving our assessments of knowledge in different counterfactual situations, are the anti-luck and ability intuitions. (See Pritchard 2012) Abilities give rise to epistemic responsibility, which seems to be crucial to explain many anti-luck intuitions in, for instance Gettier problems where the epistemic agent has justified true beliefs, not because of her epistemic abilities, but because of some other accidental factor.

Another source of disagreement with traditional epistemic accounts of knowledge that fuels the interest in virtue epistemology is the pragmatic dimension of knowledge attributions. Jason Stanley (2005) calls these pragmatic accounts 'anti-intellectualist' views of knowledge. As Stanley says, the main idea behind these accounts is that knowledge is to be *constitutively connected to action*, in the sense that one should act only on what one knows. This notion is related to the virtue epistemology principle that agents and their epistemic achievements must be the main subjects of epistemic evaluation. Thus, it seems clear that there are

C. Montemayor (✉)
San Francisco State University, San Francisco, CA, USA
e-mail: montema@sfsu.edu

A. Fairweather (ed.), *Virtue Epistemology Naturalized: Bridges Between Virtue Epistemology and Philosophy of Science*, Synthese Library 366, DOI 10.1007/978-3-319-04672-3_5,

powerful theoretical motivations to study virtue epistemology in detail, in order to achieve a better understanding of its implications and eventually be in a position to assess if it really provides the best account of knowledge.

A central problem that any virtue epistemology must confront concerns the *type* of agency that is required for epistemic achievements and knowledge attributions. For the sake of conciseness, I shall focus on reliabilist versions of virtue epistemology. The type of problem I will raise extends to non-reliabilist accounts of epistemic virtue, because it is based on psychological findings on action selection and motor control that *any* naturalized version of epistemic virtues must account for. Although the findings on motor control and action selection have important implications for naturalistic accounts of virtues, they have received very little attention in the philosophical literature. This paper aims at solving this problem, by demonstrating the relevance of these findings while providing an account of minimal agency that adequately satisfies the constraints of a naturalized virtue epistemology for knowledge.

Some reliabilist theories of epistemic virtues require a reflective component for knowledge. For instance, Ernest Sosa distinguishes between two types of epistemic agency (a non-reflective and a reflective type), which yield two types of knowledge attributions: animal and reflective knowledge (See Sosa 2009). According to Sosa, it is only when agents achieve reflective knowledge that they display the highest, most reliable and apt form of knowledge (i.e., full knowledge). Reflective knowledge is a meta-competence (which presumably involves, at least, metacognition and metarepresentation) that allows for a refined epistemic achievement in which object-based, first level animal knowledge is assessed in terms of the way in which such knowledge was produced, taking into consideration the overall epistemic situation of the agent. Thus, reflective knowledge is supposed to be based on a reflective kind of *justification* concerning the aptness of the belief given the epistemic situation of the agent, which animal knowledge cannot provide.

In contrast, John Greco (2004, 2010) argues that a reflective requirement on knowledge imposes unrealistic psychological demands on epistemic agents. One can understand this claim as follows: Sosa's account of full knowledge introduces distinctions where there are none (or where there *should* be none). For instance, when a child accurately perceives a red object and forms the belief that there is a red object in front of her, she seems to *know* that there is a red object in front of her and we *should* attribute such knowledge to *her*, because this knowledge is an achievement of her reliable perceptual skills. The same holds for animals with the capacity to form basic beliefs. Sosa grants that this *is* a case of knowledge (animal knowledge), but the question is, why would a reflective component be required for her to *fully* know that there is a red object in front of her when she seems to know everything there is to know about the case (i.e., the color of the object)?

Clearly, this challenge concerns the *empirical plausibility* of the reflective requirement on full knowledge. Greco seems to be suggesting not only that this requirement introduces unnecessary distinctions, but also that it is empirically inadequate, in the sense that it *may* be incompatible with findings in psychology, because it imposes unrealistic demands on knowledge. This would mean that virtue epistemology for full knowledge may not be susceptible of being naturalized. However, this challenge does not seem decisive because one may point at psychological findings on complexity or perceptual monitoring in many basic epistemic processes, which may *resemble* reflection, for instance, construed as memory based or navigational metacognitive constraints on object-level perception. Obviously, for this rejoinder to work, one must have a very clear notion of the type of agency that reflection demands.

It seems that reliably detecting features of the environment may be one of many types of epistemic achievements that are required for full knowledge. Sosa illustrates this with his example of Diana, the huntress, who not only has to be successful in hitting targets (and reliably so), but also *selective* with respect to the type of target she *should* hit. So not any metacognitive process will do. For instance, just monitoring perceptual processes will not do. Selection based on *criteria* seems to be an important epistemic achievement, which is fundamental to assess evidence and withhold judgment. This is what Sosa calls reflective justification. For the purposes of this paper, I shall call the reliable object-based knowledge that Diana uses to succeed in hitting targets her 'motor control' skills, and the reflective knowledge that she uses to selects appropriate targets her 'action-selection' skills.[1]

Irrespectively of how one defines 'reflection,' it is clear that it involves control, selective criteria and a higher degree of voluntary involvement on the part of the epistemic agent (certainly more involvement than successful behavior based on motor control ability requires). One may insist that reflective justification may occur implicitly, unconsciously or without full conscious access to normative criteria for selection. Requiring full conscious control of the agent imposes very constraining requirements on knowledge (requirements that cognitive psychology has proven to be not only excessive, but simply *false* about human cognition, particularly concerning perception). So Greco seems to be right in his criticism that reflective justification at least *seems* excessive. Can one make this criticism more compelling, in a way that it is clear that reflection, even of an implicit or unconscious kind, is *incompatible* with the main findings in cognitive psychology? If successful, the challenge I present here will demonstrate this incompatibility.

Before proceeding, I shall distinguish the problem I will raise here from a recent criticism against reflection that also tries to demonstrate its empirical inadequacy. Hilary Kornblith (2010) argues that reflection is a problematic requirement for knowledge because instead of increasing reliability and success (as Sosa thinks) it actually *reduces* or *interferes* with them. I partly agree with Kornblith, but I will argue that this is just one aspect of a larger problem, which is that the reliability of motor control skills is largely *dissociated* from action selection and, thereby, from any form of reflection. Thus, the problem is more severe because the empirical evidence suggests that reflection in action selection either interferes with reliability or is epistemically irrelevant.

The psychological findings I am about to present suggest a tradeoff: the more minimal the sense of agency, the less plausible the postulation of a

[1] I am borrowing these terms from the psychological literature (see Rosenbaum 2002).

reflective-requirement for knowledge; and the more enriched the sense of agency, the less epistemically relevant the characterization of virtues for knowledge. So unlike Kornblith's challenge, which concerns decreased *reliability* produced by introspective-like processes, this challenge shows that besides diminished reliability, motor control and action selection (even of an implicit kind) cannot be consistently integrated into an empirically plausible virtue epistemology for knowledge. Or, at the very least, that such integration seems to be highly problematic in the light of the evidence.

Section 2 explains the contrast between animal knowledge and reflection, describes the theoretical role that it is supposed to play, and highlights the central issues that a naturalized theory of epistemic virtue needs to address. Section 3 explains the distinction between motor control and action selection, which underlies the tradeoff between robust conscious and reflective agency, on the one hand, and reliable agency for motor control, on the other. This is the challenge that naturalized versions of reflection must confront. Finally, Sect. 4 presents objections and replies.

2 Agency and Success

Many cases of knowledge involve success from epistemic ability in a way in which the connection with action is very straightforward. When one is driving to an important meeting and one's copilot insists that she knows how to get there, but keeps getting it wrong, it seems one should conclude that she does not know how to get there. In such a case, it is best to rely on a computerized navigational system, like GPS. Standard process reliability about knowledge has no problem accounting for the type of knowledge one achieves by following the information from the GPS. For a reliabilist virtue epistemology, however, knowledge must be attributable to the agent, based on her epistemic ability, such as being a reliable perceiver and interpreter of GPS information.

How much involvement of the agent is required for an empirically and theoretically plausible *reliabilist* version of virtue epistemology? This is a critical question. It seems that paradigmatic cases of perceptual belief involve *very little* involvement of the agent in the sense that her perceptual abilities give her immediate knowledge of the environment without further reflective or conscious effort on her part. The same is true about the navigational abilities that produce spatiotemporal knowledge. The repertoire of perceptual-like reliable abilities of an agent seem to satisfy the requirements for knowledge, even if no conscious or reflective effort of the agent is involved. These *stable dispositions* produce successful behavior across relevantly similar environments, and are attributable to the agent because they are *her* perceptual abilities. These are the abilities that constitute what I am calling her *motor control* skills.

As mentioned, Diana the huntress uses motor control skills to hit targets. She forms immediately justified beliefs, or at least epistemic entitlements, about the position of the targets (basic perception) and through skillful motor control, she deploys her knowledge of how to best balance the tension of the bow, taking into account atmospheric conditions, distance, etc. (specialized knowledge based on memory, experience, etc.). She can do this almost mechanically and be very reliable in hitting her targets successfully. All this knowledge about archery is attributable to her because it is non-accidentally based on her skills, and not dependent on luck or other contingencies of the environment. If Diana were a good archer because someone else is always helping her, or because she has been so far incredibly lucky by having favorable winds every time she shoots, then she would *lack* such knowledge (and we should not attribute such knowledge to her).

Psychologists have found that perceptual and motor control skills are indeed very reliable. Perceptual knowledge is based on stable dispositions to respond to stimuli across relevantly similar environments, which play the important role of satisfying goals and meeting accuracy conditions for veridical content. The same is true about procedural knowledge and other forms of implicit memory, which require a very minimal type of agency (for instance, this type of knowledge can be achieved unconsciously). I will explain this in more detail in the next section. What is important to highlight now is that knowledge for motor control is reliably produced by abilities attributable to agents (at least to a high degree), without the requirement that those agents be consciously controlling these epistemic processes. Thus, knowledge for motor control is a *first level* type of luck elimination for epistemic goals, which is perfectly compatible with a fully naturalized theory of epistemic virtues.

But Diana is not only an extremely reliable archer; she is also a fantastic hunter. Her performance as a hunter cannot be fully captured by her first level reliability, as a good archer, because formidable hunting performance must manifest her *selective judgment* regarding adequate targets. Hitting any target would be detrimental, waste energy and reduce first level reliability by introducing an eagerness to hunt that may be incompatible with precise archery. She must hit specific prey and *refrain* from hitting anything else, thereby reflecting on the adequacy of her decisions. We would not attribute this reflective knowledge to Diana if she decided to hit prey by flipping a coin or by basing her decisions on the instructions of someone else. Thus, her knowledge for action selection is a *second level* type of luck elimination for epistemic goals, attributable to her because of her selective skills.

Two questions arise for the project of giving a naturalistic account of epistemic virtues in terms of first and second level abilities. First, how can the minimal agency required for successful first level goals be *cognitively integrated* with a more enriched type of agency that is relevant for different types of goals? It seems clear that the reflective skills that Diana manifests, regarding her overall epistemic situation with respect to prey, depend critically on her voluntary intervention. It is true that one could think of these skills as highly procedural ones (she may detect targets and reflect on the adequacy of shooting prey quite quickly and by means of implicit processing). But a naturalized account of the cognitive integration of motor control and action selection depends on empirical evidence, rather than intuition alone.[2]

[2] For the importance of cognitive integration to solve problems that standard reliabilism faces see Greco (2010).

In the next section, I argue that the evidence strongly suggests that these two types of skills cannot be cognitively integrated into stable, systematic or coherent epistemic virtues.

A second and related question is, how "far up" one needs to go to find the type of agency required for reflection? This is not a trivial question, since most examples of reflection involve *conscious deliberation*. Take the case of Diana. She can, presumably, hit targets very rapidly relying on epistemic processes of which she is unaware, which constitute her motor control knowledge. But this is not a plausible characterization of her reflective, *action selection* processes. She is clearly involved in the selection of actions (hitting adequate targets) in a much more robust and conscious way than when she is just hitting targets based on her perceptual knowledge.

It is useful to think of this problem in terms of knowledge attribution. The case of Diana is helpful to draw analogies with epistemic processes, but it is not very clear how knowledge attributions would work in general. It is one thing to say that Diana clearly knows how to hit targets because of her first level knowledge, and that without the second level knowledge her hunting would be deficient (knowledge for hunting should not be attributed to her in the absence of reflection). But in cases of perceptual knowledge, why would *reflection* be relevant? The idea is that, like in the case of Diana, in standard perceptual knowledge attributions one is attributing animal knowledge to agents. But only those who reflect on the reliability of their senses, as well as the adequacy of their overall epistemic situation, deserve full knowledge attributions.

Why should full knowledge attributions require reflection? Any naturalized account of virtue epistemology must answer this question by appealing to findings in psychology, in order to verify if these standards for attributions are realistic. As mentioned, Sosa suggests that we do not need to go very far up to find the type of reflection required for full knowledge, since reflection could be unconscious. But is this compatible with the empirical findings?

Kornblith (2010) suggests that when there is cognitive integration between these types of processes the result is diminished reliability. Kornblith is referring to processes that concern introspection, and forms of reflection about one's own epistemic status, which have been proven to be highly unreliable.[3] The proponent of reflection may respond by saying that these processes are *too high up* the scale of reflection, and that more minimal forms of reflection would do the job. The challenge I present in the next section shows that this response is problematic because animal knowledge and reflection cannot be fully integrated into a repertoire of stable epistemic virtues. Thus, a coherent and naturalized account of epistemic reflective virtues for knowledge does not seem forthcoming. Any form of action selection, no matter *how minimal*, is epistemically dissociated from motor control and perceptual-like knowledge, which impedes the kind of cognitive integration required for epistemic virtues.

[3]The classic findings on the unreliable nature of introspection are by Nisbett and Wilson (1977), which have inspired decades of voluminous research concerning unconscious biases. Findings concerning the limits of introspection also abound, for instance, with respect to blindsight, see Weiskrantz (2009).

3 Success and Cognitive Dissociation

The distinction between motor control and action selection is crucial for the challenge I am about to present for reliabilist-naturalistic accounts of reflection (which will also question naturalistic versions of responsibilist accounts). The problem is that any minimal construal of reflection, which for instance may be procedural and automatic, would have to account for reflective selection. Action selection for basic tasks (for instance, choosing one option over the alternative, even though one is *inclined* to do both) seems to be the most adequate candidate for reflection. Obviously, the epistemic processes required for reflection would involve minimal (not fully conscious and available to introspection) selective choices, such as withholding judgment over endorsing a belief, which would, presumably, increase reliability.

However, the findings on action selection suggest something very different: the more minimal the sense of agency, the less plausible the postulation of a reflective-requirement for knowledge, even if construed as procedural or automatic action selection; on the other hand, the richer the sense of agency, the less epistemically relevant the characterization of cognitive virtues for knowledge. Also, a rich sense of conscious agency generates problems with reliability, as Kornblith (2010) points out. In any case, the problem is not just one of diminished reliability, which could be fixed by finding other forms of cognitive reflection with less introspective constraints and more connection with perceptual processing. The real problem is, as mentioned, that any form of action selection seems to be epistemically dissociated from reliable knowledge for motor control.

To illustrate the distinction between motor control and action selection, it is useful to start with a non-perceptual case (or at least not *strictly* perceptual), in order to demonstrate the generality and importance of this distinction, and then move to more distinctively perceptual cases, with which reliabilists are much more familiar. Take for instance the knowledge of language. Knowing how to speak a specific language requires, fundamentally, knowledge of its syntax. It also requires knowledge of how to gesticulate, generate the specific sounds associated with letters, words, etc. Syntax processing and other motor control aspects of linguistic representation are beyond our conscious reach. One easy way to see that this is the case is that linguists cannot analyze the syntax of a language merely by introspection. It is true that intuitions about grammar guide their research, but it is also true that the knowledge of language structures that speakers manifest when they speak is beyond their conscious grasp, and this includes expert linguists.

Syntax is the systematic manipulation of information in terms of the strictly formal characteristics of linguistic stimuli. A string of symbols or sounds are processed according to formal distinctions such as subject, predicate, noun, adverb, etc. Meaning is embedded, modified, and composed according to these rules. But this is not only unconscious cognitive processing, it is also strictly sensory-motor knowledge, and no form of reflection *must* occur for speakers to have this knowledge. In the case of speech production, one is conscious of the meaning of words, but not of the syntactically driven articulatory code that is involved in gesticulation and sound intonation. Similarly, one is consciously aware of the meaning of sentences, but not of how their syntax is processed by the brain. This distinction holds for *any kind of action,* or behavior. Actions have an unconscious motor control component and a conscious action-selection component with phenomenal perceptual features, such as the meaning of words.

Now think of knowledge attributions of language. Syntax processing, articulatory codes for gesticulation and other formal skills for language manipulation are certainly abilities of speakers, and speakers know a language *because* of these abilities. Children certainly have these skills and when they manifest them we attribute knowledge of language to them. Knowledge of language requires syntax, but knowledge of syntax does not seem to require any kind of reflection or selection of choices that are evaluated by speakers. A vast amount of psychological findings actually show that the motor control aspect of language, including syntax processing, *cannot* depend on any type of reflection or semi-conscious monitoring.

Two aspects of syntax processing are crucial to understand the importance of motor control for *reliability*, which have implications for the challenge that it cannot be cognitively integrated with selective processes that require reflection. First, knowledge of syntax concerns the type of reliable process that leads to immediate perceptual beliefs. Just as seeing a red object in the vicinity leads to the kind of perceptual knowledge that requires minimal reflective agency (i.e., knowledge about the color of objects), knowledge of syntax is dependent upon reliable processes that require minimal agency, and *no kind of reflection or monitoring*. Thus, it seems that knowledge of syntax should qualify, oddly, as *animal* knowledge (even though it seems to be a *uniquely human* trait).

This shows that the distinction between conscious or unconscious processing is *not* the main issue at stake for naturalized virtues that include reflection. Conscious perceptual knowledge is produced with a very minimal sense of reflective or selective agency, while knowledge of syntax is produced unconsciously. Thus, the main problem is that conscious or unconscious *reflection* is not susceptible of cognitive integration with reliable minimal agency for motor control. This is why the present challenge is much broader and decisive than challenges that depend on conscious reflection construed as introspection (such as Kornblith's 2010).

The second noteworthy aspect of syntax processing is, as mentioned, that it seems to be one of the most distinctive and uniquely human epistemic processes, regardless of whether one is evaluating children or adults. Unconscious syntax processing is a formal, complex, and systematic cognitive achievement that is completely dissociated from any type of introspective reflection, no matter how minimal. Syntax processing is particularly interesting because it is partly responsible for the language capacity that makes us humans and distinguishes us from most species on earth. It may not be reflective knowledge (it does not even seem to involve metacognition), but it is certainly a type of knowledge that is based on highly sophisticated cognitive skills, which do not depend in any way on reflection.

One may object that this challenge to reflection is theoretically biased towards a modular, unconscious, sub-personal and mechanical account of knowledge that most epistemologists would find unpalatable. Moreover, one may also reject this challenge on the grounds that it is unrealistically minimal, even for exclusively psychological purposes. With respect to the unpalatable consequences of the challenge, one must keep in mind that a naturalistic account of epistemic virtues must be based on psychological research, such as the research concerning syntax processing, which may conflict, like in this case, with the commitments and desires of epistemologists. But the challenge has merit because of the apparent theoretical bias towards massive modularity.

However, this objection loooses its force once one considers carefully the distinction between different types of cognitive integration. It is true that modular-like processes are inadequate to account for the type of unity characteristic of conscious integration, or higher order processes that involve metarepresentations of a unified self, although even these processes do not fall under a unique type of integration. (See Proust 2007) But these processes are sensitive to information that is *epistemically irrelevant*, such as how to best interpret information to achieve phenomenal integration, or how to best satisfy strictly pragmatic goals, rather than achieving epistemic goals, such as producing true beliefs.

As Greco (2010, 166) says, cognitive integration for epistemic virtues must be *sensitive only* to those parts of the cognitive system that produce reliable information, and *insensitive* to those that produce unreliable information or play a different, not necessarily epistemic, role. His example is that perceptual beliefs are insensitive to highly theoretical beliefs (such as philosophical beliefs about the existence of the external world, or beliefs about particle physics). This kind of epistemic sensitivity is captured by the psychological notion of *cognitive impenetrability*. In many standard cases of color, shape, size and depth perception, as well as cases concerning other properties of perceived objects, including gestalt effects, background knowledge and emotion seem to affect very little how one perceives these constancies. So in standard cases (not illusory cases) of perception, perceptual beliefs (reliably produced) are insensitive to highly theoretical beliefs.

But even in the case of perceptual belief one must be careful. Not *all* the cognitive integration for perception is epistemically sensitive. Perceptual illusions illustrate this point. One consciously *sees* the difference in length of two lines in the Müller-Lyer illusion, even though one *knows* (and therefore truly believes) that they are the same length. Our conscious visual perception is in this particular case, impervious to reliable epistemic influence. But surprisingly, and this is a finding that has received almost no attention in the philosophical literature, motor control is epistemically sensitive to such reliable information, *even in cases of perceptual illusion*.

For instance, in the Müller-Lyer illusion, although the subjects' conscious self-report is *inaccurate* and reflects the illusion's cognitive influence, their motor control (specifically their unconscious manual behavior for grasping) *is accurate* and *not influenced* by the illusion. This finding seems to suggest that conscious perception has little influence on action. However, Stottinger and Perner (2006) showed that although motor control is not influenced by the illusion, cognitive processes that involve agency for *action selection*, just as conscious perception, *are influenced by the illusion*.

In their experiment, Stottinger and Perner presented subjects with vertical lines grouped in two sets (one with open brackets and the other with closed brackets, as in the standard Müller-Lyer illusion). When asked 'which gang of lines would you fight?' subjects chose the "smaller" lines although their motor control in the absence of this question did not distinguish between the sets of lines, because it was not influenced by the illusion. This finding demonstrates the dissociation between action selection and motor control. Morsella and Bargh (2010, 7) say that this dissociation occurs because inborn or learned information from the ventral stream (which is associated with conscious urges) constrains action selection but not motor control.[4]

Conscious inclinations about fighting are clearly not fundamentally associated with *epistemic reliability* and, in this particular case, the conscious decision to fight the longer lines is based on false information. This information for action selection may lead to good practical decisions, but not to reliably produced true belief. Accurate motor control concerning length, on the other hand, is a precondition for *successful navigation*. So it makes sense that the epistemically relevant information that allows agents to succeed, based on their knowledge of the environment, ignores, or is insensitive to, the epistemically *irrelevant* conscious information concerning who to fight.

Success (achieving true belief) from epistemic virtue seems to be guaranteed only at the motor control level, at least in the case of the illusion just mentioned, but the dissociation between motor control and action selection extends to many forms of action. Crucially, cognitive integration for motor control processes that lead to success in a reliable fashion is insensitive to epistemically irrelevant inclinations, or highly sophisticated theoretical or philosophical beliefs, in spite of the fact that those inclinations may underlie practical interests.[5] However, as the example just mentioned shows, cognitive integration for conscious processes and action selection *is* sensitive to epistemically irrelevant information. So motor control knowledge complies with the right kind of cognitive integration required for *stable epistemic virtues.*

Epistemic virtues are generally described as stable dispositions attributable to an agent. The more stable the disposition, the more successful the agent. Using the language of cognitive psychology, the less sensitive epistemic virtues are to practical or highly theoretical considerations, the more stable they will be, and vice versa. This is what explains the basic tradeoff I described before: the more minimal the sense of agency, the less plausible the postulation of a reflective-requirement for knowledge; and the more enriched the sense of agency, the less epistemically relevant the characterization of 'virtue' for knowledge. This is because rich agency for action selection makes epistemic capacities *less stable* across different epistemic situations and more sensitive to epistemically irrelevant information.

[4] See also Goodale (2010).

[5] This may question the empirical adequacy of the semantics for knowledge attributions based on practical interests, defended, for example by Stanley (2005), but I shall not comment on this issue here.

Motor control knowledge is firmly associated with *facts* about the environment and the success of agents is contingent upon these facts. True beliefs about environmental features are formed reliably because of these virtues, thereby allowing agents to avoid errors and lucky guesses across a large variety of situations. In other words, there is counterfactual dependency between the success of agents and these stable epistemic virtues that reliably form beliefs about facts. But this does not mean that practical reasons, conscious action selection and introspection are epistemically irrelevant *in general*. As Greco says, they may explain epistemic virtues of a different kind (not associated with knowledge, but with other epistemic goals). In the next section, I consider objections to the present proposal, and address this issue in more detail.

4 Objections and Replies

Although the tradeoff between reflective agency for action selection and motor control for perceptual knowledge is based on very well confirmed psychological evidence, one may have serious doubts about accepting these findings as a constraint on a satisfactory account of epistemic virtue. It is true that the most reliable cognitive skills (which are certainly epistemic virtues that produce a higher number of true beliefs than false beliefs) are the motor control skills that are largely insensitive to introspective, theoretical or practical information. But, the objector would say, a theory of knowledge based exclusively on these (motor control) epistemic virtues would be deprived of the most interesting types of knowledge, such as mathematical and scientific knowledge.

There are two replies to this objection, which are not based on the empirical evidence, which is abundant, but on strictly epistemic considerations. First, one must be careful with distinctions such as 'low and high level cognition,' 'sophisticated and basic reasoning' or even 'animal and reflective knowledge.' The example concerning knowledge of syntax, which, as far as scientists can tell, seems to be one of the few exclusively human traits, shows that caution is required because this (motor control) knowledge is largely unconscious, stable and insensitive to theoretical or practical information. But non-illusory conscious and unconscious *perception* is also stable and insensitive to practical or theoretical information. It is having the right kind of epistemic sensitivity to relevant information, and reliably manifesting epistemic insensitivity to any other information, that is crucial for a naturalized account of epistemic virtues for knowledge.

Maybe mathematics requires knowledge of syntax, which is insensitive to theoretical considerations (say about geometry or topology) and a more theoretical component (which may depend not on reliable manifestations of the syntax-like component, but on coherence with other mathematical truths and inferential reasoning) that is sensitive to these considerations. It is analogous to language, after all, and there is evidence of basic arithmetic-like abilities in some animals.[6] The motor

[6] See for example, Dehaene (2001), Brannon et al. (2001), and Montemayor and Balci (2007).

control epistemic virtues for mathematical knowledge lead to successful behavior that depends on accurate counting, across many different situations, while the theoretical abilities depend on assumptions and background information concerning points, lines, etc. If one wants to know how many cookies are left or how many people are chasing us, it is the stable and insensitive knowledge that will allow us to succeed. If one stops and wonders about theoretical issues concerning number theory, the real numbers and the mathematical continuum, things may get quite complicated and, in the second case, potentially dangerous.

Second, epistemic virtues are *stable dispositions* of agents. This means that these dispositions manifest in true belief in *most* situations (or at least in more situations than other cognitive dispositions). Whether an agent has these virtues or not is something that can be found by experimental manipulation, by exposing the agent to different situations, and finding out whether she always manifests true beliefs, as is the case with experiments that involve illusions. Decades of research in psychology show that motor control is remarkably stable and reliable. This is the knowledge we use to walk around, count objects on tables, parse the syntactic structure of a sentence, assess the length of things, etc. It is the kind of knowledge that is less susceptible to luck or environmental variability.

Suppose now that one tries to be more careful with the characterization of reflective knowledge by appealing to specific aspects of psychological processes. An obvious candidate for the disambiguation of this term is conscious content, but reflection need not be conscious, as mentioned before. It seems that reflection must be at least metarepresentational, but this is also problematic and requires further distinctions that proponents of reflection have not addressed carefully. For instance, metacognition (thinking about one's own mental dispositions and inclinations) is partly unconscious, and psychologists have found out that there is an important *dissociation* between metacognition and metarepresentation: some animals engage in metacognition without metarepresenting, and metacognitive processes lack the recursive characteristics of metarepresentational ones. (See Proust 2007).

Moreover, the notion of 'self,' required for metacognition is problematic because there are implicit and explicit forms of metacognition, and some of these happen without representations of the self. It seems that the type of reflection required for meta-aptness and meta-virtues constitutive of reflective knowledge need to involve metarepresentation, metacognition and *explicit* representations of the self. This is an area where proponents of reflection need to spell out in more detail how the dissociations between these three different types of cognitive processes are brought together into a naturalized account of reflection. In any case, naturalized virtues for knowledge must be the most stable dispositions across different scenarios and their manifestations must be true beliefs that lead to success. The best candidates, given the psychological evidence, are motor control epistemic processes.

A different kind of objection to the present account of naturalized virtues for knowledge (based on motor control) is that without reflection one cannot achieve any epistemic goal regarding the *adequacy* of our beliefs with respect to the situation one is in. More specifically, the challenge is that traditional problems in the theory of knowledge are blatantly dismissed by this proposal. One may want to know, for instance, how could a naturalized account of virtues addresses issues about withholding judgment or doubting, when these are the *correct epistemic attitudes* that a virtuous agent *should* deploy. A naturalized theory of knowledge that dismisses these virtues is, therefore, inadequate.

My response to this objection is that epistemic normativity does not *only* include (and should not focus exclusively on) virtues for knowledge. It is correct to think of doubt and withheld judgment as manifestations of epistemic virtue, but these virtues may not be constitutive of knowledge. Rather, they may be constitutive of different epistemic achievements, such as the overall coherence of one's own beliefs, the quality of one's understanding of a problem, the creativity with which one solves difficult tasks, etc. The empirical findings show that these virtues are not cognitively integrated with the virtues that reliably produce true belief, which are constitutive of knowledge. Actually, our cognitive system *requires* virtues for knowledge to be insensitive to these other virtues in order to work properly, which is the main reason why reflection is a problematic requirement for knowledge.

Interestingly, the need for knowledge and implicit reasoning (at least of a perceptual kind) to be insensitive to theoretical information has already been made in the philosophical literature independently of the findings on motor control and action selection. For instance, Kent Bach (1984) argues that a repertoire of *snap judgments*, either conscious or unconscious, are crucial to successfully perform tasks, and that one should question these judgments only under *very specific* circumstances. Motor control knowledge is certainly 'default reasoning' in the sense that it is impervious to most theoretical questions, and this makes sense because motor control is proven to be extremely reliable (more likely to be true than false). Knowledge is linked to success based on epistemic abilities or stable dispositions to produce true belief, according to reliabilist virtue epistemology. Epistemic virtues for knowledge must be insensitive to information that jeopardize or interfere with such success, even if this information is relevant for other epistemic goals.

Thus, it seems that there are powerful theoretical and empirical reasons against the inclusion of reflective processes in a naturalized account of epistemic virtues *for knowledge*. First, if these reflective processes are introspective and aimed at specifying reasons for action, evidence shows that they are unreliable. Second, if reflection is construed as action selection then it is dissociated (not cognitively integrated) from reliable motor control. Finally, there are problems, both theoretical and empirical, with respect to the relationship between metacognition and metarepresentation, such that there seems to be a systematic ambiguity in contemporary uses of the term 'reflection.'

5 Conclusion

I have argued that motor control is reliable, based on abilities of epistemic agents, and deeply associated with their success. Motor control abilities have the right kind of sensitivity for producing beliefs that achieve the normative status of knowledge:

they are not merely sub-personal, because they can be cognitively integrated with memory, cross-sensorial information etc., at the *organism level*, and they are valuable because they guarantee the success of the agent in performing a panoply of tasks.[7] In any event, it seems that any theory of epistemic virtue should take into account the distinction between motor control and action selection.

However, it is important to emphasize that reflection is valuable, and may actually be indispensible, for achieving epistemic goals, other than knowing. Understanding, for instance, seems to be a very important epistemic achievement, which could radically differ from knowledge because one can understand something that is false and not understand the truth. The phenomenal content of subjective experiences, introspection and metarepresentation are crucial to many debates in epistemology, such as skepticism and other theoretical issues involving withheld judgment. But maybe this is because of how much value we attribute to genuine understanding, rather than automatic success, even if attributable to the abilities of agents.

So I am not suggesting that understanding should be eliminated from a broader theory of naturalized epistemic virtues that could explain the relationship between knowledge and reflective understanding. Rather, the main point is that, based on solid psychological evidence, motor control seems to capture the main characteristics of knowledge because it is bullet proof when it comes to irrelevant information, which makes it remarkably stable across many epistemic situations. This is what guarantees success across many possibilities the agent may encounter.

Maybe the term 'knowledge' has been used loosely and applied to cases that are not really cases of knowledge. If this were the case, the situation would not be unfamiliar. Take for instance Ned Block's (1995) distinction between phenomenal and access consciousness. Uses of the term 'consciousness,' if one accepts this distinction, turned out to be ambiguous because some of them referred to global access, while others referred exclusively to the qualitative character of subjective experiences. This distinction has, according to Block, important theoretical consequences. For example, Block (2003) suggests that physicalism and functionalism about consciousness may not be rival theories, but answers to different questions. Physicalism, he says, tries to answer the question concerning the neural basis of experience (phenomenal consciousness), while functionalism tries to answer the question of what makes neural representations available for thought, decision, reporting and action (access consciousness).

Similarly, traditional views in epistemology concerning reflection (doubt, withheld judgment, etc.) may be answers to questions that concern understanding and other epistemic achievements, while views concerning immediate justification and reliable true belief concern knowledge. What is important for future accounts that try to give a general theory of these different epistemic virtues is to consider the tradeoff between robust reflective agency and minimal-reliable agency for motor control.

[7] For the importance of the requirement of cognitive integration at the organism level in order to account for mental representation, see Burge (2010). For its epistemic relevance see Greco (2010).

References

Bach, K. 1984. Default reasoning: Jumping to conclusions and knowing when to think twice. *Pacific Philosophical Quarterly* 65: 37–58.

Block, N. 1995. On a confusion about the function of consciousness. *Behavioral and Brain Sciences* 18: 227–247.

Block, N. 2003. Consciousness. In *Encyclopedia of cognitive science*, ed. L. Nadel. New York: Nature Publishing Group.

Brannon, E., C.J. Wusthoff, C.R. Gallistel, and J. Gibbon. 2001. Subtraction in the pigeon: Evidence for a linear subjective number scale. *Psychological Science* 12(3): 238–243.

Burge, T. 2010. *Origins of objectivity*. New York: Oxford University Press.

Dehaene, S. 2001. Subtracting pigeons: Logarithmic or linear? *Psychological Science* 12: 244–246.

Goodale, M.A. 2010. Transforming vision into action. *Vision Research* 10(1016): 7–27.

Greco, J. 2004. How to preserve your virtue while losing your perspective. In *Ernst Sosa and his critics*, ed. J. Greco. Oxford: Blackwell.

Greco, J. 2010. *Achieving knowledge: A virtue-theoretic account*. New York: Cambridge University Press.

Kornblith, H. 2010. What reflective endorsement cannot do. *Philosophy and Phenomenological Research* 80(1): 1–19.

Montemayor, C., and F. Balci. 2007. Compositionality in language and arithmetic. *Journal of Theoretical and Philosophical Psychology* 27(1): 53–72.

Morsella, E., and J.A. Bargh. 2010. What is an output? *Psychological Inquiry* 21: 354–370.

Nisbett, R.E., and T.D. Wilson. 1977. Telling more than we can know. *Psychological Review* 84: 231–259.

Pritchard, D. 2012. Anti-luck virtue epistemology. *Journal of Philosophy* 109: 247–279.

Proust, J. 2007. Metacognition and metarepresentation: Is a self-directed theory of mind a precondition for metacognition? *Synthese* 159: 271–295.

Rosenbaum, D.A. 2002. Motor control. In *Stevens' handbook of experimental psychology: Vol. 1. Sensation and perception*, 3rd ed, H. Pashler (Series ed.) and S. Yantis (Vol. ed.), 315–339. New York: Wiley.

Sosa, E. 2009. *Reflective knowledge: Apt belief and reflective knowledge*, vol. II. Oxford: Oxford University Press.

Stanley, J. 2005. *Knowledge and practical interests*. Oxford: Oxford University Press.

Stottinger, E., and J. Perner. 2006. Dissociating size representation for action and for conscious judgment: Grasping visual illusions without apparent obstacles. *Consciousness and Cognition* 15: 269–284.

Weiskrantz, L. 2009. *Blindsight: A case study spanning 35 years and new developments*. Oxford: Oxford University Press.

Towards a Eudaimonistic Virtue Epistemology

Berit Brogaard

1 Introduction

Virtue theories of knowledge maintain that knowledge requires cognitive success attained because of the exercise of intellectual virtue. Virtue epistemologists normally take cognitive success to be true belief. But one could take an approach to cognitive success that allows that veridical mental states other than true belief can count as cognitive success. On one such approach set forth by Timothy Williamson (2000: Introduction), a wide range of mental states, including belief states, memory states, states of visual seeming, seeings, and so on, can count as knowledge states, provided that they satisfy certain further constraints. I have defended this position in earlier work (Brogaard 2011a). Here I shall take it for granted. But the views articulated below remain valid without this assumption in place.

What counts as intellectual virtues has been the subject of fierce debate (Code 1984; Montmarquet 1987; Zagzebski 2003; Greco 2003; Baehr 2006; Baehr 2011; Sosa 2007; Kelp 2011). Virtue responsibilists take intellectual virtues to be character traits that are beneficial to cognitive success, such as patience, intellectual honesty and curiosity (Montmarquet 1987; Zagzebski 2003; Axtell 1997). Virtue reliabilists, on the other hand, take intellectual virtues to be reliable cognitive faculties, such as the memory system and the visual system (Greco 2003; Sosa 2007; Kelp 2011). Some take a mixed approach requiring both reliability and responsibility (Greco 2000, 2003; Greco and Turri 2011).[1] Here I shall subsume the mixed approaches under virtue responsibilism. Though defenders of the mixed approach

[1] Sosa's approach can also be considered a mixed approach. Sosa distinguishes between animal knowledge and reflective knowledge. Animal knowledge requires the exercise of reliable cognitive faculties or abilities, whereas reflective knowledge requires a more active, or reflective, second-order stance (see e.g. Sosa 2007). Since Sosa does not require responsibility as a condition on knowledge, I shall here treat his approach as a kind of virtue reliabilism.

B. Brogaard (✉)
Philosophy Department, University of Miami, St. Louis, MO, USA
e-mail: brogaardb@gmail.com

A. Fairweather (ed.), *Virtue Epistemology Naturalized: Bridges Between Virtue Epistemology and Philosophy of Science*, Synthese Library 366, DOI 10.1007/978-3-319-04672-3_6,

only take exercise of virtuous character traits to be a necessary condition on knowledge, my criticisms will apply equally to these approaches.

I think all three approaches are fraught with difficulties, and even if the hurdles could be overcome, I doubt that any of these theories would be superior to a conventional reliabilist account of knowledge. The main problem facing virtue responsibilism is that it must be limited to special forms of knowledge. It cannot account for the appearance that agents can attain knowledge even when they do not possess any virtuous character traits. Virtue responsibilism furthermore fails to accommodate important forms of implicit knowledge, for example, some forms of knowledge-how. The knowledge-how that precedes spontaneous action forms on a subpersonal level and so cannot in any way be considered the result of exercising virtuous character traits. The view also fails to account for the knowledge possessed by individuals who have a superior cognitive skill that reliably produces true belief without requiring the exercise of virtuous traits.

The main obstacle for virtue reliabilism is that of accounting for knowledge that is the result of shared effort or that does not arise primarily from the exercise of reliable cognitive faculties. Virtue reliabilists typically respond to this kind of difficulty by allowing cognitive skills to be assessed for credibility even if they require assistance from external sources. I fear, however, that once the view is extended in this way, it will be indistinguishable from a traditional form of reliabilism.

After identifying the main problems with both types of virtue theoretical approach, I argue that traditional reliabilism avoids the problems facing virtue theoretical accounts and solves the perplexities virtue epistemology was introduced to untangle. I then sketch the basics of a eudaimonistic virtue epistemology. This theory takes our ultimate intellectual goal to be to proceed intellectually in ways that do not hinder intellectual flourishing. Intellectual flourishing is the epistemic equivalent of Aristotle's eudaimonia (well-being, flourishing, happiness). Though one can sometimes flourish intellectually by being intellectually virtuous in the traditional sense, being intellectually virtuous in this sense can be at odds with intellectual flourishing. Or so I will argue.

2 The Main Problems with Virtue Responsibilism

Virtue responsibilism takes the exercise of intellectual virtues to be the main source of knowledge. Intellectual virtues, on this view, are those personality traits a rational person who desires the truth would want to have. This view is feasible only if mental states with knowledge status actually are the result of the exercise of intellectual virtues. Cognitive success is a result of an exercise of character traits just when the agent would be similarly successful in a broad enough range of conditions in which he exercises the same character traits. The two main reasons to question that cognitive success is a result of exercise of this kind are (1) that what may look like the exercise of intellectual virtues often is really the expression of attitudes, social conditioning and a desire for the world to be a certain way, and (2) that many mental states with knowledge status do not require any form of exercise of virtuous character traits, as the processes that underlie them are not within the agent's conscious control.

2.1 Agents Do Not Exercise Intellectual Virtues

In his article "Do You Know what You're Doing?" and his forthcoming book *A Natural History of the Self* John Doris reviews social-psychological studies that show that we have limited conscious access to our reasons for acting the way we do.

Studies revealing our limited rational behavior are not hard to come by. Dan Ariely (2008: 177), for example, reports on studies showing that in countries in which people applying for a driver's license are asked to check a box if they want to join an organ donor program, they don't check the box and they don't join. In countries in which people applying for a driver's license are asked to check a box if they don't want to join an organ donor program, they don't check the box and they join. The reason people make these decisions is that they don't want to deal with the complex issue of organ donation and so remain content with the option already chosen for them.

To test whether experts make similarly bad decisions, Ariely ran a parallel study with physicians. The physicians were divided into two groups. Both groups were presented with a case of a patient with hip pain. The physicians were told that they had sent him off to hip replacement surgery after experimenting with several medications. The first group was then asked what they would do if they suddenly realized that they had not tested the efficacy of ibuprofen. The majority of participants in this group chose to cancel the hip replacement surgery in favor of experimenting with this medication. The second group was asked what they would do if they realized that two medications hadn't been tested. The majority of participants in this group chose to let the patient have the hip replacement surgery. Apparently, the increased complexity of cancelling the surgery made them choose not to do so.

The real reasons for acting the way we do need not be accessible to introspection. When we introspect it seems that we were acting for rational reasons. The reasons we cite as reasons for action, however, are in many cases pure constructs. In one study, cited by Doris, Johansson and colleagues presented pairs of images representing female faces to participants. The participants were asked to pick the face they found more attractive (Johansson et al. 2005; cf. Johansson et al. 2006). In three of the 12 trials the researchers contrived to treat the photo that the participants did not pick as though they did pick it. When people were asked to explain their choices for three of the non-manipulated trials and the three manipulated trials, the explanations were nearly indistinguishable.

These types of studies cast doubt on the extent to which we act on the basis of reasons to which we have conscious access. The perhaps most impressive case of apparent failure of being the agent of our actions is the case of slow walkers. In one study, participants were given a puzzle and asked to construct grammatical sentences out of randomly ordered words. One group received a version containing words associated with stereotypes of the elderly, such as wrinkled, gray and Florida. The other group received a test containing only age-neutral words, such as private, thirsty and clean. The study showed that participants who were given the puzzle with the geriatric words were walking significantly slower afterward than the participants who were provided with the age-neutral vocabulary (Bargh et al. 1996: 236–7; Doris 2011: chap. 4).

Doris concludes from cases like these that our standard notion of agency is in need of revision. Though Doris' primary concerns lie within the area of practical reason, he indicates in several places that his trepidation carries over to the realm of epistemology. I don't know exactly how Doris plans to adapt his situationalism to the epistemic realm.[2] But there are, no doubt, lots of situations in which even the most seemingly virtuous agents are influenced by irrelevant factors. For example, studies show that unseen (happy or unhappy) faces can affect subsequent ratings of unrelated characters (Murphy and Zajonc 1993) and sodas (Winkielman and Berridge 2004). An unseen happy face makes us rate faces and sodas more positively than an unseen unhappy face. So even if we are unbiased in rating the faces and sodas in neutral situations, we fail on this virtue in affectively charged situations. These kinds of cases raise a question of whether people's character traits, being so susceptible to irrelevant influences, can count as intellectual virtues.

The situationalist critique, as it stands, is not fully convincing though. It remains a theoretical possibility that there are enough situations in which we are unbiased and otherwise good and admirable intellectual agents and in which we manage to achieve cognitive success by exercising our intellectual virtues.

Furthermore, to say that our beliefs are influenced by factors beyond our control is not to say that virtuous character traits play no role in belief-formation. Seeing a smiley face before making a judgment about a person or a drink may make our ratings more positive than they otherwise would have been. Suppose I assign a 9 to a person I would otherwise have rated an 8. You assign a 6 to the same person because you don't like the color of his skin but had you not seen the smiley face, you would have assigned a 5. Though our decisions are influenced by the smiley face, the differences in our ratings may be due to differences in our personality traits or how we choose to use what we have.

There is, however, a different way of making the same point. There are myriads of examples of beliefs not grounded in the exercise of intellectual virtue. People believe in improbable conspiracies, the healing powers of crystals, the diagnostic value of auras and psychics who are in contact with dead relatives. Some deny that the holocaust took place, others that evolution did.

Why do people believe these things? Some believe implausible tings because the world would be a more exciting place if their beliefs were true. The world

[2]Others have expressed similar misgivings about virtue responsibilism. See e.g. Mark Alfano (Forthcoming). Alfano's focus is on showing that none of us possesses virtuous character traits, except in particular situations, whereas I am content with showing that even clearly non-virtuous agents possess knowledge. I will not be dealing explicitly with Alfano's critique, as I had already completed this paper when I became familiar with his. Thanks to Abrol Fairweather for drawing my attention to it.

would be far less mundane if Kennedy's assassination were part of a plot that the CIA conjured up rather than a one-man show.

Others believe what they do because it fits prejudices or fears they already have. Anti-Semitics believe the holocaust never took place because they hate Jews, and creationists believe evolution never took place because they fear that admitting that evolution did happen will put them in hell. Yet others believe what they do because it makes them feel better. On December 10, 1997, on the Larry King Live show, James Van Praagh, who specializes in being contacted by people's dead relatives, claimed that he could feel Larry's dead parents and pointed to a location where the spirits were located. Hordes of believers called in on the show. They believed van Praagh because he told them what they wanted to hear.

In many of the just-cited cases people believe falsely. But it is not hard to conjure up cases in which people believe truly without exercising intellectual virtue. Inmate Jay Lewis Biggs' wife believed in her husband's innocence long before there was evidence suggesting that he might have been wrongly convicted. Conversely, many people who heard about the Biggs case believed Jay was guilty before his case went to trial. No one of a sound mind would ascribe knowledge to the agents in the preceding examples. But even clearly non-virtuous agents possess knowledge when nothing is at stake and when it takes no effort to get there. All the world's creationists, anti-Semites, misogynists, followers of Deepak Chopra, and so on, possess knowledge, for example, basic sensory knowledge. But they lack important intellectual virtues and therefore cannot have exercised these in arriving at their beliefs about basic matters. Suppose an exceptionally slothful, uncaring, apathetic and dishonest detective is put on a murder case. His virtuous son feels sorry for his idle dad, solves the case and gives the evidence to his father. Using very simple sensory and cognitive skills, the lazy detective now reliably forms a belief that so-and-so is the murderer and takes credit for the discovery. Intuitively the indolent detective knows that so-and-so is the murderer but he did not attain cognitive success by exercising intellectual virtues. He attained cognitive success by exercising basic sensory and cognitive faculties.

The problem that these cases pose for virtue responsibilism is clear. We are not very intellectually virtuous on a whole. We believe many of the things we do because of ignorance, desires, wishes, love, fear or issues of control. Non-virtuous people have what appears to be knowledge of basic matters on the basis of the proper function of visual pathways and cognitive domains but most of their cognitive successes fail to be the result of exercising virtuous character traits.

2.2 *Implicit Knowledge*

Another big hurtle for virtue responsibilism is to find a way to accommodate certain basic forms of knowledge-how associated with spontaneous action. Prior to any spontaneous action, we form neural representations of our bodies and our environment that do not correlate with conscious awareness. These neural representations form in the dorsal stream, a pathway in the brain that starts in the primary visual cortex and runs upward through parietal cortex and ends before the motor cortex. These dorsal stream representations are necessary for spontaneous action (Goodale and Milner 1992; Milner and Goodale 2008). If the action is that of reaching to and grasping an object in front of you, the dorsal stream representations represent the absolute size of the object, so you can adjust the aperture of your hand to fit the object. It also represents your body position relative to the object and the trajectory your hand must take from where it is currently located to where the object is located.

These dorsal stream representations are the main components of vision-based knowledge-how (Brogaard 2011b). We rely on them when we walk, run, swim, eat and so on. However, these representations do not correlate with conscious awareness. Studies show that conscious awareness of changes in our environment precedes updates of representations in the dorsal stream. For example, if an object unexpectedly changes location, subjects adjust arm velocity and trajectory in less than 100 ms. 100 ms is not enough time for the human brain to consciously represent a change in object location or a corresponding change in velocity and trajectory (Paulignan et al. 1991). Further, when subjects are asked to use a minimally demanding vocal response (Tah!) to indicate awareness of a change in an object's location, they correct their movements considerably faster than the vocal response. Corrections of trajectory and hand aperture occur within 100 ms, the vocal response happens after 420 ms (Castiello et al. 1991; Castiello and Jeannerod 1991).

Studies of pointing and saccadic eye movement further indicate that we can correct saccadic eye and pointing movements faster than we can consciously perceive a change in an object's location (Goodale et al. 1986; Pelisson et al. 1986). In one study, the researchers asked participants to point as quickly and accurately as possible to stimuli occurring in the dark (Pelisson et al. 1986). In the first series of trials, the target leaped from an initial position to a randomly selected position. In the second series, the target made a second jump in the same direction as the initial leap. The participants reported that they were unaware of the second jump, and that they were unable to predict its direction, but while saccadic eye and pointing movements were initially aimed at the target's position after the first jump, both were instantly adjusted to fit the target's new location. Even though the subjects had no awareness of the two jumps, they were evidently seeing and acting on both. The findings indicate that the subjects updated movement trajectory and target location without awareness.

Similar results were reported by Jakobson and Goodale (1989). They first showed that subjects could not detect a three-degree shift in vision through wedge prisms. They then monitored the participants' movements. Despite no reported awareness of the shift in vision, the shift in vision produced a modified hand-path curvature. Together these findings indicate that dorsal stream representations are inaccessible to consciousness. Exercising virtuous character traits requires a conscious effort. Since dorsal stream representations do not correlate with conscious awareness, they cannot be the result of an exercise of virtuous character traits. In fact, the brain can generate dorsal stream representations in individuals acting automatically, such as sleepwalkers.

One of the most talked-about sleepwalking trials in US history was R. v. Parks, [1992] 2 S.C.R. 871. Kenneth James Parks, a 23-year-old man from Toronto with a wife and baby, was suffering from severe insomnia and anxiety owing to unemployment and gambling debts.[3] Kenneth had repeatedly placed bets on horses the previous summer and this had caused him great financial problems. To obtain money for gambling he stole $32,000 from his employer Revere Electric. Kenneth kept losing money and when the company found out about his embezzlement in March 1987, he was fired. Court proceedings were brought against him, and his marriage was in trouble.

In the early morning of May 1987 Kenneth got out of bed and drove 23 km from Pickering to his in-laws' house, Barbara Ann and Denis Woods, in the Toronto suburb of Scarborough. After collecting a tire iron from the trunk of his car, he entered the house using his key. He continued to the bedroom and choked his father-in-law unconscious. He then beat his mother-in-law with the tire iron and stabbed her repeatedly with a kitchen knife. He also stabbed his father-in-law.

Barbara was later found in a room 5–6 ft from the bedroom. She had been stabbed in the chest, the shoulder and the heart. She had furthermore sustained blunt-force injuries to her eye, nose and skull that caused a subarachnoid hemorrhage. Denis was unconscious after the assault but his wounds were less severe. That night Kenneth also picked up the phone in the kitchen and set it down again, off the hook. He ran upstairs to the teenage daughters' bedrooms. But he stopped, just stood there, then ran down again and left.

After the killing Kenneth then drove to the police station. He arrived at 4:45 A.M., covered in blood, and said "I just killed someone with my bare hands; oh my God, I just killed someone; I've just killed two people; my God, I've just killed two people with my hands; my God, I've just killed two people. My hands; I just killed two people. I killed them; I just killed two people; I've just killed my mother- and father-in-law. I stabbed and beat them to death. It's all my fault." The police reported that he was shaking and seemed distressed. Despite having cut tendons in both hands, he did not appear to be in pain. This is an example of dissociative analgesia, a profound blunting of pain sensation in the absence of pain killers. Dissociative analgesia can occur during states of sleepwalking but also after drug use and in states of shock or great distress.

After careful examination of the case, the experts couldn't find a cause of the attack, except sleepwalking. Kenneth underwent a series of sleep tests and psychological tests. The electroencephalography (EEG) scans showed that Kenneth had some abnormal brain activity during deep sleep, periods of partial awakenings, which is indicative of parasomnia, and since there is no way to fake one's own EEG results, and Kenneth had not felt pain when he arrived at the police station, it was concluded that he was sleepwalking when he assaulted his in-laws.

The experts described Kenneth's actions as the result of many circumstances converging: he had promised to fix his in-laws' furnace, he was familiar with the route to their house, and he was worried about his upcoming trial. The experts thought that it suddenly occurred to Kenneth in his sleep that he should repair his

[3] http://csc.lexum.umontreal.ca/en/1992/1992scr2-871/1992scr2-871.html

in-laws' furnace. He then got up and drove to the house but was startled by the in-laws. He attacked both of them without awareness of what he was doing.

Sleepwalking doesn't automatically lead to full acquittal. An involuntary act entitles an accused to an unqualified acquittal only if his condition did not originate in "a disease of the mind" that has made the person insane. In the latter case, the accused is not entitled to a full acquittal, but only to a verdict of insanity. "Disease of the mind" is a legal term, not a medical term. Because it is a legal term, a judge cannot rely only on medical opinion but must also give weight to the likelihood of recurrence and the cause of the offence. A condition likely to present persistent danger should be treated as insanity. A condition originating in the internal constitution of the accused, rather than external factors, should also lead to a verdict of insanity. These two conditions might seem sufficient to justify less than full acquittal of sleepwalkers who kill. But the defense at Kenneth's trial argued that a combination of external factors caused the killing and that it was unlikely that a similar pattern of factors would occur again in the future. In the medical review it was concluded that

> the legal defense was, therefore, one of homicide during noninsane automatism as part of a presumed episode of somnambulism… the defendant did not have any preexisting "disease of the mind" within the meaning of… the Canadian Criminal Code. There was no evidence for psychosis or other mental pathology. Moreover, it was believed that the clustering of such a number of triggering factors was extremely unlikely to occur again, so that the possibility of recurrence of sleepwalking with aggression was considered extremely remote.

Accordingly Kenneth was acquitted of the killing of his mother-in-law and the assault on his father-in-law.

Sleepwalking occurs in the deep stage of sleep when slow brain waves (50 % + delta waves) begin to appear. Because of the slow brain waves, people who are asleep are not normally consciously aware of sensory input from their surroundings. During sleep there is also a gating mechanism that blocks input from the cognitive brain to the motor system. The chemical messenger gamma-aminobutyric acid (GABA) acts as an inhibitor that suppresses the activity of the motor system (Navarro 2008; Barlow and Durand 2008). However, in parasomnia the gating mechanism fails. Accordingly, there is considerable input to the motor system. Owing to the failure of the gating mechanism the brain issues commands to the muscles during sleep. In children the neurons that release this neurotransmitter are still developing and have not yet fully established a network of connections to keep motor activity under control. Sometimes the gating mechanism remains underdeveloped. In other cases, sleep deprivation, fever, anxiety, or drugs cause the gating mechanism to function less effectively. In those cases sleepwalking can persist into adulthood. People can do amazing and complex things in their sleep. Former chef Rab Wood described cooking spaghetti bolognaise and fish and chips while sleeping.

While slow delta brain waves occur during sleepwalking, a significant amount of high oscillation waves occur in areas related to movement. Sleepwalkers have their eyes open, they can see their environment but not consciously, they are unaware of what they see. While sleepwalkers are in a state of deep sleep, the part of the brain in charge of motion is awake. Only the part of the brain that correlates with awareness and cognition remains asleep. Sleepwalkers are essentially awake and asleep at the same time. As the cortex, which is the part of the brain that controls thinking and voluntary movement, is asleep during slow wave sleep, the movements sleepwalkers make are controlled by other parts of the brain and are more or less reflexive.

Sleepwalking adds evidence to the hypothesis that the parts of the brain that control thinking and voluntary movement and the parts that control motion are dis-associated. The latter parts are just regions in the dorsal stream pathway. While Kenneth Parks was not conscious of the actions he performed while asleep, Kenneth engaged in quite complex behavior while sleepwalking. He drew 23 km in a car. He had to make the drive in relatively unfamiliar surroundings in the dark at night. During his drive he encountered several major intersections which he had to maneuver. He fetched a tire iron from his car and used his key to enter the house. At some point he got a kitchen knife, and he then beat and stabbed his in-laws. Yet he was acquitted because his actions were automatic. He was not responsible for what he did and could not be blamed.

A lot of knowledge how and associated *dorsal stream representations* went into producing the actions Kenneth completed while asleep. These knowledge representations were not the result of exercising responsibilist virtue. They were completely independent of any such character traits. Whether we possess virtuous character traits makes no difference to the functionality or structure of these kinds of representations. Even if we are non-conscientious, dishonest and don't care about the truth, our dorsal streams can still reliably compute accurate representations of our bodies, our environment and our position relative to the our environment. These forms of implicit knowledge representations are unaccounted for by virtue responsibilism.

2.3 Extreme Intellectual Abilities

Some people have extremely reliable and impressive cognitive abilities. They can perform what may seem like impossible mathematical, linguistic or artistic tasks. These are the so-called 'savants'. According to Darold A. Treffert, savant abilities occur in a very narrow range of abilities (Treffert 2009). Far the most typical ones are:

Music: Piano performance or composition with perfect pitch. Example: Blind Tom, a blind autistic slave in Georgia in the nineteenth century, was an amazing pianist and performer.

Art: Drawing, painting, or sculpting. Example: Stephen Wiltshire drew an extremely accurate sketch of a four square mile section of London, including twelve major landmarks and 200 other buildings after a 12 min helicopter ride through the area.

Calendar calculation: The ability to name the day of the week that a certain date or event will or did occur in any particular year or to name all the years when a specific holiday will fall on a specific date. Example: for any chosen calendar day, the human computers and autistic twins Kay and Flo Lyman can report what they ate for dinner, what they did on that day, what weekday it was, what their favorite TV-host wore on that day, and so on.

Mathematics: Lightning calculation, geometrical acumen or computation of multi-digit prime numbers, in the absence of other special arithmetic abilities. Example: Oliver Sacks' autistic twins John and Michael computed prime numbers with more than 6 digits (Sacks 1995).

Spatial skills: Distance measurements or construction of complex structures with painstaking accuracy. Example: the real rain man Kim Peeks was able to provide map directions between any two cities (Peek 1996).

Savant skills tend to be right-brain or bilateral skills. For example, mathematical skills likely involve bilateral processing in the intra-parietal sulcus (Dehaene 1999, 2001, 2007; Dehaene et al. 1998, 2004; Piazza et al. 2007; Eger et al. 2009; Hubbard et al. 2009), and spatial reasoning skills involve right-hemisphere processing in superior temporal cortex, the regions on the right that corresponds to the language area on the left (Karnath et al. 2001).

The leading hypothesis is that savant syndrome is caused by a lesion or birth defect in the left hemisphere that results in overcompensation by the right hemisphere (Pesenti et al. 2001). An alternative, but related, hypothesis is that we all have the skills of savants but that they are dormant because of the dominance of the left hemisphere in most people (Snyder et al. 2003; Young et al. 2004). In some people with savant syndrome the dominance is weakened by an absence of information transfer between hemispheres. For example, an MRI scan of the artistic savant, Kim Peek, who lent inspiration to the fictional character Raymond Babbitt, played by Dustin Hoffman, in the movie Rain Man, revealed an absence of the corpus callosum, the anterior commissure and the hippocampal commissure, the parts of the neurological system that transfer information between hemispheres (Wisconsin Medical Society, Islands of Genius). The brain can also transfer information indirectly through subcortical areas. It is unknown whether any information was transferred between hemispheres in Kim Peek's brain.

However, in most cases the dominance of the left hemisphere is weakened by a lesion to the left hemisphere. Savant syndrome is typically accompanied by severe developmental disorders, usually autism. In the largest study of savant syndrome today, 41 out 51 subjects had been diagnosed with autism (Treffert 2009). But there are also cases in which savant syndrome occurs without any associated disability and cases in which it is acquired later in life, following central nervous system injury or disease (Lythgoe et al. 2005; Treffert 2009). For example, Daniel Tammet (DT), a young man with savant syndrome, can perform mathematical calculations faster than most people can on a calculator (Bor et al. 2007). Though there is some evidence that DT has some degree of autism, this is far from obvious, and it certainly is not the cause of any disability. DT is capable of living a quite normal life with his male partner, while also appearing on television and participating in science experiments. He can speak 10 languages, some of which he learned in the course of a few days. DT may be an example of a person with acquired savant syndrome, as he reports that his extreme mathematical abilities kicked in after a series of seizures he had when he was four.

Many individuals with savant syndrome no doubt possess virtuous character traits, such as curiosity, creativity and intellectual honesty. But some of them have extreme intellectual abilities without possessing any evident virtuous character traits. Sacks' twins would spend all day taking turns citing prime numbers. Kay and Flo have spent most of their lives watching television. Stephen Wiltshire could draw accurate pictures of landscapes he had seen but could not talk or carry out simple intellectual tasks. Many of these amazing people have virtually none of the character traits traditionally considered intellectually virtuous. But it would be hard to deny them knowledge, even knowledge of a quite impressive kind.

3 Virtue Reliabilism

Unlike virtue responsibilists, virtue reliabilists take intellectual virtues to be a reliable cognitive faculty or ability, for instance, vision, memory, introspection, and reason. On this approach, knowledge requires the attainment of cognitive success because of an exercise of a cognitive faculty or ability. As I argued in my (2006), the main problem with virtue reliabilism is that it cannot easily count cognitive success that results from assisted cognitive or electronic processing as knowledge. The main players in this field defend their approach by saying that almost all types of knowledge acquisition use assistance that is not strictly part of the virtuous system and that credit or partial credit can accrue nonetheless (Sosa 2007; Greco 2007). But if this is their reply to my concern, then I fail to see how their approach differs from more traditional forms of reliabilism.

3.1 Cognitive Success with Assistance

Memory recovery has been associated with false memory syndrome, a condition in which false memories are induced in a person using drug, hypnosis, literal dream interpretation, and so on. Because of a series of cases of false memories of childhood abuse in the 1980s, psychiatrists were advised to avoid engaging in any memory recovery techniques to uncover memories of past sexual abuse (Brandon et al. 1997).

However, memory recovery techniques have been successfully used for other purposes. For example, they have been successfully used to help sex offenders who have no memory of the offense they committed recover memory (Serran and Marshall 2005; Marshall et al. 2005). Sex offenders in this group often do not deny that they committed the offense. They agree that the evidence overwhelmingly show that they committed a crime. But they cannot remember what happened. Part of the technique is to use contextual cues that occurred on the day of the offense, including anything they can remember from when they woke up that morning to their feelings throughout the day. This approach is repeated over several days. Sex offenders who have recovered memories with the assistance of a therapist can be said to have attained cognitive success. However, the sex offenders' cognitive success is as much a result of the therapist guiding them through the memory recovery techniques as it is a result of their own cognitive effort.

Often cognitive faculties and abilities are highly unreliable without the assistance of other people. Recent studies in personality psychology, for example, reveal that we cannot attain full self-knowledge without assistance from others. There are certain areas in which we can predict how we will behave in the future, but in other areas other people are better predictors. We are our own best judges of neuroticism-related traits (e.g., self-esteem). Friends, on the other hand, are the best judges of intellect-related traits (e.g., openness, creativity and intelligence) and people of all perspectives are equally good at judging extraversion-related traits (e.g., talkativeness, leadership and dominance) (Vazire 2010). The studies indicate that full self-knowledge cannot be obtained as a result of exercise of our own sensory and cognitive abilities.

Individuals with Alzheimer's disease undergo progressive memory loss due to tangles and plaque formation in the hippocampus. Some of these individuals benefit from electronic devises that help trigger memories in their diseased brains. One such devise is MemeXerciser, which was developed by Carnegie Mellon doctoral candidate Matt Lee. MemeXerciser requires uploading photo collages and sounds from events at early stages during the disease. At later stages the device can determine importance of photos and sounds for each event, and by pressing a button and surveying the presented material, the Alzheimer's patient can recall events within a few seconds. The Alheimer's patient attains cognitive success by exercising his cognitive abilities but only in combination with electronic devises.

Andy Clark and David Chalmers (1998) have made a convincing case for what they call the 'extended mind hypothesis'. The extended mind hypothesis is the view that a mental state can extend out into the world. For example, if I routinely rely on information stored in my iPhone, then my iPhone can become part of my belief system. Alva Noe (2006) argues for a similar thesis for the case of visual experience. If the extended mind hypothesis is right, then it is plausible that our mental states do indeed include the memory states of assisting therapists, friends and fancy electronic devices.

However, the problem with this sort of response as a way of defending virtue reliabilism is that even if the extended mind hypothesis is correct, it is doubtful that our mental states do indeed extend out into the world in the kinds of cases cited above. The mental states of the therapist who assists a sex offender in recovering his memories are not constituent parts of the sex offender's mind.

The examples just cited make it clear that while we no doubt need cognitive faculties to produce knowledge, cognitive success very often is the result of joint effort. The joint effort it takes to reach cognitive success does not make cognitive faculties irrelevant but it does cast doubt on the hypothesis that knowledge requires an attainment of cognitive success *because of* the exercise of cognitive faculties or abilities. An analogy may be helpful here. Suppose I help you clean your house before your guests arrive. Knowing what a slop you are, the guests comment on the neatness of your house upon entering. It would be wrong for you to triumphantly declare that you managed to keep the dirt away because of your cleaning skills.

3.2 *Why Did We Need Virtue Epistemology in the First Place?*

There may well be ingenious ways in which virtue reliabilists can reply to the above criticisms. It is not an unfair response to maintain that almost all types of knowledge acquisition use assistance that is not strictly part of the virtuous system. But this sort of response makes it difficult to see the difference between virtue reliabilism and traditional reliabilism.

I am also doubtful of attempts to rescue virtue reliabilism for the sake of preserving it as a theory of knowledge. For virtue reliabilism to be preferable to traditional reliabilism, it must offer solutions to problems that are unsolvable on a conventional approach.

Virtue-theoretical approaches were indeed introduced because it was thought that they would solve several otherwise impenetrable problems in epistemology. One notorious problem in the recent history of epistemology is that of how to analyze knowledge. The traditional analysis of knowledge as justified true belief familiarly failed because of the Gettier problem. There have been many spirited attempts to solve the Gettier problem. Virtue epistemologists claimed to have the ultimate solution (Greco 2003; Sosa 2007; Zagzebski 1996; for discussion of the 'because of' relation, see Greco 2003; Levin 2004). According to them, Gettiered subjects do not have a true belief *because of* their exercise of intellectual virtues. Imagine that Hank sees what appears to be a sheep on a hill. What Hank actually sees is a rock. It happens, though, that behind this rock, and out of Hank's view, is a sheep. In this case, Hank does indeed enjoy a justified true belief that there is a sheep on the hill, but he does not know that there is a sheep on the hill, since it is merely accidental that one is present. According to virtue epistemologists, the reason Hank fails to know that there is a sheep on the hill is that his true belief does arise out of exercising intellectual virtues. He attained cognitive success because of the coincidental presence of a sheep not because of efforts on his part.

However, the problem with this reason for keeping virtue epistemology in the running is that anyone could adopt a similar solution to the Gettier problem. A traditional reliabilist can truly say that our Gettiered subject fails to have knowledge because her cognitive success is due to the coincidental presence of a sheep and not to the implementation of reliable mechanisms. Likewise, an evidentialist can truly say that our Gettiered subject fails to have knowledge insofar as she attained cognitive success because of a coincidence and not because of the evidence in her possession.

Virtue epistemologists also claim to hold the key to the value problem in epistemology (Riggs 2002; Greco 2003, 2007; Zagzebski 2003; Brogaard 2006; Sosa 2007). The value problem is that of explaining the intuition that knowledge is more valuable than mere true belief. A common way for virtue reliabilists to reply to this problem is to say that we value knowledge more than mere true belief because when true belief adds up to knowledge the agent is to credit for her true belief. We admire the great skill involved in producing a true belief for much the same reason that we admire the great skill of talented athletes.

I shall return to the question of whether the value problem really is a problem below. Regardless of whether it is, the virtue epistemologist does not succeed in solving it. Take Jennifer Lackey's example of the Chicago tourist (Lackey 2007, 2009; Pritchard 2008). If one comes to know that the Sears Tower is two blocks west of here by asking the first person who happens to pass by, then one really doesn't deserve much credit for one's achievement. There is hardly any great and admirable skill involved in asking an arbitrary person on the street whether she knows how to get to the Sears Tower. When considered in isolation of the efforts put into attaining it, states of knowledge aren't all that admirable.

4 Intellectual Flourishing

I think Kvanvig (2003) was right when he said that there has been too much focus on knowledge in epistemology. Kvanvig wanted to grant a higher status to understanding than knowledge. I don't think turning our attention to understanding will make a big difference. I have argued elsewhere that understanding just is a form of knowledge (Brogaard 2005). A more important observation is that there are epistemic goods that are more valuable than knowledge. Consider the following case (Brogaard 2011a):

> *Brain Damage*
> A has a brain condition that causes him to intend to keep track of truths about leaves. He believes that he can achieve this only if he intends to count the leaves on the trees in his garden every day.

If A does what he believes is necessary for him to intend to keep track of truths about leaves, and he is a good counter, his intention is likely to maximize true belief and minimize false ones. If he didn't intend to count leaves, he would go about his everyday business forming a lot more false beliefs than he does if he is just counting leaves all day. Although A has a brain condition that causes him to intend to keep track of truths about leaves, he nonetheless has excellent cognitive faculties. So his intention also ensures the greatest extent of knowledge. But we would not advice A to engage in this inconsequential and repetitious task every day. There are other things that are intellectually more important.

In precious work I argued that when we value knowledge, it is because it makes us flourish intellectually (Brogaard 2011a). If we focus too narrowly on knowledge, we lose track of the big picture. What we ought to do intellectually is avoid hindering intellectual flourishing. I suggested the following instances of this overall principle.

Intellectual Flourishing (belief)
You should believe p only if believing p does not hinder intellectual flourishing.

Intellectual Flourishing (assertion)
You should assert p only if asserting p does not hinder intellectual flourishing.

Intellectual Flourishing (action)
You should treat p as a reason for action only if treating p as a reason for action does not hinder intellectual flourishing.

Intellectual Flourishing is an epistemic norm just like principles such as 'Don't form beliefs while drunk, 'Be open-minded' and 'Avoid the gambler's fallacy'. But there is good reason to think that, unlike 'Don't form beliefs while drunk', 'Be open-minded' and 'Avoid the gambler's fallacy', Intellectual Flourishing is the fundamental epistemic norm. If we were to give epistemic advice to A in the Brain Damage case, we would instinctively recommend that he stop counting leaves and start using his time in a more futile way. So, in the envisaged scenario, we would attach less epistemic value to intentions to engage in activities that maximize true belief and minimize false ones than we would to activities that do not do this.

If Intellectual Flourishing is the fundamental norm, then all other epistemic norms are derivative. Generally it is good to aim at the truth because doing so often doesn't hinder intellectual flourishing. Likewise, asserting only what you know is normally a good norm to obey because obeying it often does not hinder intellectual flourishing. But the principle that you should aim for knowledge as well as the principle that you ought to maximize true belief and minimize false belief are derivative epistemic norms, or what we might call 'ceteris paribus laws', and so can come into conflict with Intellectual Flourishing.

To say that intellectual flourishing is an ultimate epistemic good is not to say that it is an ultimate goal that we work toward and receive only at the end of our life. Intellectual flourishing is a continuous process of living a good intellectual life. It is the epistemic equivalent of Aristotle's eudaimonia (well-being, flourishing, happiness). For Aristotle, eudaimonia requires having a virtuous character, being loved and having close friends. If we extend this idea to intellectual flourishing, then intellectual flourishing might involve such things as avoiding intellectual bigotry, seeking to expand on one's knowledge, making wise intellectual choices, being respected and admired intellectually and having good intellectual cohorts. Just as we cannot flourish, in Aristotle's sense, in solitude, so we cannot flourish intellectually outside of a community. Intellectual flourishing differs in this respect from knowledge acquisition. While a brain in a vat that is not properly connected to other individuals could, in principle, acquire knowledge as well as you and me, it cannot flourish intellectually.

While virtuous character traits and well-functioning cognitive faculties and abilities can lead to a good intellectual life, there are many cases in which true belief flows from virtuous character traits or well-functioning cognitive faculties and abilities but in which the agent is not on the right track intellectually speaking. Each individual is unique and thus possesses a particular set of personality traits and mental abilities and is situated in her own social and historical context. Needs, mental acumen and circumstances affect an individual's eudaimonia. An activity that can contribute to one individual's eudaimonia may not be relevant to another's. For example, it's intellectually admirable if a 5-year old writes a 20-page short-story about a hamster that goes to the dentist because the achievement outruns our expectations for 5-year olds. But writing a 20-page short-story about a hamster that goes to the dentist may not contribute to the intellectual flourishing of a grown man or woman, regardless of how intellectually virtuous he or she is and regardless of how well his or her cognitive faculties and abilities function.

Being intellectually virtuous may also be insufficient for intellectual flourishing if intellectual achievements that flow from the virtues are not admirable by public measures. Suppose you are an ambitious philosopher with generally well-functioning cognitive faculties and abilities and many good personality traits.[4] You invest a great deal in writing articles and books. You have what seems to be a perfectly successful career. Your articles and books regularly win prizes and public praise. In fact, however, unbeknownst to you, all your papers and books are published and assessed by people hired by your rich uncle who took pity on your because you are such a bad philosopher. Needless to say, in this case you do not flourish intellectually despite believing that you do, as your intellectual achievements fail to meet public measures of greatness.

Being intellectually virtuous in the conventional sense may preclude flourishing intellectually. Suppose A hears of a new proof that God does not exist. A knows that if he sees the proof and the proof is correct, he will become terribly depressed and will spend the rest of his life in isolation from intellectual cohorts. To ensure that he flourishes intellectually, A must refrain from looking at the proof, even if this move does not involve the exercise of intellectual virtue.

We can, of course, correctly say of an agent who flourishes intellectually that he or she is 'intellectually virtuous', he or she just isn't virtuous in the classical sense. There is no one set of character traits that an agent who flourishes intellectually must have. What can be a positive trait in one situation may be a bad trait in a different situation. For example, you should not be intellectually honest while carrying out an experiment that involves deceit. Truth-telling in this situation would ruin the experiment. Even intellectual justice can counteract eudaimonia. To be unjust in the intellectual domain is to do something that could potentially hinder the intellectual flourishing of others. Destroying other people's intellectual property, preventing others from developing their mental abilities, rewarding unworthy rather than worthy intellectual achievements, obstructing intellectual amity and camaraderie, disrespecting the intellectual work of others on irrelevant grounds, such as gender or skin color, are all prima facie intellectually unjust activities. The very possibility of eudaimonia presupposes justice. However, what counts as unjust in one situation may count as just in another. In general, it is unjust to prevent people from posting their thoughts on their website. However, it may be just to prevent people from posting bigoted content.

The epistemic theory that forms from the above considerations is best understood as a eudaimonistic virtue epistemology. But it is not one that gives center stage to knowledge. Nor is it one that ranks, weighs or balances intellectual virtues independently of each situation. In some situations, virtue in the traditional sense is

---

[4] This example is adapted from Brogaard and Smith (2005).

overridden by a concern for the other dimensions of eudaimonia, including having good intellectual cohorts and being admired and respected.

5 Conclusion

We considered the prospects of two types of virtue theories of knowledge, virtue responsibilism and virtue reliabilism. Both types of virtue theories hold that knowledge requires attaining cognitive success because of the exercise of intellectual virtue. However, they disagree about the nature of intellectual virtues. Virtue responsibilism maintains that intellectual virtues are virtuous personality traits, such as creativity, curiosity and intellectual honesty, whereas virtue reliabilism holds that intellectual virtues are cognitive faculties and abilities, such as vision, memory, introspection and reason.

I have argued that virtue responsibilism falls short of accounting for everyday knowledge that non-virtuous individuals appear to possess. It also fails to account for certain kinds of knowledge-how that do not require making any conscious choices. Finally, it cannot accommodate the knowledge of individuals who fail to be intellectually virtuous but who possess extremely impressive cognitive mechanisms.

Virtue reliabilism fares better in these respects. However, virtue reliabilism has no straightforward way of accounting for knowledge that ensues owing to shared effort. A standard reply offered by virtue reliabilists is that we cannot deny that many forms of knowledge require more than just reliable cognitive faculties and abilities. But once they make this concession, it is unclear how their view differs from traditional reliabilism.

I concluded by making a case for a eudaimonistic virtue epistemology that rejects the idea that knowledge is central to the enterprise of epistemology. Knowledge may be what we aim for in some situations but not all kinds of knowledge are equally good. Some kinds promote intellectual flourishing, other kinds hinder it. Because our search for the intellectually good life can override our pursuit of knowledge, intellectual flourishing is a more fundamental intellectual good than knowledge.

References

Alfano, M. Forthcoming. Expanding the situationist challenge to respinsibilist virtue epistemology. *Philosophical Quarterly*.

Ariely, D. 2008. *Predictably irrational: The hidden forces that shape our decisions*. New York: Harper.

Axtell, G. 1997. Recent work in virtue epistemology. *American Philosophical Quarterly* 34: 410–430.

Baehr, J. 2006. Character, reliability and virtue epistemology. *The Philosophical Quarterly* 56(223): 193–212.

Baehr, J. 2011. *The inquiring mind: On intellectual virtues and virtue epistemology*. Oxford: Oxford University Press.

Bargh, et al. http://www.yale.edu/acmelab/articles/bargh_chen_burrows_1996.pdf.
Barlow, D.H., and V.M. Durand. 2008. *Abnormal psychology: An integrative approach*. Belmont: Cengage Learning.
Bor, D., J. Billington, and S. Baron-Cohen. 2007. Memory for digits in a case of synaesthesia and Asperger syndrome is related to hyperactivity in the lateral prefrontal cortex. *Neurocase* 13: 311–319.
Brandon, S., J. Boakes, D. Glaser, R. Green, J. MacKeith, and P. Whewell. 1997. Reported recovered memories of child sexual abuse: Recommendations for good practice and implications for training, continuing professional development and research. *Psychiatric Bulletin* 21(10): 663–665.
Brogaard, B. 2005. I know. Therefore, I understand. Unpublished manuscript. Available at: https://sites.google.com/site/brogaardb/.
Brogaard, B. 2006. Can virtue reliabilism explain the value of knowledge? *Canadian Journal of Philosophy* 36: 335–354.
Brogaard, B. 2011a. Intellectual flourishing as the fundamental epistemic norm, University. In J. Turri and Clayton Littlejohn. Oxford: Oxford University Press, 2012.
Brogaard, B. 2011b. Knowledge-how: A unified account. In *Knowing how: Essays on knowledge, mind, and action*, ed. J. Bengson and M. Moffett. Oxford: Oxford University Press.
Brogaard, B., and B. Smith. 2005. On luck, responsibility, and the meaning of life. *Philosophical Papers* 34: 443–458.
Castiello, U., and M. Jeannerod. 1991. Measuring time to awareness. *Neuroreport* 2: 797–800.
Castiello, U., Y. Paulignan, and M. Jeannerod. 1991. Temporal dissociation of motor responses and subjective awareness. A study in normal subjects. *Brain* 114: 2639–2655.
Clark, A., and D.J. Chalmers. 1998. The extended mind. *Analysis* 58: 7–19.
Code, L. 1984. Toward a 'responsibilist' epistemology. *Philosophy and Phenomenological Research* 45(1): 29–50.
Dehaene, S. 1999. *The number sense*. Oxford: Oxford University Press.
Dehaene, S. 2001. Précis of the number sense. *Mind & Language* 16: 16–36.
Dehaene, S. 2007. Symbols and quantities in parietal cortex: Elements of a mathematical theory of number representation and manipulation". In. *Sensorimotor foundations of higher cognition, volume XXII of Attention and Performance, chapter 24* (ed.) Patrick Haggard, Yves Rossetti, and Mitsuo Kawato, 527–574. Oxford: Harvard University Press.
Dehaene, S., G. Dehaene-Lambertz, and L. Cohen. 1998. Abstract representations of numbers in the animal and human brain. *Trends in Neuroscience* 21: 355–361.
Dehaene, S., N. Molko, L. Cohen, and A. Wilson. 2004. Arithmetic and the brain. *Current Opinion in Neurobiology* 14: 218–224.
Doris, J. 2011. A natural history of the self. forthcoming.
Eger, E., V. Michel, B. Thirion, A. Amadon, S. Dehaene, and A. Kleinschmidt. 2009. Deciphering cortical number coding from human brain activity patterns. *Current Biology* 19: 1608–1615.
Goodale, M.A., and A.D. Milner. 1992. Separate visual pathways for perception and action. *Trends in Neurosciences* 15: 20–25.
Goodale, M.A., D. Pelisson, and C. Prablanc. 1986. Large adjustments in visually guided reaching do not depend on vision of the hand or perception of target displacement. *Nature* 320: 748–750.
Greco, J. 2000. *Putting skeptics in their place*. New York: Cambridge University Press.
Greco, J. 2003. Knowledge as credit for true belief. In *Intellectual virtue: Perspectives from ethics and epistemology*, ed. M. DePaul and L. Zagzebski. Oxford: Oxford University Press.
Greco, J. 2007. The nature of ability and the purpose of knowledge. *Philosophical Issues* 17: 57–69.
Greco, John, and John Turri. 2011. Virtue epistemology *The Stanford Encyclopedia of Philosophy* (Winter 2011 Edition), Edward N. Zalta (ed.), forthcoming. http://plato.stanford.edu/archives/win2011/entries/epistemology-virtue/
Hubbard, E.M., M. Piazza, P. Pinel, Dehaene S. Numerical, and Spatial Intuitions. 2009. A role for posterior parietal cortex? In *Cognitive biology: Evolutionary and developmental perspectives on mind, brain and behavior*, ed. L. Tommasi, L. Nadel, and M.A. Peterson, 221–246. Cambridge: MIT Press.

Jakobson, L.S., and M.A. Goodale. 1989. Trajectories of reaches to prismatically-displaced targets: Evidence for 'automatic' visuomotor recalibration. *Experimental Brain Research* 78: 575–587.
Johansson, P., L. Hall, S. Sikström, and A. Olsson. 2005. Failure to detect mismatches between intention and outcome in a simple decision task. *Science* 310: 116–119.
Johansson, P., L. Hall, S. Sikström, B. Tärning, and A. Lind. 2006. How something can be said about telling more than we can know. *Consciousness and Cognition* 15: 673–692.
Karnath, H.-O., S. Ferber, and M. Himmelbach. 2001. Spatial awareness is a function of the temporal not the posterior lobe. *Nature* 411: 950–953.
Kelp, C. 2011. In defence of virtue epistemology. *Synthese* 179: 409–433.
Kvanvig, J. 2003. *The value of knowledge and the pursuit of understanding*. Cambridge: Cambridge University Press.
Lackey, J. 2007. Why we don't deserve credit for everything we know. *Synthese* 158: 345–361.
Lackey, J. 2009. Knowledge and credit. *Philosophical Studies* 142: 27–42.
Levin, Michael. 2004. Virtue epistemology: No new cures. *Philosophy and Phenomenological Research* 69(2): 397–410.
Lythgoe, M., T. Pollak, M. Kalmas, M. de Hann, and W.K. Chong. 2005. Obsessive, prolific artistic output following subarachnoid hemorrhage. *Neurology* 64: 397–398.
Marshall, W.L., G. Serran, L.E. Marshall, and Y.M. Fernandez. 2005. Recovering memories of the offense in "amnesic" sexual offenders. *Sexual Abuse* 17: 31–38.
Milner, A.D., and M.A. Goodale. 2008. Two visual systems re-viewed. *Neuropsychologia* 46: 774–785.
Montmarquet, J. 1996. Epistemic Virtue. *Mind* 96:384 (1987), pp. 482–497; *Epistemic virtue and doxastic responsibility*. (Lanham, Md.: Rowman & Littlefield, 1993); L. Zagzebski, *Virtues of the Mind*. Cambridge: Cambridge University Press, 1996.
Murphy, S.T., and R.B. Zajonc. 1993. Affect, cognition, and awareness: Affective priming with optimal and suboptimal stimulus exposures. *Journal of Personality and Social Psychology* 64: 723–739.
Navarro, C. 2008. Why do some people sleepwalk? Scientific American, 31 Jan 2008.
Noë, A. 2006. Experience without the head. In *Perceptual experience*, ed. Tamar S. Gendler and John Hawthorne. Oxford: Oxford University Press.
Paulignan, Y., C.L. MacKenzie, R.G. Marteniuk, and M. Jeannerod. 1991. Selective perturbation of visual input during prehension movements. 1. The effect of changing object position. *Experimental Brain Research* 83: 502–512.
Peek, F. 1996. *The real rain man: Kim Peek*. Salt Lake: Harkness.
Pelisson, D., C. Prablanc, M.A. Goodale, and M. Jeannerod. 1986. Visual control of reaching movements without vision of the limb. II. Evidence of fast unconscious processes correcting the trajectory of the hand to the final position of a double-step stimulus. *Experimental Brain Research* 62: 303–311.
Piazza, M., P. Pinel, D. Le Bihan, and S. Dehaene. 2007. "A magnitude code common to numerosities and number symbols in human intraparietal" cortex. *Neuron* 53: 293–305.
Pritchard, D. 2008. Radical scepticism, epistemic luck and epistemic value. *Proceedings of the Aristotelian Society* 82(suppl): 19–41.
Riggs, W. 2002. Reliability and the value of knowledge. *Philosophy and Phenomenological Research* 64: 79–96.
Sacks, O. 1995. *The man who mistook his wife for a hat, and other clinical tales*. New York: Summit Books.
Serran, G.A., and V.L. Marshall. 2005. The "memory recovery technique" a strategy to improve recall of offense-related details in men who commit sexual assaults. *Clinical Case Studies* 4: 3–12.
Snyder, A.W., E. Mulcahy, J.L. Taylor, D. Mitchell, P. Sachdev, and S.C. Gandevia. 2003. Savant-like skills exposed in normal people by suppressing the left fronto-temporal lobe. *Journal of Integrative Neuroscience* 2: 149–158.
Sosa, E. 2007. *A virtue epistemology. Apt belief and reflective knowledge*, vol. 1. Oxford: Oxford University Press.
Treffert, D.A. 2009. The savant syndrome: An extraordinary condition. A synopsis: Past, present, future. *Philosophical Transactions of the Royal Society B* 364: 1351–1357.

Vazire, S. 2010. Who knows what about a person? The self–other knowledge asymmetry (SOKA) model. *Journal of Personality and Social Psychology* 98: 281–300.
Williamson, T. 2000. *Knowledge and its limits*. Oxford: Oxford University Press.
Winkielman, P., and K.C. Berridge. 2004. Unconscious emotion. *Current Directions in Psychological Science* 13: 120–123.
Wisconsin Medical Society, Islands of Genius. http://www.wisconsinmedicalsociety.org/savant_syndrome/
Young, R.L., M.C. Ridding, and T.L. Morrell. 2004. Switching skills by turning off part of the brain. *Neurocase* 10: 215–222.
Zagzebski, L. 1996. *Virtues of the mind: An inquiry into the nature of virtue and the ethical foundations of knowledge*. Cambridge: Cambridge University Press.
Zagzebski, L. 2003. The search for the source of epistemic good. *Metaphilosophy* 34: 12–28.

Expanding the Situationist Challenge to Reliabilism About Inference

Mark Alfano

1 Introduction

Following Lorraine Code's landmark article, "Toward a 'responsibilist' epistemology" (1984), virtue epistemology today is typically taken to cleave into three families of views:

- *Reliabilism*, exemplified by Ernest Sosa, Alvin Goldman, and John Greco, which sees the intellectual virtues as non-motivational capacities, dispositions, or processes that tend to lead their possessors to increase the balance of truths over falsehoods in their belief sets (e.g. sound deduction, good eyesight, capacious memory, etc.)[1]
- *Responsibilism*, exemplified by Lorraine Code, James Montmarquet, and Linda Zagzebski, which views the intellectual virtues on analogy with the neo-Aristotelian moral virtues as motivational, reasons-responsive dispositions to act and react in characteristic ways (e.g., open-mindedness, curiosity, intellectual courage, etc.)
- *Mixed virtue epistemology*, exemplified by Jason Baehr, Christopher Hookway, and Christopher Lepock, which countenances the virtues of both reliabilism and responsibilism.

Meanwhile, the last 13 years have seen the rise of the so-called *situationist challenge to virtue ethics*. John Doris (1998, 2005) and Gilbert Harman (1999, 2000, 2001, 2003, 2006) have been the primary proponents of this challenge, with further arguments made by Christian Miller (2003, 2009), Peter Vranas (2005) and me (2011a, 2013a, b). Virtue ethicists in the neo-Aristotelian tradition such as

[1] It might seem that Goldman does not belong on this list, but in his (1992) he embraced the virtue epistemology label.

M. Alfano (✉)
Center for Human Values, Princeton University, Princeton, NJ, USA

Department of Philosophy, University of Oregon, Eugene, OR, USA
e-mail: mark.alfano@gmail.com

A. Fairweather (ed.), *Virtue Epistemology Naturalized: Bridges Between Virtue Epistemology and Philosophy of Science*, Synthese Library 366, DOI 10.1007/978-3-319-04672-3_7,

Alasdair MacIntyre (1984) have been wont to argue that their view provides an empirically adequate moral psychology, one that presupposes that many people have such traits as honesty, temperance, and courage. Unlike consequentialists and deontologists, whose theories focus on the purely evaluative, virtue ethicists use a vocabulary that is simultaneously evaluative and explanatory. To say that someone acted compassionately is both to praise her (or her action) as manifesting a virtue and (partially) to explain that action as flowing from her character. This means that virtue ethicists have empirical skin in the game: if virtue ethics is explanatory, then the virtues had better be psychologically real. Philosophical situationists argue, however, that most people do not possess traits that resemble the virtues as traditionally conceived. Both in our behavior and in our thought, feeling, and deliberation, we are astonishingly susceptible to seemingly trivial and normatively irrelevant features of our situations, such as mood elevators, mood depressors, ambient sounds, ambient smells, social distance cues, and even the weather. If exceedingly few people are virtuous, then explaining human conduct in terms of the virtues is a hopeless endeavor.

It seems only natural that eventually we would see the convergence of the twain: *the situationist challenge to virtue epistemology.*[2] In my (2011b), I argued that the challenge straightforwardly extends to responsibilist varieties of virtue epistemology. In this paper, I extend that argument, claiming that reliabilist virtue epistemology must come to terms with a situationist challenge of its own. I begin by spelling out in more detail the reliabilist approach to epistemology. Next, I outline the contours of the situationist challenge to virtue ethics. I then show, using evidence drawn from cognitive psychology, that the challenge can be expanded to reliabilism about inference.

2 Virtue Epistemology

Code's distinction between reliabilism and responsibilism remains the touchstone of contemporary taxonomies such as Heather Battaly's (2008), though some virtue epistemologists have begun to argue that only a mixed theory that countenances the virtues of both reliabilism and responsibilism can do justice to the panoply of epistemically important dispositions. This section briefly sketches the reliabilist and mixed.

2.1 Reliabilist Virtue Epistemology

Reliabilist theories aim to resolve well-known epistemological puzzles such as Gettier cases, lottery paradoxes, and skeptical arguments. Solutions to these puzzles

[2] John Doris (2005, p. 110) briefly gestures in this direction, as does Jason Baehr (2006, p. 8 fn. 15).

are framed in terms of non-motivational traits, abilities, capacities, and processes such as perception, intuition, and memory. Although such dispositions are not Aristotelian moral virtues, it is generally agreed that they are sufficiently virtue-like (because they are stable dispositions to think and reason in characteristic ways) that it is not a misnomer to class them with the virtues (Sosa 1991, p. 271). Three of the most influential virtue epistemologists are Alvin Goldman (1992), John Greco (1992a, b, 1993, 2000, 2009), and Ernest Sosa (1980, 1985, 1991, 2001, 2007, 2011), whose views I will treat as representative throughout this article.

The key to this approach to justification and knowledge is the direction of analysis. One starts with notions of the various intellectual virtues, then uses them to define more traditional epistemic concepts, such as justification and knowledge. Roughly, reliabilists define epistemic *justification* in terms of the epistemic *virtues* and define *knowledge* in terms of *truth* and epistemic *justification* (and hence indirectly in terms of the epistemic virtues).[3] Beliefs acquired through the exercise of these faculties are justified (Goldman 1992, p. 157). Or, in Sosa's words, "A belief *B* is justified if and only if it is the outcome of a process of belief acquisition or retention which is reliable, or leads to a sufficiently high preponderance of true beliefs over false beliefs" (1992, p. 80).[4] True beliefs so acquired count as knowledge. As Greco puts it, "S knows that p *if and only if* S's believes [sic.] the truth (with respect to p) because S's belief that p is produced by intellectual ability" (2009, p. 18). Or, as Sosa says, "knowledge is true belief out of intellectual virtue, belief that turns out right by reason of the virtue and not just by coincidence" (1991, p. 277).

Thus for the reliabilist, someone has a justified belief that the cat is on the mat if he comes to believe that the cat is on the mat because he *sees* the cat on the mat, whereas someone has an unjustified belief that the cat is on the mat if he comes to believe that the cat is on the mat because he *hopes* the cat is on the mat. Moreover, the belief is justified only if the agent is stably disposed to see well. If he is legally blind but occasionally identifies cats correctly after much squinting, we would be less inclined to admit that his belief is justified. If he is good at identifying tabbies as cats but only mediocre at identifying other breeds as cats, or if he only identifies mats as mats when he is in a good mood, we would be reluctant to attribute justification. Reliabilists explain this reluctance in terms of the epistemic virtues: "When we categorize a belief [as justified or reasonable] we speak directly of the belief but also, indirectly, of the believer, whose intellectual reliability is also under evaluation" (Sosa 2001, p. 58; see also Sosa 2007, p. 29). If we recognize that the agent is

[3] Note that epistemic justification and other types of justification may come apart, as Driver (2003, p. 110) and Sosa (1991, p. 165) have argued. For example, it seems that an ill person has practical but not epistemic justification for believing that he will recover. So believing would tend to trigger a placebo effect, which would in turn facilitate recovery, but that is not in itself warrant for the belief.

[4] It should be clear that I'm dealing with what Sosa refers to as "reflective knowledge," not "animal knowledge." As far as I can tell, animal knowledge is impervious to situationist critique.

unreliable, then – even if he gets it right this one time – we will be disinclined to say that his belief is justified.

Another example, this time having to do with knowledge: someone knows the cat is on the mat if he concludes that the cat is on the mat because he knows the cat is either on the mat or in the box, knows the cat is not in the box, and makes an inference using disjunctive syllogism. He would not know that the cat is on the mat (even if it is) if he concludes that the cat is on the mat because he knows that George Washington was the first American president, makes an inference using *tonk*-introduction to <George Washington was the first American president *tonk* the cat is on the mat, then makes another inference using *tonk*-elimination to <the cat is on the mat>. Moreover, as before, the agent has knowledge only if he is stably disposed to use disjunctive syllogism (and other sound deductive rules) and stably disposed not to use *tonk* and its unsound ilk. If he uses the rules of classical deductive inference when it's cloudy but *tonk*y deductive inference when it's fair, we would find it difficult to attribute knowledge to him even during stormy weather. As before, reliabilists would explain this reluctance in terms of intellectual character. To say that a certain belief rises to the level of knowledge is to say something about the belief, but it's also to say something about the believer. According to Sosa (2001, p. 58), "What one cares about in oneself and in one's epistemic fellows is a relevantly stable, dependable character."

To sum up, then, reliabilist virtue epistemology starts with a subset of the stable, counterfactual-supporting traits of intellectual character – namely, the virtuous ones, which tend to increase their possessors' balance of truth over falsehood. It then defines both the epistemic justification of beliefs and the epistemic justification of agents in terms of intellectual character. An agent has justification if and only if she is exercising her intellectual virtues and not exercising any intellectual vices, and any given belief of hers is justified just in case it was acquired and retained through the exercise of her intellectual virtues (and not through the exercise of any of her intellectual vices). Finally, reliabilism defines knowledge as true belief acquired and retained through the exercise of intellectual virtues (and the absence of intellectual vices).

2.2 *Mixed Virtue Epistemology*

More recently, epistemologists in both the reliabilist and the responsibilist camps have groped towards a consensus on mixed virtue epistemology.[5]

From the responsibilist side, Zagzebski has not resisted saying that her view presupposes reliabilist virtues. For one's desire to believe the truth and avoid falsehood to be satisfied, one must possess many cognitive capacities, abilities, and dispositions. She goes so far as to claim that even the moral virtues involve perceptual and

[5] See, among others, Baehr (2006), Greco (1992a, 2000), Lepock (2011), Sosa (2011, pp. 15–16), and Zagzebski (1996, p. 149).

cognitive states, saying, "No one has the virtue of fairness or courage or compassion or generosity without generally being in cognitive contact with the aspect of reality handled by the respective virtue. Otherwise, one would not be reliably successful." Furthermore, she continues, "Being reasonably intelligent within a certain area of life is part of having almost any moral virtue" (1996, p. 149).

3 The Situationist Challenge to Virtue Ethics

Like responsibilists, virtue ethicists typically construe virtues as motivational traits of character. To be generous is to have a complex disposition to *notice* when others are in need, to *construe* ambiguous social cues charitably, to *want* to help others whom one takes to be in need, to *deliberate* soundly about what would in fact help a given person in particular circumstances, and to *succeed* in helping when one tries. Such traits are "thick" in the sense that they are descriptively rich (knowing that someone is generous tells you a lot about her), explanatorily powerful (the generous person not only *does help* but *would help* in circumstances where the same reasons were in force), and evaluatively laden (it's admirable to be generous). As Alasdair MacIntyre (1984, p. 199) puts it, "to identify certain actions as manifesting or failing to manifest a virtue or virtues is never only to evaluate; it is also to take the first step towards explaining why those actions rather than some others were performed." Furthermore, virtues are reasons-responsive traits; the generous person helps not because it makes her feel good (though ideally it does make her feel good) but because someone is in need.

What's more, virtue ethicists are rather egalitarian; they take virtues (and vices) to be instantiated in lots of ordinary people. MacIntyre goes so far as to say that "without allusion to the place that justice and injustice, courage and cowardice play in human life very little will be genuinely explicable" (p. 199).

The situationist challenge to virtue ethics is an attack primarily on this conjunction of consistency, explanatory power, and egalitarianism. According to the situationist critique, most people do not respond – or do not respond robustly and directly – to moral reasons. Seemingly trivial and normatively irrelevant features of our environments predict and explain our behavior better than such traits. Though consistency requires that people respond the same way whenever they have the same reasons (e.g., generously when they have decisive reason to help), psychologists and other social scientists have found that they respond differently depending on weather conditions, presence of bystanders,[6] social distance,[7] ambient smells, and ambient sounds,[8] among other things.

[6] For more on the power of bystanders, see Latané and Darley (1968, 1970), Latané and Nida (1981), Latané and Rodin (1969), Schwartz and Gottlieb (1991).

[7] For more on the power of social distance, see Bohnet and Frey (1999a, b) and Hoffman et al. (1996).

[8] For more on the power of ambient sensory stimuli, see Baron (1997), Baron and Thomley (1994), Boles and Haywood (1978), Cohen and Lezak (1977), Donnerstein and Wilson (1976), Geen and

While it is possible for virtue ethics to retreat into purely normative territory by giving up egalitarianism, such a move is distasteful for a number of reasons. One of the supposed strengths of virtue ethics is an empirically adequate moral psychology; such a retreat would abandon that strength. Additionally, the move toward moral elitism threatens to violate the "ought"-"can" implication. If people ought to be virtuous, presumably they can. But if virtue is rare and exceedingly difficult to attain, it might be that they really cannot.

A less extreme retreat endorses something like Doris's (2002, p. 62) theory of local virtues, which are individuated much more finely than the traditional virtues. Doris seems inclined, for instance, to distinguish a large variety of local traits that fit within the global virtue of courage, individuating traits as finely as required for them to actually support counterfactuals and confident predictions. Instead of *courage* or even *physical courage*, he would have us speak of *battlefield physical courage*, of *storms physical courage*, of *heights physical courage*, and of *wild animals physical courage*. Indeed, he even seems willing to differentiate between *battlefield physical courage in the face of rifle fire* and *battlefield physical courage in the face of artillery fire*. Though it might seem that he is being flip by cutting the fabric of traits so finely, he claims that this principle of individuation "is the beginning of an empirically adequate theory" (p. 62).

It's beyond the scope of this paper to discuss the intricacies of the dialectic between situationists and virtue ethicists. In this Section I aim only to limn the outlines of the situationist view. In a nutshell, the view is that traditional virtues are individuated coarsely by their characteristic reasons (e.g., others' need and one's ability to help), but that most people's actual thought, feeling, deliberation, and behavior are better characterized by finely individuated dispositions that make reference to seemingly trivial and normatively irrelevant situational features. In the next section, I marshal empirical evidence previously uncited in the present context to show that a similar argument applies to reliabilism about inference.

4 Expanding the Challenge to Reliabilism

At first blush, empirical evidence about what sorts of cognitive dispositions people actually possess would seem to be welcome news to reliabilists because it would help to solve the so-called *generality problem*. Recall that, on this view, knowledge is true belief acquired and retained through the exercise of intellectual virtues. In other words, someone knows that *p* just in case her belief that *p* was acquired and retained through a reliable process. However, any event of acquiring a belief could be classed under indefinitely man headings, some of which are reliable, others of which are not. Suppose that Susie comes to believe that the cat is on the mat, and that the cat really is on the mat. If we describe her

O'Neal (1969), Grimes (1999), Konecni (1975), Korte and Grant (1980), Korte et al. (1975), Matthews and Cannon (1975), and Page (1974).

belief-formation process as *seeing a cat on a mat*, then of course it is reliable. If, however, we describe it as *seeming to see a cat on a mat*, then it's not so obvious. All seeings of cats are seemings as of cats, but not all seemings as of cats are seeings of cats. Furthermore, not all seemings as of cats are veridical, but all seeings of cats are. And if we describe her belief-formation process as *seeing or hoping that the cat is on the mat*, then it's downright unreliable.

The problem is one of picking out the right principle of individuation for belief-formation processes. Should cognitive virtues be coarsely individuated, so that *inference* makes the cut, or should they be finely individuated, so that disjunctive syllogism makes the cut? Various answers have been proposed to the individuation question. James Beebe (2004) argues that cognitive processes should be individuated by the problems they solve, the algorithms they instantiate, and the cognitive architecture backing them. I've argued (2009) that they should be individuated more finely than the sense modalities (where two tokens fall under the same heading if and only if they are both cases of vision, or both cases of hearing, etc.) but less finely than doxastic equivalence (where two tokens fall under the same heading if and only if they would lead to all and only the same beliefs). Empirical evidence about what cognitive processes people actually use would help in sorting out such proposals. Instead of sitting in our armchairs wondering how to individuate cognitive dispositions, we could consult the psychological literature to find out what kinds of cognitive dispositions people actually have.

However, my interpretation of the empirical literature bodes ill for reliabilism, for it's hard to construe the cognitive dispositions we do in fact possess as intellectual virtues. A first-pass elaboration of the situationist challenge to reliabilism can be framed as an inconsistent triad:

(*inferential non-skepticism*) Most people know quite a bit inferentially.
(*inferential reliabilism*) Inferential knowledge is true belief acquired and retained through inferential reliabilist intellectual virtue.
(*inferential cognitive situationism*) People acquire and retain most of their inferential beliefs through heuristics rather than intellectual virtues.

The thesis of inferential *non-skepticism* is near-orthodoxy. In a recent PhilPapers survey of philosophers around the globe, 81.6 % of philosophers and 84.3 % of epistemologists rejected skepticism.[9] I will therefore treat *inferential non-skepticism* as unrevisable for the purposes of this article. At the very least, it would take impressive argumentative acrobatics to convince most epistemologists that they should abandon *inferential non-skepticism* instead of *inferential reliabilism*; if one of the three propositions must go, it's unlikely to be *inferential non-skepticism*.

The crucial question is therefore whether to accept *inferential cognitive situationism* and reject *inferential reliabilism* or, conversely, to reject *inferential cognitive situationism* and accept *inferential reliabilism*. Though I could not hope to summarize even a substantial plurality of the relevant research, my interpretation of the cognitive psychology literature is that both ordinary people and experts utilize a

[9] Data available at http://philpapers.org/surveys/results.pl.

motley of fine-grained heuristics to arrive at their inferential beliefs. These heuristics are surprisingly accurate in some ordinary circumstances, but they can easily lead to gross error. In this section, I will describe in some detail just a few illustrative studies of the *availability* and *representativeness heuristics*.

Ask yourself the following question: "In four pages of a novel (about 2,000 words), how many words would you expect to find that have the form _ _ _ _ *ing* (seven-letter words that end with 'ing')?" If you're like the subjects in Amos Tversky and Daniel Kahneman's (1973) study, you probably guessed 13 or 14. Now ask yourself a slightly different question: "In four pages of a novel (about 2,000 words), how many would you expect to find that have the form _ _ _ _ _ *n* _ (seven-letter words whose second-to-last letter is 'n')?" If you're like those subjects, then you probably guessed 4 or 5. Note, however, that any word of the form _ _ _ _ *ing* is necessarily of the form _ _ _ _ _ *n* _. Hence, to be consistent, your answer to the second question must be at least as high as your answer to the first. The availability heuristic leads people to expect that the probability of an event or the proportion of a property in a population is directly correlated to the ease with which examples can be brought to mind. It's easy enough to think of an example of a seven-letter word ending in 'ing': just start with a four-letter verb, then form a participle. It's not so easy to think of a seven-letter word whose penultimate letter is 'n'. Thus, our use of the availability heuristic leads us to estimate inconsistently in this case. The point here is not about estimating *correctly* but about the *internal consistency* of one's estimates. The same phenomenon occurs when subjects are asked to guess the number of words of the form _ _ _ _ _ *ly* and the number of words of the form _ _ _ _ _ *l* _ in a stretch of 2,000 words of prose. Average estimates in Tversky and Kahneman's study were 8.8 and 4.4 respectively. As in the previous example, every seven-letter word ending in 'ly' is also a seven-letter word whose penultimate letter is 'l', but because it's easier to think of an example of the former than the latter, people inconsistently estimate that the latter is more common than the former.

The availability heuristic uses availability or ease of recall as an index of probability or frequency. Because words ending in 'ing' are easier to conjure up than words whose penultimate letter is 'n', people take them to be more common. Another, closely related, cognitive process is the representativeness heuristic, in which the representativeness of a token is treated as an index of probability or frequency. For Tversky and Kahneman, "*Representativeness* is an assessment of the degree of correspondence between a sample and a population, an instance and a category, an act and an actor, or more generally, between an outcome and a model." They elaborate further, saying that representativeness "can be investigated empirically by asking people, for example, which of two sequences of heads and tails is more representative of a fair coin or which of two professions is more representative of a given personality" (2002, p. 22). If x is more representative of F than of G, then people typically say that it's more likely that x is F than that x is G, even when it's logically impossible because F is the property of being G and H. The upshot is that the representativeness heuristic leads to the conjunction fallacy.

Consider the now-infamous case of Linda: "Linda is 31 years old, single, outspoken and very bright. She majored in philosophy. As a student, she was deeply concerned with issues of discrimination and social justice, and also participated in anti-nuclear demonstrations." In a preliminary survey, Tversky and Kahneman had participants rate the degree to which Linda was representative of the following classes: elementary school teachers, bookstore employees who take yoga classes, feminists, psychiatric social workers, members of the League of Women Voters, bank tellers, insurance salespeople, and feminist bank tellers. 85 % said that Linda was more representative of feminists than feminist bank tellers. This is unobjectionable. The representativeness heuristic means, however, that people will therefore say that it's also *more likely* that Linda is a feminist bank teller than a bank teller. This is objectionable.

Tversky and Kahneman (2002) went on to investigate the effects of the representativeness heuristic using the case of Linda with a series of increasingly direct studies. In the least direct study, participants read the description of Linda, then ranked for probability (1 = most probable, 2 = second most probable, etc.) the following statements:

(T) Linda is a teacher in elementary school.
(S & Y) Linda works in a bookstore and takes yoga classes.
(F) Linda is active in the feminist movement.
(P) Linda is a psychiatric worker.
(L) Linda is a member of the League of Women Voters.
(B) Linda is a bank teller.
(I) Linda is an insurance salesperson.
(B & F) Linda is a bank teller and is active in the feminist movement.

In this study, either B and F were dropped from the list or B & F was dropped from the list. Therefore, no participants were afforded the opportunity to rank the conjunction as more probable than one of its conjuncts. Pooling the results, however, showed that statistically naive subjects on average ranked B & F at 3.3 and B at 4.4. That is, they considered it more likely that Linda was a feminist bank teller than that she was a bank teller. Statistically sophisticated subjects fared no better: they ranked B & F at 3.1 and B at 4.3. This is troubling, but only minimally so. After all, we don't expect people to be infallible in their inferences. Perhaps when given the opportunity to stare the conjunction fallacy in the eye, people would not make a mistake.

In a more direct test, all eight items were included in the list for subjects to rank. The results were even more troubling. 89 % of naive subjects and 85 % of sophisticated subjects committed the conjunction fallacy by ranking B & F higher than B. Worse still, the directness of the test had no measurable effect in helping participants to avoid the conjunction fallacy. The average rank of B & F was 3.3 for naïve and 3.2 for sophisticated subjects, while the average rank of B was 4.4 and 4.3, respectively. Bovens and Hartmann (2003) attempt to explain this and related results away by saying that people quite reasonably trust a source more when it tells them things they expected or already knew. Hence, when one source says that Linda is a feminist and the other doesn't, they are inclined to trust the former source more than the latter, which in turn means that they are rational in placing more confidence in that source when it also tells them that Linda is a bank teller than the other source which tells them that Linda is a bank teller out of the blue. This is a clever work-around, but I fear it doesn't work. For one thing, in this study the propositions B, F, and B & F are all simply listed; they aren't presented as if they came from different sources. For another, subjects explicitly put their trust in representativeness, not the reliability of the source, as the studies I cite below demonstrate.

Even so, the reliabilist might push back by saying that it's unsurprising that people have trouble keeping track of eight propositions simultaneously. Most people never undertake such a task, so their inability to do so without committing the conjunction fallacy is consistent with modest reliabilism about simpler inferential knowledge. The implicit assumption here is that, when the task is simplified in the appropriate way, the conjunction fallacy will go away. Tversky and Kahneman set out to test this assumption, embarking upon what they describe as "a series of increasingly desperate manipulations designed to induce subjects to obey the conjunction rule" (2002, p. 26). The filler items were deleted, and a new batch of subjects was asked to rank just B and B & F. Even with the distractions removed, 85 % of respondents committed the conjunction fallacy. Defenders of inferential reliability cannot shrug this result off as easily. Deciding which of two similar propositions is more probable is actually something that people do on a regular basis, and the propositions being compared related to personal and vocational characteristics, properties that we often attribute and reason about. The familiarity defense has no purchase here.

Still, one could respond that Tversky and Kahneman were merely exploiting a trick. Surely, one might think, participants would realize their mistake and make the correct judgment if the reason for choosing B over B & F were made clear. Were this so, it would show that the representativeness heuristic is easily triggered but corrigible. To test this hypothesis, Tversky and Kahneman designed another study in which participants read the description of Linda, then were asked to indicate which of the following arguments they found more persuasive:

(A1) Linda is more likely to be a bank teller than she is to be a feminist bank teller, because every feminist bank teller is a bank teller, but some women bank tellers are not feminists, and Linda could be one of them.

(A2) Linda is more likely to be a feminist bank teller than she is likely to be a bank teller, because she resembles an active feminist more than she resembles a bank teller.

A solid majority of 65 % chose A2. This is an improvement, but it provides only cold comfort. The representativeness heuristic overpowered the natural light of reason in a large majority of participants.

Furthermore, this study indicates an important difference between cognitive situationism and moral situationism. Subjects in this study actually reasoned in terms of the situational factor of representativeness. In the moral studies, subjects typically do not reason in terms of their moods or the level of ambient noise.[10] This feature of cognitive situationism suggests that, even if the challenge is fended off, higher-order knowledge on the reliabilist model faces a distinct challenge. Ordinary people not only deploy the heuristic but also reflectively endorse it over the sound

[10] Thanks to Jonathan Adler for emphasizing this point to me.

inference rule. If that's right, then even if it could be shown that heuristics are reliable, we should conclude that most people don't have second-order knowledge of their heuristically derived knowledge. Suppose I arrive at knowledge of *p* based on a heuristic inference: K*p*. Do I know that I know that *p*? I.e., KK*p*? Probably not, because I have a false belief about the reliability of my way of arriving at K*p*: I think that it's more reliable than it really is, and I think that it's more reliable than a relevant alternative rule of inference.

Proponents of ecological validity might retrench further, claiming that when we make comparative judgments of likelihood, it's usually with an eye to the potential payoffs for being right (or wrong). Moreover, one might worry that the thorny semantics and pragmatics of terms such as 'probability' and 'likelihood' make the results cited so far difficult to interpret. What would really show that people employ the representativeness heuristic is the introduction of stakes for being right or wrong, as well as the elimination of troublesome vocabulary. To ensure that their results could not be explained away in this fashion, Tversky and Kahneman conducted a study in which participants read the description of Linda, then answered the following question: "If you could win $10 by betting on an event, which of the following would you choose? [B or B & F]" This time, 56 % of respondents committed the conjunction fallacy – a lower proportion than in the other studies but still more than half.

The question now arises whether the representativeness heuristic is used only in making person-level judgments. If its use is highly circumscribed, the argument for epistemic situationism would also be highly circumscribed. In a series of follow-up studies, Tversky and Kahneman (2002) investigated whether the representativeness heuristic would induce the conjunction fallacy in other content areas (medicine, sports, and betting) and whether it would be undercut by expertise and monetary incentives. The results are a source of genuine consternation.

In the medical study, internists enrolled in Harvard postgraduate courses or with admitting privileges at the New England Medical Center responded to surveys like the following:

> A 55-year-old woman had pulmonary embolism documented angiographically 10 days after a cholecystectomy. Please rank order the following in terms of the probability that they will be among the conditions experienced by the patient (use 1 for the most likely and 6 for the least likely). Naturally, a patient could experience more than one of these conditions.

Dyspnea and hemiparesis	Syncope and tachycardia
Calf pain	Hemiparesis
Pleuritic chest pain	Hemoptysis

Consulting physicians had determined that hemiparesis (partial paralysis) was highly unrepresentative of pulmonary embolism (blood clots in the lungs) while dyspnea (shortness of breath) was highly representative. The question, then, was whether statistically sophisticated physicians would commit the conjunction fallacy even in a content area where they were experts. In all five versions of the case, the conjunction was ranked more probable (2.7) than its conjunct (4.6). In the best case, 73 % of the physicians committed the conjunction fallacy; in the worst, 100 % did. Unlike ordinary participants, the doctors were quick to revise their judgments when the fallacy was pointed out. This suggests that statistical expertise does make people's use of heuristics corrigible. However, it's unclear how comforting this should be. It seems to indicate that experts are just as likely as ordinary people to have unreliably formed beliefs except in those rare circumstances where their errors are pointed out to them. Such an outcome would still lead to a kind of skepticism.

In other studies, the pattern of responses was just as dismaying. Adult Oregonians said it was more likely that Bjorn Borg would win the finals at Wimbledon after losing the first set than that he would lose the first set.[11] In another study, undergraduates at the University of British Columbia and Stanford responded to the following prompt:

> Consider a regular six-sided die with four green faces and two red faces. The die will be rolled 20 times and the sequence of greens (G) and reds (R) will be recorded. You are asked to select one sequence, from a set of three, and you will win $25 if the sequence you chose appears on successive rolls of the die. Please check the sequence of greens and reds on which you prefer to bet.

1. RGRRR
2. GRGRRR
3. GRRRRR

In this study, the words 'probability' and 'likelihood' made no appearance. In addition, subjects had monetary skin in the game: they would be rewarded with a non-trivial sum if they predicted correctly. Although sequence 2 is more representative of the die (2 greens, 4 reds), sequence 1 is necessarily more likely than sequence 2. In the language of decision theory, choosing 1 weakly dominates choosing 2. Nevertheless, 65 % of participants chose sequence 2. Even when sequence 2 was replaced with 'RGRRRG' in a follow-up study to make it clearer that it contained sequence 1, 63 % of respondents chose it.

It should be clear that I could go on *ad nauseum* about the representativeness heuristic and the conjunction fallacy, but I want to step back now to draw a tentative philosophical conclusion. The robustness of the representativeness heuristic throws a pall of doubt over the notion that most people possess the intellectual virtues related to even rudimentary deductive and inductive reasoning. The process used to arrive at beliefs about likelihood in no way resembles sound inferential practice; rather, people follow a heuristic that treats representativeness as an index of probability. My claim is that the same holds true for other heuristics, and that this creates trouble not only for the cognitive virtues related to deductive and inductive reasoning,

---

[11] Michael Levin has emphasized to me that this result could be explained by assuming not that participants used the representativeness heuristic but that they compared the unconditional probability P(Borg loses the opening set) to the conditional probability P(Borg wins the match | Borg loses the opening set). I agree, but it seems to me that the more parsimonious explanation appeals to the representativeness heuristic, since that seems to figure in many other judgments as well. In any event, conditionalizing explanations of this sort will not salvage most of the other violations of the conjunction fallacy canvassed in this article.

but for many of the other cognitive virtues related to inference, such abduction. If this is right, reliabilists face a dilemma. If they say that such heuristics are not intellectual virtues, skepticism looms: if most people use non-virtuous heuristics to arrive at their inferential beliefs, then most people have unjustified beliefs, which do not count as knowledge even when true. If, however, reliabilists say that these heuristics are intellectual virtues, then they need to explain how these dispositions are to be construed as reliable.

The first horn of the dilemma faced by reliabilists is to admit partial defeat in the sphere of inference. This admission could take three forms. First, one could adopt a mixed theory of knowledge, according to which a non-inferential belief counts as knowledge only if it was formed by a reliable process while an inferential belief counts as knowledge only if it satisfies some to-be-specified criterion. This would save inferential non-skepticism at the cost of circumscribing the applicable domain of the reliability criterion of knowledge. I am unfamiliar with any extant mixed theories of knowledge, so I'm not sure how appealing this option might be. Hybrid theories do have their appeal in ethics. Perhaps epistemologists should try them on for size.

Alternatively, reliabilists could maintain their theory even in the realm of inference. This leads to the second way of admitting partial defeat, which is to give up on virtue egalitarianism. Many virtue ethicists have made this move in responding to the situationist challenge to virtue ethics, claiming that virtue is rare, so empirical studies of groups should not be expected to turn up a great deal of virtuous behavior. I don't find this response appealing in the case of virtue ethics, and it seems even less appealing for virtue epistemology, for it would entail a great deal of skepticism. Yes, some people have knowledge, but they're an elite epistemic minority. This flies in the face of the Moorean platitude of non-skepticism.

The third way of admitting partial defeat attempts to partition off a class of knowledge claims rather than a class of knowers. In the face of the evidence for inferential situationism, reliabilists might want to conclude that people who use these heuristics really *don't* have inferential knowledge but that they could still have knowledge derived from many other sources, such as perception, memory, testimony, etc. The Moorean platitude of non-skepticism may not extend to the far reaches of inference. Perhaps it really only includes more mundane things like knowledge that the external world exists, that I have a hand, and so on.

To bolster this reply, reliabilists might try to argue that really, when you think about it, most human beliefs are not arrived at through inference. So even if we are forced to concede skepticism about inference, that would not impugn a critical mass of our beliefs. And when people do use sound inferential practices, such as Bayes's Law or *modus tollens*, their beliefs can still count as knowledge. This would rescue scientific inference, which is typically much more careful than everyday inference because scientists presumably avoid using heuristics when they can.

There is some precedent for localized skepticism of this sort. The reply in some ways resembles Nozick's admission that the negations of skeptical hypotheses are not known, even when truly believed. Of course, extending ignorance from the negations of skeptical hypotheses to most inferential beliefs is a step, but it may not be such a big step as to be unacceptable.

If the first horn of the dilemma doesn't tempt you, perhaps you'd prefer to argue that, despite what I've said so far, heuristics such as availability and representativeness really *are* reliable. After all, I did mention above that heuristics can be surprisingly accurate. Perhaps the cases where they break down – the cases that I harped on above – are quite rare. They can be constructed *ad infinitum* in laboratory contexts, but in the real world perhaps they don't crop up too often. If they are sufficiently rare, then heuristics may not be so bad after all.

A related argument holds that heuristics are reliable in the appropriate domain; it's only when they're used outside that domain that they lead to systematic error. In a slogan, "heuristics are reliable except when they aren't."[12] Well, bully for them. So is flipping a coin to decide whether it's raining: if you only flip the coin, or only infer based on the result of the flip, when it's right, then of course it's reliable. Heuristics, along with every other decision procedure one can imagine, satisfy that condition of reliability. The idea must be something more like this: heuristics are reliable *when we're disposed to use them*, and when they're unreliable *we stop using them*.

This suggests an interesting difference between heuristics and sound inference rules. If you input truths to *modus ponens*, it outputs truths. If you input truths to disjunctive inference, it outputs truths. If you input good evidence to Bayes' Law, it outputs rational probabilities. If you input triggering conditions into a heuristic – well, it depends. Sound inference rules are context-neutral. Heuristics aren't. So, instead of talking about whether heuristics are categorically reliable, perhaps it would be better to talk about the contexts in which heuristics are reliable or unreliable. Take availability: it's pretty good when you've had a large, unbiased sample of the domain, but not otherwise. So instead of talking about whether the availability heuristic is reliable, we should talk about whether large-unbiased-sample-availability is reliable (maybe), whether small-unbiased-sample-availability is reliable (no), whether small-biased-sample-availability is reliable (no), etc.

This move parallels Doris's theory of local moral virtues. Instead of asking whether someone is generous, he thinks it's more fruitful to ask whether she's good-mood-generous, bad-mood-generous, etc. If you relativize to context, you can start to make supportable attributions. There's an important difference, though. On the moral side, global dispositions are normatively adequate but empirically inadequate: they're what you'd want, but most people don't have them. On the epistemic side, global heuristics are normatively inadequate but empirically adequate: they're unreliable, and most people use them. Relativizing moral virtues to contexts makes them empirically adequate but threatens to leave them normatively uninspiring. It might be seen as damning with faint praise to say that someone is loyal to his male friends. Relativizing heuristics to contexts makes (some of) them reliable, but empirically inadequate. Why empirically inadequate? Because, if the data cited above on the representativeness heuristic is any guide, that's not how (most) people deploy them.

One might retreat even further and say that, while people *don't* curb their use of heuristics in the right way, they *could learn to*. My initial response to this is to

[12] Thanks to Guy Axtell for emphasizing this argument to me.

shrug. Here's how I see the dialectic: the defender of reliabilism wants to fend off the skeptical challenge, according to which the processes people actually use to arrive at their inferential beliefs are unreliable. It doesn't help to say that, though the processes people actually use to arrive at their inferential beliefs are unreliable, *people could use reliable processes.* That wouldn't show that people have knowledge; what it would show is that they *could* have knowledge. Furthermore, it remains to be shown that people can actually use heuristics responsibly. The place to look would be the Kahneman-Gigerenzer controversy, but my own reading of that controversy is that Kahneman's side prevails.

Consider instead the following claim in favor of the reliability of heuristics, which has the flavor of a transcendental argument:

> Let's grant that people often arrive at their inferential beliefs via heuristics. It follows that heuristics must be reliable. Furthermore, there's good evolutionary reason for this supposition. People who routinely make unreliable inferences are less fit than people who routinely make reliable inferences, so the fact that most people are strongly disposed to use heuristics means that heuristics must be adaptive and, hence, reliable.[13]

I say the argument has a transcendental flavor because it takes for granted some empirical claim (widespread use of heuristics), then articulates what would have to be the case for that empirical claim to be possible. Consider first the argument without the evolutionary backstory. The granted claim is that heuristic use is widespread. The conclusion is that heuristics must be reliable. There's a missing premise here: *inferential non-skepticism.* If we add that in, the granted claim is that, although heuristic use is widespread, most people know quite a bit on the basis of inference. And the conclusion of the argument remains that heuristics must be reliable. Even as amended, the transcendental argument is clearly invalid. What needs to be added is that knowledge is arrived at by reliable processes. Otherwise, it remains open to say that people acquire inferential knowledge on the basis of *un*reliable heuristics.[14] While it is of course possible to add this further premise to the argument, to do so would beg the question. Reliabilism is precisely what is at stake in this debate. The defender of reliabilism is not entitled to use it as a premise.

Though the purely transcendental form of the argument is invalid, perhaps the evolutionary backstory will help. I take it that a fleshed-out version of the argument would go like this:

> Suppose that some of our ancestors tended to make inferences using processes *p, q,* and *r*, and that other of our ancestors tended to make inferences using processes *s, t,* and *u*. Suppose further that *p, q,* and *r* are more reliable than *s, t,* and *u*. Then the *pqr-ancestors* would have ended up with more reliable beliefs than the *rst*-ancestors, which in turn means that they would outcompete the *rst*s. We are the offspring of the *pqr*s, so they must have used reliable decision processes, which were passed on to us. Hence, the heuristics we use must be reliable.

Consider again the availability heuristic, which treats ease of recall as an index of frequency and probability. Certainly, one might argue, it's got to be better to use

[13] Jennifer Lackey and Guy Axtell have both made versions of this argument in conversation.

[14] John Turri (forthcoming) holds such a view.

the availability heuristic than many other rules of inference. For example, it's clearly better to use the availability heuristic than the *unavailability* heuristic, an artificial monstrosity that treats difficulty of recall as an index of frequency and probability. People wander around the world, encountering objects of various sorts. If their encounters are sufficiently like a random sample, and if accessibility depends largely on the number of encounters with things of a given type, then the easier it is to recall something, the more frequently one encountered it in the past and, hence, the more common it is. Yes, the heuristic can go haywire sometimes, but it does a solid job in most ordinary circumstances. Presumably a similar argument could be made in defense of the recognition heuristic. And then, the reliabilist might point out, heuristics may often be the only or the best thing we have to go on. People lack the time, processing power, memory, and skill to apply sound deductive and inductive rules in all cases. Sometimes they have to slum it by using heuristics. But when they do, they tend to use heuristics that work well enough to be adaptive.

There are several problems with this argument. First, at best it shows that we are descendants of our most epistemically reliable ancestors, and hence that we tend to use the most epistemically reliable heuristics available to the species tens of thousands of years ago. But that is plainly irrelevant to whether the heuristics we use are reliable *enough* to lead to knowledge. Maybe the available decision rules were all pretty bad; then the ones that survived would merely be the best of a bad lot: more reliable than some, but not reliable enough to yield knowledge.

Second, even assuming that the *pqr*s used outright reliable heuristics, and not just the best of a bad lot, the argument assumes that the contemporary inferential setting is relevantly similar to that of our ancestors. Since it's best to talk about the reliability of heuristics relative to some context, it's quite possible that using *p*, *q*, and *r* was reliable relative to the context of hunter-gatherer nomads in the African savannah, but that using *p*, *q*, and *r* is unreliable relative to the context of modern humans navigating highways, cities, and online media. Since we've changed our environment so much in the last ten millennia, the fact that something used to work is moot on the question of whether it still works.

Third, the argument crucially assumes that reliability is adaptive. This is not obvious. In recent years, the so-called *value problem* for epistemology has loomed large: why is knowledge better than mere true belief? The answer, as Plato already understood in the *Meno*, is not that knowledge is more practically useful: someone who has a true belief about how to get from point A to point B will arrive there just as surely as someone who knows the way from A to B. What I take to be the best proposed solution to the value problem is that knowledge is an achievement, and achievements are intrinsically valuable. But are achievements intrinsically adaptive? I see little reason to think so.

But, one might argue, if the *pqr*s use more reliable decision rules than the *stu*s, surely they will end up with more true beliefs (or a higher proportion of true to false beliefs), so even if their having more knowledge isn't adaptive, their having more verisimilar beliefs is. At this point I think it is essential to distinguish reliability, which is a purely epistemic notion, from adaptiveness, which is both epistemic and practical. The adaptiveness of a decision rule isn't identical to its reliability. It is, roughly speaking, the product of reliability and average payoff. An example illustrates this point. Compare two decision procedures, P1 and P2, used over ten cases. P1 leads to 8 true beliefs, while P2 leads to 6. So P1 is 80 % reliable, while P2 is only 60 % reliable. Surely, one might think, P1 is more adaptive than P2. On the contrary, it depends on what happens when the agent gets things right and what happens when he gets them wrong. For if P1 goes astray when it would be disastrous to be wrong, while P2 goes astray when it doesn't hurt too much to be wrong, then it may well be the case that P2 is more adaptive than P1. So the adaptiveness of a belief-formation process isn't just its reliability; it's reliability *when it matters*.

This just shows that it's possible for reliability and adaptiveness to come apart, which casts doubt on the evolutionary argument, but perhaps not too much. It would be more persuasive to show that, for the heuristics we actually use, reliability and adaptiveness diverge. There's reason to think that, for many of them, this is the case. Consider the so-called fundamental attribution error: the tendency to attribute others' behavior to dispositional factors rather situational ones, even when it should be clear that situation is importantly operative. Presumably the pattern of judgments identified by this error stems from the use of a heuristic: When someone does something of type *t*, infer that she is a *t*er. For example, if someone lies, infer she's a liar. If someone cheats, infer she's a cheater. If someone helps, infer she's a helper. This isn't a particularly reliable heuristic, but when it goes wrong (i.e., when it leads to the fundamental attribution error), it's often self-confirming. If you think that someone is a helper, you tend to signal that expectation to her, which in turn will make her more inclined to help. If a waiter thinks that someone is a low tipper, he'll tend to give them bad service, which in turn will lead to a low tip.[15] It looks, then, as though the heuristic that leads to the fundamental attribution error might be adaptive but unreliable.

5 Conclusion

In this paper, I've attempted to argue that the situationist challenge to virtue ethics has an epistemic cousin: the situationist challenge to reliabilism about inference. I documented evidence that people are disposed to use a variety of heuristics, such as availability and representativeness, to arrive at their inferential beliefs, and argued that these heuristics are not reliable enough to lead to knowledge – at least not in all of the circumstances in which people tend to deploy them. A number of potential counterarguments were canvassed, but it was unclear whether any of them succeed. If the arguments developed in this paper are on the right track, reliabilists need to come to grips with a serious challenge to their view.

[15] I discuss this phenomenon, which I call 'factitious virtue' in more detail in my (2013b).

References

Alfano, M. 2009. Sensitivity theory and the individuation of belief-formation methods. *Erkenntnis* 70(2): 271–281.

Alfano, M. 2011a. Explaining away intuitions about traits: Why virtue ethics seems plausible (even if it isn't). *Review of Philosophy and Psychology* 2(1): 121–136.

Alfano, M. 2011b. Extending the situationist challenge to responsibilist virtue epistemology. *Philosophical Quarterly* 62(247): 223–249.

Alfano, M. 2013a. Identifying and defending the hard core of virtue ethics. *Journal of Philosophical Research* 38: 233–260.

Alfano, M. 2013b. *Character as moral fiction*. New York: Cambridge University Press.

Baehr, J. 2006. Character, reliability, and virtue epistemology. *The Philosophical Quarterly* 56: 193–212.

Baron, R. 1997. The sweet smell of … helping: Effects of pleasant ambient fragrance on prosocial behavior in shopping malls. *Personality and Social Psychology Bulletin* 23: 498–503.

Baron, R.A., and J. Thomley. 1994. A whiff of reality: Positive affect as a potential mediator of the effects of pleasant fragrances on task performance and helping. *Environment and Behavior* 26: 766–784.

Battaly, H. 2008. Virtue epistemology. *Philosophy Compass* 3(4): 639–663.

Beebe, J. 2004. The *generality problem*, statistical relevance and the tri-level hypothesis. *Noûs* 38(1): 177–195.

Bohnet, I., and B. Frey. 1999a. Social distance and other-regarding behavior in dictator games. *The American Economic Review* 89(1): 335–339.

Bohnet, I., and B. Frey. 1999b. The sound of silence in prisoner's dilemma and dictator games. *Journal of Economic Behavior and Organization* 38: 43–57.

Boles, W., and S. Haywood. 1978. The effects of urban noise and sidewalk density upon pedestrian cooperation and tempo. *Journal of Social Psychology* 104: 29–35.

Bovens, L., and S. Hartmann. 2003. *Bayesian epistemology*. Oxford: Oxford University Press.

Code, L. 1984. Toward a 'responsibilist' epistemology. *Philosophy and Phenomenological Research* 45: 29–50.

Cohen, S., and A. Lezak. 1977. Noise and inattentiveness to social cues. *Environment and Behavior* 9: 559–572.

Donnerstein, E., and D. Wilson. 1976. Effects of noise and perceived control on ongoing and subsequent aggressive behavior. *Journal of Personality and Social Psychology* 34: 774–781.

Doris, J. 1998. Persons, situations, and virtue ethics. *Noûs* 32(4): 504–540.

Doris, J. 2002. *Lack of character: Personality and moral behavior*. Cambridge: Cambridge University Press.

Doris, J. 2005. Replies: Evidence and sensibility. *Philosophy and Phenomenological Research* 71(3): 656–677.

Driver, J. 2003. The conflation of moral and epistemic virtues. In *Moral and epistemic virtues*, ed. M. Brady and Duncan Pritchard, 101–116. Malden: Blackwell.

Geen, R., and E. O'Neal. 1969. Activation of cue-elicited aggression by general arousal. *Journal of Personality and Social Psychology* 11: 289–292.

Goldman, A. 1992. Epistemic folkways and scientific epistemology. In *Liasons: Philosophy meets the cognitive and social sciences*. Cambridge, MA: MIT Press.

Greco, J. 1992a. Agent reliabilism. *Noûs* 13: 273–296.

Greco, J. 1992b. Virtue epistemology. In *A companion to epistemology*, ed. J. Dancy and E. Sosa. Oxford: Blackwell.

Greco, J. 1993. Virtues and vices of virtue epistemology. *Canadian Journal of Philosophy* 23: 413–432.

Greco, J. 2000. *Putting skeptics in their place: The nature of skeptical arguments and their role in philosophical inquiry*. Cambridge: Cambridge University Press.

Greco, J. 2009. Knowledge and success from ability. *Philosophical Studies* 142: 17–26.

Grimes, M.B. 1999. Helping behavior commitments in the presence of odors: Vanilla, lavender, and no odor. National Undergraduate Research Clearinghouse, 2.
Harman, G. 1999. Moral philosophy meets social psychology: Virtue ethics and the fundamental attribution error. *Proceedings of the Aristotelian Society*, New Series 119: 316–331.
Harman, G. 2000. The nonexistence of character traits. *Proceedings of the Aristotelian Society* 100: 223–226.
Harman, G. 2001. Virtue ethics without character traits. In *Fact and value*, ed. Byrne, Stalnaker, and Wedgwood, 117–127. Cambridge: MIT Press.
Harman, G. 2003. No character or personality. *Business Ethics Quarterly* 13(1): 87–94.
Harman, G. 2006. Three trends in moral and political philosophy. *The Journal of Value Inquiry* 37: 415–425.
Hoffman, E., K. McCabe, and V. Smith. 1996. Social distance and other-regarding behavior in dictator games. In *Handbook of experimental economics*, vol. 1, ed. Plott and Smith, 429–435. Amsterdam: Elsevier.
Hookway, C. 2006. Epistemology and inquiry: The primacy of practice. In *Epistemology futures*, ed. Heatheringon, 95–110. Oxford: Oxford University Press.
Konecni, V. 1975. The mediation of aggressive behavior: Arousal level versus anger and cognitive labeling. *Journal of Personality and Social Psychology* 32: 706–716.
Korte, C., and R. Grant. 1980. Traffic noise, environmental awareness, and pedestrian behavior. *Environment and Behavior* 12: 408–420.
Korte, C., A. Ypma, and C. Toppen. 1975. Helpfulness in Dutch society as a function of urbanization and environmental input level. *Journal of Personality and Social Psychology* 32: 996–1003.
Latané, B., and J. Darley. 1968. Group inhibition of bystander intervention in emergencies. *Journal of Personality and Social Psychology* 10: 215–221.
Latané, B., and J. Darley. 1970. *The unresponsive bystander: Why doesn't he help?* New York: Appleton-Century-Crofts.
Latané, B., and S. Nida. 1981. Ten years of research on group size and helping. *Psychological Bulletin* 89: 308–324.
Latané, B., and J. Rodin. 1969. A lady in distress: Inhibiting effects of friends and strangers on bystander intervention. *Journal of Experimental Psychology* 5: 189–202.
Lepock, C. 2011. Unifying the intellectual virtues. *Philosophy and Phenomenological Research* 83(1): 106–128.
MacIntyre, A. 1984. *After virtue: A study in moral theory*. Notre Dame: University of Notre Dame Press.
Matthews, K.E., and L.K. Cannon. 1975. Environmental noise level as a determinant of helping behavior. *Journal of Personality and Social Psychology* 32: 571–577.
Miller, C. 2003. Social psychology and virtue ethics. *The Journal of Ethics* 7(4): 365–392.
Miller, C. 2009. Empathy, social psychology, and global helping traits. *Philosophical Studies* 142(2): 247–275.
Page, R. 1974. Noise and helping behavior. *Environment and Behavior* 9: 311–334.
Schwartz, S., and A. Gottlieb. 1991. Bystander anonymity and reactions to emergencies. *Journal of Personality and Social Psychology* 39: 418–430.
Sosa, E. 1980. The raft and the pyramid: Coherence versus foundations in the theory of knowledge. *Midwest Studies in Philosophy* 5(1): 3–26.
Sosa, E. 1985. Knowledge and intellectual virtue. *The Monist* 68(2): 226–245.
Sosa, E. 1991. *Knowledge in perspective*. New York: Cambridge University Press.
Sosa, E. 2001. For the love of truth. In *Virtue epistemology: Essays on epistemic virtue and responsibility*, ed. A. Fairweather and L. Zagzebski, 49–62. Oxford: Oxford University Press.
Sosa, E. 2007. *A virtue epistemology*. Oxford: Oxford University Press.
Sosa, E. 2011. *Knowing full well*. Princeton: Princeton University Press.
Turri, J. forthcoming. Unreliable knowledge. *Philosophy and Phenomenological Research.*
Tversky, A., and D. Kahneman. 1973. Availability: A heuristic for judging frequency and probability. *Cognitive Psychology* 5: 207–232.

Tversky, A., and D. Kahneman. 2002. Extensional versus intuitive reasoning: The conjunction fallacy in probability judgment. In *Heuristics and biases: The psychology of intuitive judgment*, ed. T. Gilovich, D.W. Griffin, and D. Kahneman. Cambridge: Cambridge University Press.
Vranas, P. 2005. The indeterminacy paradox: Character evaluations and human psychology. *Noûs* 39: 1–42.
Zagzebski, L. 1996. *Virtues of the mind*. Cambridge: Cambridge University Press.

Inferential Abilities and Common Epistemic Goods

Abrol Fairweather and Carlos Montemayor

1 From Moral to Epistemic Situationism

A wide range of psychological research on trait attribution and rationality has chipped away at what appeared to be a solid empirical footing for virtue ethics, thereby challenging the adequacy of virtue ethics on the very point that appeared to be a primary strength.[1] Philosophers such as Gilbert Harman (2000) have been led to question the very existence of character traits, and others like John have denied their robustness and explanatory value. Character trait attributions enjoin predictive and explanatory commitments that simply fail too often to meet norms of epistemic success that require the manifestation of epistemic virtues. Doris argues that traditional forms of virtue ethics cannot be empirically adequate and normatively adequate at the same time. Only recently has the situationist challenge been applied to virtue epistemology. In two recent papers, Mark Alfano (this volume) defends the first thorough application of situationism to responsibilist virtue epistemology, and (2013, this volume) develops the first thorough challenge to virtue reliabilism. Alfano's challenge to virtue reliabilism is based a diverse range of empirical results in social psychology on rationality, inferential abilities and trait attribution.

Alfano nicely frames the psychological challenge to virtue epistemology as an inconsistent triad:

(a) *inferential non-skepticism*: most people know quite a bit inferentially
(b) *inferential reliabilism*: inferential knowledge is true belief acquired and retained through inferential reliabilist intellectual virtue
(c) *inferential cognitive situationism*: people acquire and retain most of their inferential beliefs through heuristics rather than intellectual virtues.

[1] The literature on relevant research is quite large, for some comprehensive treatments see Miller (2003, 2014), Alfano (2011, 2012, and 2013), Alfano and Fairweather (2013).

A. Fairweather (✉) • C. Montemayor
Department of Philosophy, San Francisco State University, San Francisco, CA, USA
e-mail: afairweather@gmail.com; montema@sfsu.edu

A. Fairweather (ed.), *Virtue Epistemology Naturalized: Bridges Between Virtue Epistemology and Philosophy of Science*, Synthese Library 366, DOI 10.1007/978-3-319-04672-3_8,

The dilemma for the virtue epistemologist is that empirical adequacy will require accommodating the empirical work presented by situationists and thus will have to accept (c). But if this empirical research shows that all too rarely will an agent meet virtue theoretic standards for epistemic success, we are now unable to account for the robustness of knowledge affirmed in (a). Failing to meet the non-skepticism desiderata would be a *normative inadequacy* in virtue epistemology because any such theory will be unable to assign positive epistemic standings to actual beliefs in a way that keeps pace with the actual frequency of human knowledge. Alternatively, if a virtue epistemologists crafts norms that assure meeting (a), such an account will now fall prey to *empirical inadequacy* because no such account will be supported by research in psychology presented in support of (c). Alfano argues that virtue epistemology must be the discarded commitment. Call this the challenge of *epistemic situationism.*

This essay will challenge Alfano's argument on many points and will defend a naturalistic account of reliable inferential abilities that not only meets Alfano's challenge virtue reliabilism, but will also illuminate the nature and norms of inference, rationality and assertion. Key to this defense will be the Bounded Rationality (BR) research program started by Herbert Simon and recently developed by Gerd Gigerenzer according to which fast and frugal heuristic reasoning often outperforms optimizing rationality for bounded rather rational agents. Gigerenzer (2008) argues that norms of rationality cannot be reduced to assessments of ideal-approximation, or how closely an agent approximates an ideal rational outcome. If Gigerenzer is right here, situationists are applying the wrong kind of norms in their interpretation of the research on rationality.

We take a closer look at the nature of inference and Alfano's triad in Sects. 2 and 3, and then get right to Gigerenzer's account of "ecological rationality" in Sect. 4. Section 5 defends a number of reliable inferential abilities that are supported by relevant empirical work on knowledge of syntax, communication, assertion and directed memory and Sect. 6 examines Ernest Sosa's virtue reliabilism in light of the situationist challenge and concludes with a novel proposal for rigidifying the normal conditions for epistemic assessment to the (psychologically normal) conditions for assertion. We argue that an empirically grounded account of normal conditions for epistemic assessment can be provided by work examining work on the psychological and semantic processes involved in assertion.

2 A Closer Look at Inference

Since Alfano restricts his argument to inferential virtue reliablism and not perceptual virtue reliablism, it will be important to get clear on what is meant by inference. Alfano should be able to show that the same considerations that count against extending his argument against responsibility virtue epistemology to perceptual knowledge do not also count against extending his argument to *perception like forms of inference*, and, independent of the specifics of Alfano's argument, it will be necessary to get as clear as possible on the nature of inference to assess the merits of any argument against inferential virtue reliabilism.

2.1 Graham on Association and Inference

In "Psychological Capacity and Positive Epistemic Standing" (2012), Peter Graham distinguishes a number of distinct abilities and capacities which underlie different kinds of positive epistemic standings. These include critical reasoning, propositional thinking, perceptual representation and sensory registration. Graham claims that only the latter is a genuinely inferential capacity, though certain forms of association can be easily confused with reasoning and inferring. On the distinction between associating and inferring, he says:

> Associating is one 'intelligent' capacity for learning about and navigating one's environment. And it is widely thought that no matter how much representation actually goes on in animals when associating, associating isn't reasoning or relying on inference. Logical reasoning is a different kind of psychological capacity. (2014, pg. 154)

Of particular interest is Graham's discussion of research by Premack and Call on inference in apes and chimpanzees. In full view of chimpanzees, researchers took two boxes and placed an apple in one and a banana in the other, and then proceeded to eat the banana out of one of the boxes, again in full view. When given the opportunity to pick from the boxes, the chimps went right to the other box containing the apple. Graham explains that the researchers concluded

> That the chimps reasoned something like this: there is an apple in box *A* and a banana in box *B*. But there is no longer a banana in *B*, so there's just an apple in *A*. That's why they went right for *A*. Animals that don't reason like this, but presented with the same information, might still look for a banana in box *B*, or might only slowly make their way to box *A*.

To show that the chimps were inferring and not just associating, Joseph Call's experiments with apes involved putting two opaque cups in full view, one full of food, the other empty, and shook both in front of the subjects. If the cup with food was shaken, the apes went right for it. If the cup without food was shaken, they went right for the other one. Call concluded that apes were reasoning something like: 'when there's no noise, there's no food in the shaking cup, so grab the other one'.

This is particularly interesting because similar research with dogs showed that they rely on associative intelligence rather than "logical guidance", reasoning or inference. With dogs searching for a ball placed behind one of three screens, the speed of their search would slow down as they went from 1 to 2, and again from 2 to 3, whereas with children performing a similar task the speed of their search will increase. This explained by the fact that inference in the child shows the failure to find the target in the first attempt as making it more likely that it is behind 2 or 3, and if not 2 then definitely 3. When the dog responds to the failure to find the ball as an 'extinction trial' (Graham, *ibid*) that signals it is less rather than more likely to be found in screens 2 and 3. Children exhibit the kind of inferential intelligence attributed to apes and chimps, rather than the associative intelligence of the dogs.

The best explanation of these findings is that very robust, stable inferential abilities exist and these provide the basis for distinguishing inferential intelligence from merely associative intelligence. Graham distinguishes the inferential reasoning of chimps, apes and children from the full blown "critical reasoning" exhibited by most adult humans, but the existence of basic inferential abilities is sufficient for our purposes here. We argue that the stability of inferential abilities for many epistemic tasks is confirmed by these psychological findings as well as many additional considerations discussed throughout this essay.

In "Epistemic Virtues and Cognitive Dispositions", Henderson and Horgan (2009) distinguish between *classically inferential processes* and *inferential processes broadly construed*. The former are much more restricted kinds of processes and, they argue, have been the dominant focus of epistemology since the modern period.

> Inferential processes broadly construed: are simply those cognitive cognitive processes in which beliefs are formed or maintained on the basis of the information. Being based on information is a causal notion, pointing to arrays of counterfactual dependencies and to dispositions. This is the broadest and most tolerant notion of an inferential process. (2009, pg. 301)

Classically inferential processes are restricted by two additional things, "the information figuring in the inference is *explicitly represented* in the cognitive system that is the agent.... Second, the causal processes whereby beliefs are fixed (formed, revised, or retained) must be *occurrently isomorphic* with the deductive and inductive support relations obtaining between the information that the agent possesses." (ibid.) Cases where content that is occurrently not represented (and thus not classically inferential) but is nonetheless causally salient in belief formation will be broadly but not classically inferential, and beliefs formed on the basis of perception rather than on the basis of other beliefs may be a large and interesting example of such cases.[2]

Henderson and Horgan argue that properly recognizing the distinction between classical and broad inferential processes supports epistemic virtue theory. In particular, they argue that epistemic virtues have the right dispositional structure to support a theory of broad and classical inference, and thus "a superior epistemological perspective will give significant attention to virtues–to epistemically good dispositions." Graham's distinction between associative and inferential intelligence is not essentially about what content is or is not represented as in Henderson and Horgan's distinction between classical and broad inferential processes, but this just further demonstrates the heterogeneity of inference and the range of human cognitive activity where stable and reliable inferential dispositions show up. Their distinction between classical inferential processes and inferential processes broadly construed will also have some resonance with Gigerenzer's distinction between optimizing rationality and ecological rationality discussed at length below.

---

[2] Henderson and Horgan distinguish a third form called "argumentative inference" to cover cases where the isomorphism between logical relations and causal relations within the information represented by the agent fails to hold.

3 A Closer Look at the Triad

Returning to Alfano, the above discussion shows that inferential virtue reliabilism can rely on the existence of cognitive capacities between mere association and full blown critical reasoning (Graham), or on cognitive dispositions that are broad and non-representational rather than classical (Henderson & Horgan 2009). In both cases, we have empirical support not only for distinguishing kinds of inferential abilities, but also for the claim that some of these are clearly stable, robust and reliable in human beings. Below, we take a much closer look at the three principles in Alfano's triad before moving to our defense in the following Section.

A. ***Non-Skepticism***: "People know quite a bit through inference". Alfano follows Moore in advancing an optimistic intuition about the frequency of knowledge in human cognition. While we are in broad agreement that "people know quite a bit through inference", (NS) contains an important ambiguity. Clearly enough, an adequate theory of knowledge must 'get it roughly right' about how often and when to attribute knowledge and other valuable epistemic achievements to actual human believers. However, it is important to note that a theory can fail to satisfy (NS) in at least two different ways; a theory might *over-attribute failure* or *under-attribute success*. We will see that, depending on how read the normal conditions constraint on abilities and virtues, some beliefs will be neither epistemic successes nor epistemic failures, and this might be used to show that virtue epistemology does not violate (NS) even if we grant Alfano's *inferential reliabilism* and *inferential cognitive situationism*. This will turn out to be an important distinction when we examine Ernest Sosa's virtue reliabilism and will ultimately require looking more closely at 'normal conditions' requirements in virtue theory to determine how to treat these cases.

Local skepticism: While (NS) is an anti-skeptical intuition, it is fully compatible with "local skepticism". It may be true that people know quite a bit through inference all told, but nonetheless perform very poorly within certain domains of inquiry. Cognitively limited creatures using fallible methods of inquiry will be expected to have certain dark areas in their full set of beliefs, even when they are reliable in their actual inferential practices. If the research he relies on only supports local virtue theoretic failures then situationism will not be enough to push virtue epistemology to skepticism. As shown below, some cognitive failures actually entail broader cognitive success. The response available here to virtue epistemologists is to broadly individuate epistemic abilities being evaluated in virtue epistemology, and narrowly individuate the failures shown in the empirical research.

Alfano presents (NS) as a widely shared intuition about knowledge. True enough, but, just as clearly, it is true that our untutored intuitions about knowledge might be slightly off the mark in any number of ways. In particular, our intuitions about the frequency of doxastic success may turn out to far outstrip the actual frequency of doxastic success. So, we add that if given sufficient reason, (NS) can be revised down. Here is one reason to revise (NS) downward. If the best interpretation of the empirical data implies that *any plausible theory of knowledge* will violate (NS), then there is good reason to revise (NS) downward. If any plausible epistemic theory will fall victim to (NS) in the face of the situationist's empirical results, then this cannot be a special problem for virtue epistemology in particular. Thus, virtue epistemology must be shown to run afoul of the non-skepticism principle in ways that other plausible epistemic theories do not.

B. Virtue Epistemology: Alfano argues that both responsibilist and reliabilist virtue epistemology are threatened by empirical findings (Alfano 2011; Alfano 2013). However, it will be difficult to establish both. Consider this: If we grant that Alfano's empirical findings show epistemic irresponsibility and thus succeed against virtue responsibilism, we must assume that the subjects accurately represented the stimulus in the cases they did not manifest the virtue-relevant outcome. This is because, if S *incorrectly* represents a given stimulus that is actually P as being P*, and subsequently fails to achieve a virtue relevant outcome (O) by performing action (A) on the basis of P*, *S can still act responsibly* by A-ing so long as the virtue in question would require S to A when P* obtains. This is actually very common – When we say "I see why you would have thought that", and while we might express disagreement with a conclusion we can also grant cognitive responsibility as intended above. Responsibility also appears consistent with misrepresentation in "new evil demon" cases. Thus, if the failures Alfano cites are to count against the attributability of responsibilist virtues, these very agents must be accurately representing the stimulus conditions. But, it then appears that *virtue reliabilism must be true if virtue responsibilism is shown false in the way Alfano proposes*. Going the other way, if we grant Alfano that virtue reliabilism is shown false, then his argument against virtue responsibilism cannot succeed because we cannot assume that the agents are correctly representing the stimulus.

Perception like inferential abilities: Alfano appears to be aware of this, and thus only targets reliabilism for 'inferential' rather than perceptual knowledge. This appears to avoid the dilemma above, since the abilities assumed in representing the stimulus accurately seem to be perceptual. However, this is also problematic. Representing an epistemic environment or cognitive task does not easily reduce to reliably perceiving one's environment, and perception itself must likely involve very inference-like cognitive actions and abilities. Also, a wide range of research in bounded rationality, language acquisition and assertion shows that there are perception-like forms of inference that are very stable and reliable. Collectively, these research programs make it extremely likely that human beings have basic inferential abilities that are stable and reliable across an impressive range of situations and environments.

C. Inferential Situationism: Alfano reports that psychological research shows that the inferences people actually make employ heuristics rather than optimizing methods of formal logic and probability theory, citing a wide range of studies from Kahneman and Tversky (1973, 2011). However, heuristics as studied in bounded rationality research present a more optimistic story. Simon, Gigerenzer and others take seriously the fact that rationality theory studies a cognitively limited creature and have flourishing research programs that suggest heuristic use is often *more reliable* for a cognitively limited agent than using an optimizing rule. Alfano states that empirical results show that inferential beliefs are typically formed by heuristics rather than intellectual virtue. At a minimum, it must also be shown that heuristics cannot be virtues. In the current context, this will be a question of whether they are reliable. Properly understood, we argue that heuristics can be sources of relevant epistemic success when properly selected in the right environments. Because of some slothfulness involved in the process, it might be difficult to argue that heuristic reasoning is a form of responsibilist epistemic virtue, although there are prospects for a 'heuristic responsibilism' in Gigerenzer's recent work. However, we only aim to defend virtue reliabilism against Alfano, so this will be an independent issue. While the appropriate conditions for using heuristics are very narrow and they can lead to mistakes in reasoning, we suggest that human beings can manifest a certain kind of epistemic virtue through the appropriate use of heuristics. These will be *frugal virtues*.

4 Gigerenzer's Ecological Rationality, Bounded Agents and Epistemic Norms

Initiated by the pioneering work of Herbert Simon (1972), research on bounded rationality takes seriously that the subjects of epistemic evaluations are *cognitively limited,* and that heuristics often play an important role in successful human reasoning. Gerd Gigerenzer (2008) has now developed Simon's early insights into a well developed naturalistic epistemic perspective he calls *ecological rationality*, and has recently presented his research to mainstream epistemologists. Our interest here is to see how Gigerenzer's work provides the basis for an empirically grounded inferential virtue reliabilism that can meet the challenge from epistemic situationism.

4.1 Less Is More, Sometimes

Perhaps the most essential point in the bounded rationality research is that limited cognitive agents will often perform *less reliably* when using an ideal or optimizing epistemic rule than when properly employing fast and frugal heuristics. Gigerenzer illustrates this with the example of an outfielder tracking a fly ball who could potentially mathematically calculate the trajectory of the ball or apply some formal method to determine its future location and a strategy for catching it. Or, they could just keep the ball held fixed at the center of their visual field and keep running. The latter is a far more reliable way to succeed in catching the ball, even though the former would yield more accurate information if allowed to run to completion. In such cases, rational agents should not do what ideal epistemic rules prescribe.

This is a very important point. Optimizing rules are not always epistemically normative for limited cognitive agents and cannot fully prescribe what a limited cognitive agent ought to do. We get problematic results about rationality when we lose sight of this, but the results are not surprising when this is kept in mind.

4.2 Reinterpreting Linda

Consider Gigerenzer's (2008) interpretation of the Linda case from Kahneman, the main example examined in Alfano (2013). To quickly review the case: Infamously, when asked whether, given a character description of Linda, it is more probable that she is (a) a bank teller or (b) a bank teller and active in the feminist movement, 85 % of the subjects answered (b), clearly committing the "conjunction fallacy" and violating basic theorems of probability calculus. Gigerenzer notes that subjects are required to use syntactic, content blind rules of reasoning where the values of the variables are not relevant to getting the answer right, agents do not have additional cues from context or a specific rule to use. However, if one asks 'how many' instead of 'how probable', research shows better results. When asked *how many* out of 100 people that satisfy Linda's description would be bank tellers and *how many* would be bank tellers and active in the feminist movement, subjects' performance significantly improves and they do not commit the conjunction fallacy. This shows that different framing of logically equivalent information gets very different results, and the framing is thus playing a big role here. Gigerenzer shows that when the same information is presented relative to certain frames, people answer quite rationally. The poor performances that worry Alfano may thus to be very *local,* and we agreed above that local failures are consistent with Non-skepticism.

The significance of cognitive limitations can be easily missed. The point is not just that our threshold standards for *approximating ideal epistemic rules* should be informed by facts about cognitive limitations. This is a reasonable enough view, but the stronger implication is that we need to use an entirely different kind of norm. This would be a major shift in normative epistemology away from 'ideal-approximation assessments'. Adam Morton (2012) puts this point very well: *from the fact that we have an ideal epistemic rule, it does not follow that non-ideal epistemic agents should be evaluated in terms of how closely they approximate the ideal epistemic rule*. Morton argues that ideal-approximation norms are not sufficient instruments for evaluating limited cognitive agents. But what will this other kind of norm look like? How is it different from an optimizing norm? Will it be anything like a virtue?

4.3 Heuristics: What Are They? Why Do We Need Them?

In "Bounded Rationality: Models for some fast and frugal heuristics", Arlo Costa and Helzner (2011) develops results in formal epistemology that nicely illustrate the structure of heuristic reasoning, with both good and bad results for epistemic normativity. Drawing on Simon's famous image of "the two great scissor blades of rationality", *heuristic reasoning implements threshold evaluations for selected criteria that exploit reliable features of task environments rather than performing computations on sets of evidence.* That is not to denigrate optimizing rationality, but rather to emphasize that optimizing rationality and ecological rationality are two distinct and equally legitimate forms of rational response, two equally good and importantly different scissor blades. Traditional thinking about rationality sees it as normatively governed by optimizing norms alone, and is thus monistic in this sense.

Ecological reasoning transitions from threshold and criteria assessments to search and stopping rules, and Arlo-Costa shows that this can be formalized as a reliable type of reasoning. He also suggests a weak and strong reading of ecological rationality, and correctly locates Gigerenzer as adovating the stronger form.

> *Weak Ecological Rationality*: heuristic reasoning can be and often is near optimal when used in appropriate circumstances. Since optimizing norms are nearly approximated, no deep revision in epistemic norms is necessary, we just expand the rational strategies for satisfying them to include heuristic inferential processes and abilities.
>
> *Strong Ecological Rationality*: It is rational to use heuristics even when doing so goes against the dictates of optimizing rationality. This is normatively revisionist compared to traditional conceptions of rationality. This is Gigerenzer's stated view.

If strong ER forces significant revision to epistemic norms, some virtue epistemologists will approach bounded rationality cautiously,[3] while others might embrace Gigerenzer as a fellow epistemic revisionary. The issue raised here is over the tenability of a certain kind of epistemic value monism. In epistemology, value monism typically refers to something like Goldman's "t-value monism",[4] which claims that truth and the reliable means to it will be the sole values in normative epistemology. Many have argued for dualist or pluralist accounts of epistemic value in the literature on the "value problem", but the question above is over the scope of optimizing norms rather than whether truth is the sole epistemic value. Are optimizing norms sufficient to provide evaluation and/or guidance for the full range of cognitive tasks and achievements relevant to epistemology?

Weak heuristic rationality does not give up the truth goal, but constrains norms for success around the *cognitive limits* imposed on real world decision-making. Yet, even weak ER is at odds with optimizing rationality in the *content of guidance norms*, since rational agents will be instructed to do very different things when they are being ecologically rational than when they are being optimally rational. Weak ER thus preserves a fundamental commitment to truth, but will have a different prescriptive content than many traditional epistemic theories.

---

[3] Greco and Pritchard both clearly endorse the "traditional epistemic project", while others like Zagzebski and Roberts and Woods (2007) and Axtell are more revisionist in how they see the epistemic project, but not in how they see they epistemic virtues.

[4] See Goldman (2002).

4.4 Ecological Virtues?

Gigerenzer (2008) proposes a novel 'ecological definition of terms' according to which elements of an epistemic theory will actually represent complex relations between an organism and its environment, rather than properties seated entirely in either. This is a significant move regarding the ontology of epistemology. The agent is now just part of a broader epistemic ecology, and this ecology is the fundamental unit of analysis for evaluating human rationality. However, *this move potentially conflicts with virtue epistemology's emphasis on the agent,* and thus may be metaphysically (rather than normatively) revisionist. Virtue epistemology is defined as being 'agent based' rather than 'belief based', but it is not clear that Gigerenzian 'ecological virtues' would still be agent-based enough and in the right way to properly constitute what virtue epistemologists have in mind.[5] Many virtue epistemologists rely on some form of *agent-based credit for success* to both answer the value problem and respond to a range of problems related to epistemic luck, including Gettier Problems (Greco, Sosa, Zagzebski, Turri). Person-level abilities are an important arrow in the quiver and this will have to be worked out in any virtue theoretic formulation of ecological rationality.

On the other hand, perhaps virtue theory has always been understood ecologically. Dispositions have the very ecological structure Gigerenzer refers to because even the most robust dispositions like fragility and solubility will only manifest with the help of "reciprocal causal partners". Abilities and dispositions are also sensitive to environmental cues through "normal conditions" requirements that account for relevant forms of masking and mimicking, only some of which will imply agent culpability for lack of success. If disposition theory can bring virtue epistemology and ecological rationality together in a single account of reliable inferential abilities, virtue epistemologists will have at least the basis of a powerful empirical response to the situationist's empirical challenge. The social dimensions involved in cultivating virtues are also ecological in the relevant sense. Since virtues most likely have an ecological structure to begin with, accommodating Gigerenzer's research will not require metaphysical revisionism in virtue epistemology.

5 Knowledge of Syntax, Directed Memory and Basic Inferential Abilities

Knowledge of Syntax: Heuristics are not the only inferential ability supported by research in psychology. Research on generative grammar, language acquisition and communication all show that human beings have very stable and robust inferential

[5] See discussions of "the direction of analysis" in Greco (2010), Blackburn (2001).

abilities, though perhaps these are *basic inferential abilities* compared to the higher order calculations that Alfano appears to be concerned with.[6]

Basic inferential abilities are critically involved in acquiring the lexicon and generative rules of a language. Knowledge of syntax requires the manipulation of information according to strictly formal rules. Children have epistemic skills that allow them to learn any language based on these rules and their modal robustness is extraordinary. A vast amount of research in neuroscience and linguistics aims at explaining this robustness.[7] Specifically, scientists have tried to understand how it is possible for infants to learn a language given the incredibly diverse contexts they are in, the impoverished stimuli they are exposed to, the complexity of the grammatical rules etc. Despite there being many open questions, it is clear that some kind of inferential abilities are essential to language acquisition and, like perceptual skills involved, are remarkably stable across different situation types and individual differences. Although knowledge of syntax is highly formal, humans manifest such knowledge at a very early age, and they do so reliably and without conscious effort or monitoring. Infants do not need classes of universal grammar and rules of syntax in order to distinguish the syntactic components of (in many cases poorly constructed) utterances of a language. They are certainly not introspecting on these rules, or accessing evidence that could justify them to parse an utterance in terms of subject and predicate. What the infant is doing is highly complex, but the infant performs this incredible epistemic task in a *perception-like fashion*. Widely accepted results in linguistics and cognitive science also show that there is something inferential going on.

Notice that Gigerenzer makes it much easier to see this case as inferential but not optimizing or computational. Without claiming too much, we can at least say that the kind of inferential ability the child is manifesting will be something like the inferential abilities manifested in Gigerenzer's account of ecological reasoning. The inferential abilities involved in language acquisition may well be among the most basic capacities that heuristics feed on. But it will be enough for our purposes here if knowledge of syntax requires a distinct type of inferential ability, because this will still be problematic for Alfano independently of this convergence with Gigerenzer. The situationist may insist that even the most robust epistemic dispositions can be easily disturbed by very easy manipulations of the stimuli, perhaps the framing effect in the Linda case and other studies on the effect of font size show precisely how fragile these abilities are. Even here, they only function well when seemingly irrelevant environmental variables are not present. This in turn might threaten the anti-luck and safety intuitions endorsed by many virtue epistemologists.

---

[6] See Bach (1984), Montemayor (2014) and Proust (2007) for accounts of basic action that may be amenable to a theory of basic abilities. If there are basic actions, there are very likely abilities to cause the actions. These abilities might themselves be inferential even if the basic action is not itself an inference.

[7] See Jackendoff (2003), Chomsky (1986 and 1987) and Hornstein (1984).

In response, we would like to provide an illustration of why although information processing may always be disturbed under laboratory settings, this by no means threatens the stability of epistemic dispositions. For instance, in the Stroop task, the interference between inclinations (the automatic inclination to read a word vs. identifying a color) does not entail that the capacities involved are unreliable because of alleged context sensitivity. The capacities to read and detect color are incredibly reliable across subjects in many conditions. Interference only shows that having two inclinations affects processing. Any virtue conceived as a stable disposition will be disturbed or "masked" under some conditions. But being disturbed in non-standard situations is just part and parcel of *being a disposition*, but this is a point we will return to below.

It is worth noting that this response to Alfano requires widely individuated abilities for color recognition. It was conceded that in some cases the very narrow disposition to 'identify color R in *disturbing conditions* C' may not be reliable, so it is the broad recognitional ability "identifying color R" that allows reliability and susceptibility to disturbing conditions.[8] This appeal to broad abilities is not an *ad hoc* move just to defeat the situationist, but rather is the most natural way of understanding epistemic dispositions in light of the most recent evidence in psychology and linguistics. This also seems supported by the evolved basic capacities Gigerenzer cites as the life blood of ecological rationality.

If Alfano insists that inference must be rule-based, formal and regimented, one can hardly think of a type of inferential process that satisfies these constraints better than knowledge of syntax. One constantly uses the rules of syntax to parse words, identify their meanings, and translate from one language to the other. Knowledge of syntax is necessary to understand and know the meanings of any expression. So it is not trivial that these robust, widespread and stable epistemic capacities are performed in a perception like fashion.

5.1 Communication and Cognitive Success: Mellor

The inferential abilities shown above in knowledge of syntax and language acquisition point to more complex inferential abilities involved in communication. What do people know when they communicate and how are inferential abilities involved? D.H. Mellor's et al. (2003) theory of communication is based on widely shared knowledge of 'utility conditions' or what a person needs to know in order to effectively communicate. Mellor's account is inferential and shows more inferential abilities involved in language acquisition above. These will not be limited to examples of children, since we increasingly all communicate all the time.

Mellor's argues that communication involves a form of indirect inferential knowledge that is analogous to indirect observational knowledge. Communication

[8] For a nice account of recognitional abilities.

is "the production in the audience of beliefs about what the speaker believes he believes". We what we assert when we communicate is that 'S believes that he believes p'. Mellor argues roughly that x gets the belief that p from what Y believes, but not directly from what Y says, but from what Y believes he believes. The cognitive abilities involved in communication are stable, reliable and inferential.

Thus, an vast array of inferential capacities build on top of another, allowing humans to engage in a complex network of epistemic exchanges, with clear implications for social epistemology. Communication, based on knowledge of syntax and language, allows for shared forms of reliable true belief production, for instance by testimony or by collective evidence gathering. The principles of effective and epistemically virtuous communication are deeply linked to the norm of assertion as knowledge: one must only assert what one knows. We expand on this issue below.

5.2 Meta-cognition and Epistemic Feelings

A recent line of research on meta-cognition suggests that there are specific brain activities that monitor and control its own cognitive operations. At a minimum, metacognitive control involves (a) self prediction (b) post-evaluation and in many cases (c) intermediate evaluation. An interesting development in understanding meta-cognition comes from work on "epistemic feelings" – the feeling of knowing, certainty, doubt. Rather than being highly reflective, computational and costly, epistemic feelings are efficient ways of achieving meta-cognitive control. This 'non-cognitivist' account pushes us closer to Gigerenzer's gut reactions than Bayesian calculation. Lepock (this volume) argues that meta-cognition does not require meta-representation, just an adequate model. Epistemic feelings are implicit assessments of our cognitive operations, and these are used in the process of meta-cognitive regulation very effectively. Proust argues that meta-cognition regulates "mental actions" in a way that is analogous to the regulation of bodily actions by the motor system. Meta-cognitive dispositions are the motor system of the mind.

For our purposes, we need to consider whether any of this is inferential, or shows reliable inferential abilities. Borrowing from Proust (2001), performing mental actions like judging, deciding, solving, active attending, looking, listening involve *self-prediction*. This involves modeling available strategies for likelihood of achieving some cognitive goal and searching these for salient features. *Strategy selection on the basis of self-prediction certainly seems inferential.* If epistemic feelings are involved in this process in the way suggested above, this will be a less costly, fast and frugal process, and it will still be inferential.

'Post evaluation assessments' of mental actions evaluate how successful the selected operations were for the task at hand. This is a rule based judgment that also looks very inference like. Intermediate assessments are more controversial, but they present interesting assessments of likelihood of success of the selected strategy *while the strategy is being implemented.* Monitoring the selected strategy occurs largely by recognizing and responding to epistemic feelings that provide implicit assessments of how well things are going in a current cognitive operation. Even if this does not sound inferential, selecting a complimentary strategy (deciding to search your address book rather than your memory) certainly does. Complimentary strategies involve an inference that another strategy will be more effective in this particular task situation then the one currently being performed. That is clearly an inference. This kind of inference is either employed or available to an agent in most rational actions, and the ability to select complimentary strategies is a reliable and robust basic inferential ability.

5.3 Knowledge of Logic

Now consider the basis for any type of formal rule of inference: knowledge of logic. We have the capacity to reason according to modus ponens and this capacity is part of a set of stable dispositions to draw deductive inferences that are truth-preserving. One may actually say that these dispositions constitute what we *mean* by deductive inference.[9] If this is the case, then one could not know the meaning of what a deductive inference is without having such stable epistemic dispositions. It is a truism that basic deductive reasoning (for example an application of modus ponens) can be achieved without explicit understanding of such rule and that these dispositions, like those underlying knowledge of syntax, are remarkably stable. Demanding an explicit *understanding* of the rules for deductive reasoning increases cognitive demands, and although we can be trained to have such explicit understanding, this is not a necessary condition to have the stable dispositions that are implicit in our capacity to identify these rules. More importantly, requiring such explicit understanding is open to traditional objections against accessibilism and deontological accounts. Thus, it seems that the best strategy is to characterize these fundamental rules for deductive reasoning in accordance with our perception-like model.[10]

The situationist seems to face a new dilemma. Either we posses stable epistemic dispositions that allow us to identify valid rules for deductive inference or we don't. If we do, then situationism is false. If we don't, it is not clear how we are able to understand what we mean when we talk about, for instance, modus ponens. For it is not clear that highly unstable and easily disturbed capacities would help us succeed in specifying what we mean *in every situation* by the fundamental rules (modus ponens, modus tollens, etc.). Thus, it would not be entirely clear that we mean the *same* fundamental rules when we characterize a piece of deductive reasoning as modus ponens or something else. The situationist needs to explain why the

[9] See Boghossian 2000.

[10] Notice that this is quite different from having a conscious-intellectual "seeming," which is one way of defining intuitions.

psychological evidence would have such a dramatic result and this strongly suggests that situationism is in trouble. Obviously, the easy way out of this dilemma is to affirm that situationism is false, which is what we propose.[11]

6 Sosa, Assertion and Normal Conditions

Consider how Sosa's view would handle Alfano's challenge. Sosa requires that a belief must be AAA in order to count as knowledge (2007). It must be accurate (true), adroit (skillful) and apt (true because skillful). In Alfano's Linda case, Ash Paradigm and others do we have a failure of accuracy, adroitness or aptness? Clearly a failure of accuracy, whatever the right answer is, it cannot be the conjunction in the Linda case. It may also be a failure of adroitness in inductive reasoning, or a failure of accuracy because of a failure of adroitness. In either case, an agent would not know on Sosa's account, but since Alfano's challenge is to reliabilism, showing a general failure to meet the accuracy condition will be enough. If a belief is not accurate, it is not apt and is thus not knowledge. If we have too many of these kinds of failures, then we have face Alfano's worry about respecting non-skepticism.

Sosa can respond here by saying that the experimental conditions take an agent outside of normal conditions because of the presence of *interfering conditions*. Sosa says that an archer may fail to hit the mark if drugged or in a tornado, but in these cases there is no relevant sense in which the archer has failed because success is only expected in normal conditions for the exercise of the relevant competence. This is a widely accepted point about dispositions – they are only expected to manifest when in normal conditions, that is when not masked or finked. Failures outside of normal conditions are not relevant failures. If the epistemic situations that agents are placed in are not a normal for the cognitive competence being tested and they get a false belief, the agent does not *fail* any more than the archer does when in Hurricane conditions. Since Sosa requires normal conditions for relevant failures, he can say that there is no (virtue theoretic) epistemic failure shown in the research. Thus, even if there are many such cases, that does not imply that there are many epistemic failures, and thus does not push the virtue epistemologist to violating the non-skepticism principle.

Alfano might respond that these are clearly "normatively irrelevant" features of the environment, so they cannot/should not be built into the stimulus conditions. Building in too many normal conditions threatens to give a vacuous account that a virtue will manifest except when anything is preventing it from manifesting. We might avoid this through with a turn to disposition theory and distinguish between *culpable and non-culpable masks*. Alfano will need to show that environmental elements introduced in the research are environmental conditions that we should

[11] This is a concrete way of making a point suggested to us by Lauren Olin in conversation, which is that relativism is much more *troubling* in the epistemic case, as compared to the moral case. If we are right, situationism is also a lot more *implausible* in the epistemic case.

expect agents with a given ability to perform well in, even when presented with masks or finks. Alfano would need to show that these are not cases of non-culpable masking or finking. Without claiming to answer this important questions, it does not appear that Alfano has made good on this additional and necessary premise.

Assertion requires the same constraints, and situationism has extremely strong implications with respect to assertion. If the situationist is right, it would be quite difficult to satisfy the norm of assertion, threatening not only inference, but also testimony, public assessment of evidence and, ultimately, basic communication. We believe that having these consequences is a form of *reductio ad absurdum* for the situationist challenge in epistemology.

References

Alfano, M. 2011. Expanding the situationist challenge to responsibilist virtue epistemology. *Philosophical Quarterly* 62(247): 223–249.

Alfano, M. 2012. Expanding the situationist challenge to reliabilist virtue epistemology. Forthcoming.

Alfano, M. 2013. *Character as moral fiction*. New York: Cambridge University Press.

Alfano, M., and A. Fairweather. 2013. Situationism and virtue theory. *Oxford Bibliographies in Philosophy*.

Arlo Costa, H., and J. Helzner. 2011. Bounded rationality: Models for some fast and frugal Heuristics. In *Games norms and reasons*, Synthese Library, vol. 353, 1–21. Dordrecht: Springer.

Bach, K. 1984. Default reasoning: Jumping to conclusions and knowing when to think twice. *Pacific Philosophical Quarterly* 65: 37–58.

Blackburn, S. 2001. Reason, virtue, and knowledge. In *Virtue epistemology: Essays on epistemic virtue and responsibility*, ed. Abrol Fairweather and Linda Trinkaus Zagzebski, 15–29. Oxford: Oxford University Press.

Boghossian, P. 2000. Knowledge of logic. In *New essays on the a priori*, ed. P. Boghossian and C. Peacocke, 229–254. Oxford: Oxford University Press.

Chomsky, N. 1986. *Knowledge of language*. New York: Praeger.

Chomsky, N. 1987. *Generative grammar: Its basis, development and prospects*, Studies in english literature and linguistics. Kyoto: Kyoto University.

Gigerenzer, G. 2008. *Rationality for mortals: How people cope with uncertainty*. Oxford: Oxford University Press.

Goldman, Alvin I. 2002. *Pathways to knowledge: Private and public*. Oxford: Oxford University Press.

Graham, Peter J. 2012. Epistemic entitlement. *Noûs* 46(3): 449–482.

Graham, P. 2014. Warrant, functions, history. In *Naturalizing epistemic virtue*, ed. A. Fairweather and O. Flanagan. New York: Cambridge University Press.

Greco, J. 2010. *Achieving knowledge: A virtue-theoretic account*. Cambridge: Cambridge University Press.

Harman, G. 1999. Moral philosophy meets social psychology: Virtue ethics and the fundamental attribution error. *Proceedings of the Aristotelian Society* 99(1999): 315–331.

Harman, G. 2000. The nonexistence of character traits. *Proceedings of the Aristotelian Society* 100: 223–226.

Henderson, D., and T. Horgan. 2009. *Epistemic virtues and cognitive dispositions. Debating dispositions: Issues in metaphysics, epistemology and philosophy of mind*. Berlin: DeGruyter.

Hornstein, N. 1984. *Logic as grammar*. Cambridge, MA: MIT Press.

Jackendoff, R. 2003. *Foundations of language: Brain, meaning, grammar, evolution*. Oxford: Oxford University Press.

Kahneman, D. 2011. *Thinking fast and slow*. New York: Farrar, Straus and Giroux.
Kahneman, D., and A. Tversky. 1973. On the psychology of prediction. *Psychological Review* 80: 237–251.
Mellor, D.H., Hallvard Lillehammer, and Gonzalo Rodríguez Pereyra (eds.). 2003. *Real metaphysics: Essays in honour of D.H. Mellor*. London: Routledge.
Miller, C. 2003. Social psychology and virtue ethics. *Journal of Ethics* 7(4): 365–392.
Miller, C. 2014. *Character and moral psychology*. New York: Oxford University Press.
Montemayor, C. 2014. Success, minimal agency and epistemic virtue. In *Virtue epistemology naturalized: Bridges between virtue epistemology and philosophy of science*, ed. A. Fairweather, 67–81. Cham: Springer.
Morton, A. 2012. *Bounded thinking: Intellectual virtues for limited agents*. New York: Oxford University Press.
Proust, J. 2001. A plea for mental acts. *Synthese* 129(1): 105–128.
Proust, J. 2007. Metacognition and metarepresentation: Is a self-directed theory of mind a precondition for metacognition? *Synthese* 159: 271–295.
Roberts, R., and J. Wood. 2007. *Intellectual virtues: An essay in regulative epistemology*. New York: Oxford University Press.
Simon, H.A. 1972. Theories of bounded rationality. *Decision and Organization* 1: 161–176.
Sosa, E. 2007. *A virtue epistemology*. Oxford: Oxford University Press.
Sosa, E. 2011. *Knowing full well*. Princeton: Princeton University Press.
Zagzebski, L. 2010. Exemplarist virtue theory. In *Virtue and vice: Moral and epistemic*, ed. H. Battaly, 39–55. Chichester/Malden: Wiley/Blackwell.

Part II
Epistemic Virtue and Formal Epistemology

Curiosity, Belief and Acquaintance

Ilhan Inan

Philosophers have paid little attention to curiosity until quite recently. There is now at least a scarce literature that discusses how curiosity relates to certain intellectual traits that we value such as inquisitiveness and open-mindedness, whether it is an essential instrument to lead us to certain epistemic achievements such as the acquisition of truth or knowledge, whether being curious is an intellectual, an ethical, or even a moral virtue, and whether curiosity is required for a good life.[1] Most of this discussion takes place in an area where epistemology overlaps with ethics and value theory, generally known as virtue epistemology. Whether curiosity is taken to be a form of virtue or not, it should be clear that there are important connections between being curious and some of our basic epistemic attitudes and achievements. Knowing, for instance, is an epistemic achievement, at least in certain cases, and curiosity is one of its basic motivators. The question of how curiosity and knowledge are related brings about a host of interesting philosophical issues, the most important of which relates to what curiosity is.[2] After all the classical "definition" equates curiosity with a desire to know. There is then the important comparative logical question: If knowledge is a propositional attitude, is curiosity so too? There are also issues concerning how curiosity relates not to knowledge, but rather its

[1] See Daston and Park (2001), Baumgarten (2001), Kvanvig (2003), Miscevic (2007), Schmitt and Lahroodi (2008), Brady (2009), Subasi (2009), Yigit (2011). Apart from this literature there has been very little discussion on some of the basic philosophical questions concerning curiosity, such as what curiosity is, what makes it possible, how it is satisfied etc. See Kvanvig (2003) and especially Whitcomb (2010). Though not directly on curiosity there is also some current relevant research on open-mindedness, inquisitiveness, love of truth and related issues: see Zagzebski (1996), Hookway (2003), Battaly (2008), Roberts and Wood (2009), Riggs (2010), Crisp (2010), Baehr (2011).

[2] I am inclined to think that curiosity based knowledge has more value than what might be called "accidental" knowledge. If so this should provide good reason for virtue epistemologists to address philosophical questions on curiosity.

I. Inan (✉)
Department of Philosophy, Bogazici University, Bebek, Istanbul, Turkey
e-mail: inanilha@boun.edu.tr

A. Fairweather (ed.), *Virtue Epistemology Naturalized: Bridges Between Virtue Epistemology and Philosophy of Science*, Synthese Library 366, DOI 10.1007/978-3-319-04672-3_9,

opposite, namely ignorance. What are the mental mechanisms we employ which allow us to become aware of our ignorance on a particular issue, and how does this motivate curiosity? Is awareness of ignorance a precondition for curiosity? There are also related issues that concern how curiosity relates to the asking of a question. If all knowing is in fact knowing the answer to a question, does it then follow that knowledge always originates from curiosity?[3] How does our curiosity relate to the asking of a question, and how does the satisfaction of our curiosity relate to the answering of our question? How does curiosity motivate inquiry into the unknown? I have dealt with these and other related issues in detail in recent work.[4] Based on some of the ideas developed there, I now wish to elaborate on topics which should be relevant not just to virtue epistemology, but to epistemology in general, and especially to formal epistemology. These involve how curiosity relates to some of our basic epistemic attitudes that come short of knowledge. Among them two stand out as being the most relevant, that is *belief* and *acquaintance*. How does curiosity relate to the holding of a belief that is uncertain and how does it relate to having partial acquaintance with an object?

Plenty of work has been done on *belief*, very little work has been done on *curiosity*, and to my knowledge there is no work, at least in the philosophy literature, that explicitly addresses the issue of how the two are related. To start off we may say that if you have a belief that is too firm, then there will be no room left for curiosity. If you are certain that Plato was a philosopher for instance, then you cannot be curious whether that really was or was not the case. Curiosity about whether a proposition is true or false can only take place under uncertainty. Here the notion of *certainty* should be taken in the "subjective" sense. It has to do with the epistemic attitude the subject takes with respect to the truth of a proposition. Being certain, in this sense, corresponds to maximum strength of a belief. Once that level is reached genuine curiosity becomes impossible. This is not a normative notion, rather it describes the mental state one is in. Being subjectively certain is not a factive mental state; that is a person may be subjectively certain that a given proposition is true, when in fact that proposition is false. If an ancient was certain that the world is flat, then he could not have been curious about whether this was or was not in fact the case. People who are certain of their beliefs may not always have the right to be certain. The evidence they have may not entitle them to be certain, but they still may. That is why people who dogmatically hold beliefs cannot bring themselves to be curious about their beliefs without giving up their dogmatism. Fortunately, not everyone is like this. There are many rational open-minded people who hold beliefs without feeling certain that those beliefs are true. The stronger your belief gets the less possible it becomes to be curious. So it does appear that curiosity is inversely propositional to the strength of one's belief, or what in the Formal Epistemology literature is called "degree of belief".[5] This is a particularly interesting notion that

---

[3] Schaffer (2007) explicitly defends the view that knowing is always knowing the answer to a question; some of Collingwood's (1940) ideas seem to imply it. I argue against this view in Inan (2012); see especially p.147.

[4] See Inan (2012).

[5] For recent work on *degrees of belief* see Huber and Schmidt-Petri (2009).

connects epistemology with other branches of philosophy, as well other scientific disciplines. That is because the degree of our belief in the truth of a proposition partially determines how we are inclined to act, as well as how we ought act in a given context. It is a central notion concerning the norms of rationality, and it is an essential concept to be utilized in our attempts to explain and model the human mind. Now just like belief, curiosity also comes in degrees. The degree of one's curiosity is one of the parameters that determines the strength of one's motivation to learn something new. It is an instance of one of the "passions of the soul",[6] as Descartes called it, which motivates inquiry. Understanding the epistemic features of the human mind, both descriptively and normatively should then require us to take into consideration curiosity. Once we integrate the notion of *curiosity* into the formal epistemology literature we will have a better chance of understanding and in effect modeling the human mind.

We enjoy curiosity partially because we are fallible beings. The evidence we have for most of our beliefs about the external world, and perhaps even for some of our beliefs about our own minds, do not guarantee that those beliefs are true. Merely the fact that we are fallible beings however is not sufficient to explain our curiosity. Curiosity can only take place when we come to realize the fallibility of our beliefs. It requires open-mindedness. And this can only take place in the absence of certainty. That is why utterances in the form "I am certain that p, but I am still curious whether p" can never express truths. Anything short of subjective certainty should then allow some room for curiosity. Even if you know that it is extremely improbable that a belief you hold might turn out to be false, you may still be curious about it. If you have a lottery ticket which you know that its chances of winning the big prize is extremely slim, you may still be curious as to whether it will. In fact people who buy lottery tickets find the motivation to check the winning numbers which indicates that they are in fact curious as to whether their ticket won. The more interesting fact is that you may be curious whether your ticket will win even if you believe that it will not. That is, utterances in the form "I believe that p, but I am curious whether p" are fine, and in fact express truths in certain contexts. Curiosity, at least in one of its forms, has to do with how much evidence one has for the truth of a proposition, and whether one takes that evidence as being conclusive: the less evidence there is, the more room for curiosity. Curiosity would then seem to have the potential of being maximized when there is no evidence on either side. I have access to no evidence for or against the truth of the proposition that there is intelligent life on other planets. I neither believe nor disbelieve it, and, of course, I am extremely curious about it. It would seem then that such cases of suspension of belief are ones which have the potential to maximize the degree of one's curiosity. So then, it initially appears as if the stronger one's belief gets the weaker the curiosity will become. Going back to the lottery case, suppose you pay one dollar for a lottery

[6] There are six primitive passions of the soul according to Descartes (1989). Among them is wonder ("admiration" in the original French) which is a "sudden surprise of the soul". Curiosity on the other hand is only a sub-species of another primitive passion, namely desire, and it is explicitly defined as "desire to understand" by Descartes.

ticket and will collect one million dollars if you win where your chances of winning is one in a million. Now you may be curious as to whether you will win, though the strength of your curiosity under normal circumstances would not be too high. That is because the degree of your belief that you will win is close to zero. But if you played another game which had the same stakes, but very different odds things would appear to be different. Suppose you again bet one dollar, and then we flip a fair coin, and if it is heads you win a million dollars and if it is tails you win nothing. All else being equal, my hypothesis is that you would be a lot more curious as to whether you will win in this case compared to the lottery case. That is because, all else being equal, your degree of belief in the same proposition is now raised to the "medium" value, mostly represented as 0.5 in the [0, 1] interval. If we raised the odds so that this time the chances of you *not winning* is one in a million, the degree of your curiosity will go down once again, all else being equal. This appears to show that the degree of curiosity is inversely proportional to the degree of belief. Now some may object to this by pointing out that at times as the degree of belief goes higher so does one's curiosity. Suppose that after investigating the crime scene, Holmes becomes curious who the murderer is. Initially there are no suspects, but then Holmes finds good evidence that a certain Ralph, whom he knows from an earlier case, might be the murderer. He may in fact come to believe, but not know that Ralph is the murderer. Initially Ralph was not on Holmes' suspects list, there was no evidence to tie him with the murder. We may assume that at this stage the degree of Holmes' belief in the proposition that Ralph is the murderer was 0.5. Nonetheless Holmes may not have been curious whether Ralph is the murderer. But then soon as he collects new evidence that makes Ralph a suspect, Holmes' degree of belief of the proposition that Ralph is the murderer now has come to be quite high. So the degree of belief has increased significantly, but contrary to what I said earlier, we may easily imagine that Holmes has now become curious whether Ralph is in fact the murderer. So then it might seem in this case that the degree of one's curiosity increases with the increase in the degree of belief. And then this will go on until the peak is reached, that is until the subject feels certain that he now knows the proposition in question or its negation. So under this account, Holmes' degree of curiosity will increase as he gathers more evidence that Ralph is the murderer; and once he comes to know that Ralph is or is not the murderer, then he will no longer be curious and the degree of his curiosity will suddenly drop to 0. This I believe is not fully accurate. When there was no evidence for or against the claim that Ralph is the murderer Holmes was not curious whether he was the one. Holmes became curious soon as he found some evidence which made Ralph a suspect. The earlier claim was that the degree of curiosity decreases as the degree of belief increases, all else being equal. What is important to note is that in Holmes' case not all else is equal. That is because at times new evidence may also increase our *interest*. Curiosity is not merely related to our degree of belief, there is another important parameter involved, namely our interest in the object of our curiosity. Initially Holmes was not interested in Ralph, or to be more precise he was not interested in the truth of the proposition that Ralph is the murderer. After collecting evidence making Ralph a suspect, Holmes then became interested. The issue of how interest and belief relate to one another is a tough one, but at least how interest relates to curiosity should be quite clear: the degree of curiosity is directly proportional to the degree of interest.

As I said anything short of complete certainty allows for curiosity.[7] This of course does not imply that we are curious about the truth of just any old proposition we entertain in our minds of which we are not certain. The proposition that the number of words in the finished version of this article will be odd is not one that I have any evidence for or against. I am not even sure whether it has a determinate truth value. My degree of belief is 0.5, meaning that it is not even a belief that I hold. And not only do I not hold a belief one way or another, I am simply not interested in the issue. It makes no difference for me, or anyone else for that matter, whether the number of words turns out to be even or odd in this article. If the editors of this issue had developed a weird policy of publishing only those articles containing odd number of words, I might have had an interest in the topic. As it stands I don't. There are also many beliefs we in fact do hold, in which we again have no interest. After hearing the weather forecast, say just by accident, suppose you come to believe that it will rain today; yet you may not be curious whether it will or it will not rain today. You may simply not be interested in the topic. Lack of certainty only when accompanied with interest motivates curiosity. This is why you may hold two separate beliefs having the same degree, though you may be curious about the truth of one, and not the other, or you may be curious about both, but with different degrees. For instance, normally one's curiosity about something as trivial as the solution to a logic puzzle will not be as strong as one's curiosity about something as vital as the result of a critical medical exam. That is because under normal circumstances we care about our health more than we care about the solution to a puzzle and therefore we have more interest in the former. The degree of one's curiosity is fundamentally linked with one's interests in general, and, as said earlier, it is directly proportional to the degree of interest in the truth of the proposition one is curious about. Just like belief and curiosity, interest also comes in degrees; the higher it gets the more room there is for curiosity. Overall we might then conclude that for any subject and a proposition that that subject grasps, the degree of curiosity in the truth of that proposition will be inversely proportional to the degree of belief in the truth of that proposition, but it will be directly proportional to the degree of interest in the truth of that proposition. That of course does not tell us how exactly these three parameters relate to one another, but it at least tells us that these are the parameters to consider. If interest and belief were independent attitudes, then there would have been a simple equation that connects them with curiosity. However they are not independent attitudes. In fact interest interacts

[7] I hold that one can even be curious about something he or she knows, as long as that piece of knowledge is fallible and thus not certain in the subjective sense. Though an utterance such as "I know that the world population is greater than seven billion, but I am not certain that this is the case and I am still curious whether it is so" does seem somewhat odd, it may very well express a truth. Given that this would appear to be a controversial issue, I do not pursue it here since my current focus is merely on how curiosity relates to belief.

with belief in its own peculiar way, and without further inquiry into this interaction we should not jump to any conclusion. It would for instance be wrong to conclude that the degree of interest in the truth of a proposition is directly proportional to the degree of belief in the truth of that proposition. One may lack interest in the truth of a proposition regardless of whether he or she has any evidence for it. Whether the number of words in this article is odd is an issue I have no interest in, and that is totally independent of my degree of belief in the truth of this proposition. Therefore we should conclude that the reason why Holmes becomes more interested in whether Ralph is the murderer soon as he collects new evidence making him a suspect, cannot be merely due to the increase of his degree of belief. We should have to bring into consideration Holmes' interest in *who the murderer is* in order to explain the increase of his interest in whether Ralph is the murderer when he gathers new evidence making him a suspect. The most that can be said here is that the degree of one's curiosity is a function of his degree of belief and his degree of interest when there is a full proposition involved.

The preceding discussion is applicable only to curiosity which has propositional content. That is not always the case. To see this, we should distinguish between two types of curiosity. If you are curious about whether there is life on other planets, your curiosity has propositional content: you wish to know the truth value of the proposition that there is life on other planets.[8] But if you are curious about what Plato's father's name was, then there is no proposition you can single out as one whose truth value you are seeking. Or when Holmes is curious who the murderer is when he has no suspects, there is no particular proposition in the form [a is the murderer] of which Holmes is curious to know. So my hypothesis is that being curious who someone is, or being curious when or where or how or why some event took place need not involve curiosity in the truth of a proposition. Though this distinction between two types of curiosity is far from being commonplace in philosophy or any other discipline, the corresponding distinction between two types of questions was made more than a couple of millennia ago. Aristotle famously distinguished between "whether-questions" that ask for whether there is a "middle term" and "what-questions" that ask for what that middle term is.[9] Today many distinguish between direct and indirect questions, where the former admit of "yes" or "no" as answers, but the latter, which are also known as "wh-questions", do not.[10] If we assume that the use of interrogative sentences is our normal linguistic tool by which we express our curiosity, then we should expect that there are two types of curiosity as well. I will call curiosity expressible by a direct question "propositional curiosity", and

[8] This is in fact an oversimplification. At times we wish to know more than just the truth value of the proposition in question; we wish to know the fact that makes the proposition true. That is why I hold that there are two ways of satisfying propositional curiosity, de re and de dicto. For a discussion of this see *Chapter 2 Asking and Answering*, and *Chapter 9 Conditions for the Satisfaction of Curiosity* in Inan (2012).

[9] Aristotle (1924), *Posterior Analytics, Book II, Chapter 1*, p.50.

[10] In contemporary philosophy the distinction was made by a number of philosophers. An early version can be found in Leonard (1957).

curiosity expressible by an indirect question "objectual curiosity". So even if we gave a satisfactory account of how degree of belief and propositional curiosity relate to one other, that will not be sufficient. We will have to account for objectual curiosity as well which cannot be reduced to a propositional attitude. This will require us to introduce at least one new epistemic parameter into our equation. The moral to be drawn from all this is that our epistemic attitudes which motivate us to act are not merely limited to the strengths of our beliefs and interests. We are intellectually a bit more complicated than that.

So I take it that propositional curiosity is what is expressible by a question in the form "is it the case that s?" where s is a full declarative sentence that expresses a proposition. If we further assume that truth and falsity are properties of propositions, then the object of propositional curiosity will be an unknown truth value. If we put this in terms of a *desire to know*, then we may say that in such cases the curious subject desires to know which of the two truth vales a proposition has. In this sense we may take this form of curiosity as a propositional attitude of a peculiar kind. This is not the case though for objectual curiosity, i.e. curiosity that is expressible by a wh-question. In such cases it is not that the degree of belief together with the degree of interest are not sufficient to account for curiosity. Rather in these cases the notion of *degree of belief* is no longer applicable. That is because objectual curiosity is not propositional. In other words being objectually curious is not a propositional attitude. We can no longer account for curiosity in terms of belief, given that there is no such thing as "objectual" belief. The difference between the logical status of belief and objectual curiosity reveals itself in surface grammar. Sentences in the form "S is curious about the F" are perfect constructions and are in fact used quite frequently, but there is no analogous construction for belief. When Holmes asks "who is the murderer of Smith?" out of curiosity, we may take that to mean that he is curious about the murderer of Smith. So "Holmes is curious about the murderer of Smith""expresses a truth, but "Holmes believes about the murderer of Smith" is ungrammatical. (There is of course one specific use of the verb *to believe* in which we say things like "Holmes believes John" and we might even say "Holmes believes the murderer", but that is obviously not an objectual attitude.) When we say that Holmes' curiosity is not propositional we do not wish to say merely that the interrogative sentence that he uses does not contain a full proposition. The claim is in fact a lot stronger than that. What we wish to say is that we cannot single out any proposition of which Holmes wishes to know whether it is true or false. There simply is no such proposition. Now some may perhaps wish to say that there is at least a certain long disjunctive proposition in which each disjunct is a possible answer to the question. This long disjunction may be along the lines of "Ralph is the murderer of Smith or Brown is the murderer of Smith or …". And then we may say that Holmes wishes to know which disjunct is true. Now this might be true in certain cases. If Holmes has, say, four possible suspects, and he knows that the murderer is among them, then he may have at his disposal a disjunction with four disjuncts. But that is on the assumption that Holmes has certain suspects to form the disjunction. What if he is totally in the dark about the identity of the murderer? It might simply be the case that the murderer is totally unknown to Holmes and neither his name nor any other information about him appears in any of Holmes' files. He has no actual suspects, and not even possible ones. Nevertheless Holmes is curious who the murderer is. Regardless of whether Holmes has suspects or not, it is important to notice here that being curious about who the murderer is, is not the same thing as being curious about which disjunct is true in a disjunction. If we can formulate a disjunction with all the possible answers to the question appearing as separate disjuncts, then it should be clear that Holmes cannot grasp this very long proposition. Of course Holmes knows very well what he is curious about; that is, *being curious* is a mental state, and Holmes has access to it. So given that he cannot single out a certain proposition that he grasps as giving the content of his curiosity, we should conclude that his curiosity does not have propositional content. It is of course true that if Ralph is the murderer and Holmes comes to know this, then his curiosity will be satisfied. But that does not imply at all that Holmes was curious about whether Ralph was the murderer. He may have never heard of Ralph before, and no information may have been available to him about Ralph initially when he was curious about the murderer. It is one thing to be curious about whether Ralph is the murderer, it is another to be curious about who the murderer is; the former is propositional the latter is not. I hold that these are very different mental states. Objectual curiosity is not propositional nor can be reduced to it.

Now even if you are convinced that objectual curiosity is not propositional, you may be inclined to think that at least its satisfaction is propositional. If Holmes is curious about the murderer, and Ralph is the one, then once Holmes comes to know that Ralph is the murderer he should have satisfied his curiosity. That is not always correct. That is because when Holmes comes to know that Ralph is the murderer, it does not immediately follow that he knows who the murderer is. Suppose that Holmes receives an anonymous phone call from a man who claims to be the murderer. Let us assume that caller is in fact the murderer and he manages to convince Holmes that this is the case by telling Holmes very specific detailed facts about the murder. Let us further suppose that Holmes now has come to know that the caller is in fact the murderer. Even so Holmes still knows very little about this guy, in fact even if the caller tells him that his name is "Ralph" it might make no difference. After all the name "Ralph" may be a made up name, and Holmes may still wonder who this person is. Under this scenario it would not be wrong for Holmes to assert that he does not know who the murderer is.[11] There is at least a strict use of the notion of *knowing who* under which this is the case. He might come

[11] It is commonplace in philosophy to hold that *knowing who* is an interest relative term. I have argued in my (2012) that the reason for this is because in many contexts the notion of *knowing who* is used elliptically for a longer notion, though there is also what I called a "strict use" of this notion that is non-elliptical and therefore not interest relative. Braun (2006) is perhaps the only one in the literature who also argues against the interest relativeness of knowing who. However the epistemic standards on Braun's view of knowing who someone is, is so low that all it takes for one to know who someone is to know a property of that person which need not even be a uniquely identifying one. Obviously I disagree with Braun, for it appears that on his view we would not be able to express genuine curiosity by asking a who-question. See my (2012, pp. 60–61) for a discussion of Braun's position.

to know that Ralph is the murderer, and if he was asked who the murderer is he could say "It is Ralph", but that does not change the fact that he does not know who Ralph is; his degree of acquaintance with Ralph is not sufficient. There have been actual cases like this. One was the famous Unabomber case. Before the suspect was caught, the police and the media had given the name "Unabomber" to the person who was responsible for a number of mail bomb incidents. Even the name all by itself aroused curiosity. People were curious about the Unabomber, given that they did not know who he was. Now go back to early 1990s when the Unabomber sent one such mail to a university office. Initially the police may have been curious as to whether the Unabomber struck again, and whether he or someone else was responsible. After investigating the evidence let us assume that they found out that it was the Unabomber again who was responsible for this latest incident. That may have satisfied their curiosity whether the Unabomber was responsible for the latest incident, but they still did not know who he was. They were still curious about this. The satisfaction of objectual curiosity requires more than learning that a certain proposition is true. It requires raising the degree of your acquaintance with the object of your curiosity to a certain level. What that level should be depends on one's interests and many other contextual factors. Reaching a certain degree of acquaintance of the object of curiosity may satisfy one but not satisfy another, and even the same person may change his standards from one context to another.[12] What is important to note here is that the police and the media and the interested public were curious about who the Unabomber was given that their degree of acquaintance with this person was too low. All that they knew of him was what they were able to gather from the evidence of the bombs he had sent. And given that there was a lot of interest in the case, there was a lot of curiosity.[13]

So we may then wish to conclude that there are two main parameters that determine the degree of one's (objectual) curiosity, namely the degree of interest and the degree of acquaintance. That would not be fully accurate. If you are curious about the colors of the Jamaican flag, that does not imply that there are certain colors in this flag of which you have a low degree of acquaintance. It is not that you wish to know more about a certain color and raise your degree of acquaintance with it. Rather given that you are already acquainted with the basic colors, you wish to know which ones appear in the Jamaican flag. In fact you may truthfully say "I am acquainted with the colors of the Jamaican flag", and then you may add "but I do not know which colors those are". If you have a particular interest in flags, then you may be

[12] For a more detailed discussion of this see *Chapter 10 Relativity of Curiosity and Its Satisfaction* in Inan (2012).

[13] I am in full agreement here with Kvanvig (2003) in his emphasis on the need to appeal to an objectual epistemic notion to explain our epistemic virtues. Kvanvig makes a further distinction between *understanding* and *knowledge*, and places *objectual understanding* at the top of the epistemic values. For the present purposes all that I am committing myself is the view that in order to account for the satisfaction of curiosity we need to appeal to some epistemic notion that forms a relation between an agent and an object. It seems to me that our common use of the verb *to know* in the objectual sense captures exactly this, though following Kvanvig we might prefer to replace it with the notion of *understanding*.

curious about this even if you know that you are acquainted with the object of your curiosity (which is a set of colors in this case). Curiosity does not always imply lack of acquaintance. As in this case, a curious subject may have a relatively high degree of interest, but also a relatively high degree of acquaintance with the object of his curiosity. If the degree of acquaintance is high, why should our subject be curious? The short answer to this question is that curiosity has conceptual content. What you lack in this case is not acquaintance with certain colors, but rather you wish to know which of those colors (that you are already acquainted with) fall under the concept *the colors of the Jamaican flag*. You are curious given that you do not know which colors this term refers to. Acquaintance is an extensional notion, whereas what we need is an intensional one, that is, we need a notion that is sensitive not only to the degree of acquaintance of the object of curiosity, but also to what concept you represent that object in your mind. I will call this parameter "the degree of ostensibility". Roughly this notion applies to how the curious subject is epistemically related to an object *under a concept*. To be curious about an object we need to be able to conceptualize it; the basic tool by which we achieve that is by constructing a definite description whose referent is unknown to us, what I have called an "inostensible term"[14] relative to a subject, that is a term whose referent is unknown for that subject. The referent may be unknown because the subject may simply not have come across it before. Holmes may be curious about the murderer even when he has no suspects, and when he has no epistemic connection to the murderer except for whatever evidence there is at the murder scene. But we may also be curious about the referent even when we do have some close epistemic connection to it, when the referent is in fact an object we are partially acquainted with, and even when we know that this is the case. If Holmes has sufficient evidence to come to know that the murderer is one of the two suspects both of whom he knows to a certain degree, he may still be curious as to which of them is in fact the murderer. Satisfaction of curiosity takes place only when we come to know that a certain object is the referent of our inostensible term. For Holmes to satisfy his curiosity, he must be able to connect his inostensible term "the murderer" with one of the two suspects and come to know this. If Ralph is the murderer, Holmes must come to know Ralph as being the murderer, where "Ralph" is an ostensible term for Holmes in that he knows that this name refers to a person with whom he has some high degree of acquaintance. We may now say that the degree of ostensibility of a term d for a subject S reaches its maximum level if there is an object o such that S is completely acquainted with o and S knows that o is the referent of d. The degree of ostensibility will be very low if there is no object that S is acquainted with which S knows to be the referent of d. And then there will be intermediate cases in which there is an object o with which S has a certain intermediate degree of acquaintance.

The degree of curiosity then is a function of two factors: degree of interest and degree of ostensibility. It is directly proportional to the former and inversely proportional to the latter. Acquaintance is by itself not one of the direct parameters that

[14] See Inan (2010, 2012) for a detailed discussion of the distinction between ostensible versus inostensible terms.

determines curiosity and its degree. Note that acquaintance is an epistemic relation between an object and a subject; it is a purely extensional relation – rather than an intensional one. That is because the *acquaintance* relation says nothing about how the subject conceptualizes the object in question. Presumably this is not true for propositional knowledge or belief. When you say Sue knows that the world is round, you do say something about how she conceptualizes a certain fact. Bu when you say that Sue is acquainted with Ben, you say nothing about how Sue conceptualizes Ben. Now it might be the case that *acquaintance* always requires a form of conceptualization, i.e. in order for Sue to be acquainted with Ben she must have some kind of mental representation of Ben which has conceptual content. Or one might follow Russell here and hold that there is direct acquaintance with some kinds of entities that is pre-linguistic having no conceptual content. Either way it is a fact that an acquaintance attribution in the form [S is acquainted with o] says nothing about how the subject conceptualizes the object. This is exactly why *acquaintance* together with *interest* are not sufficient to explain curiosity. I claim that curiosity always requires the representation of an unknown object. That kind of representation for us has conceptual content. If there are other forms of representation that we, or some animals, or some extra-terrestrial beings employ, then there are other forms of curiosity that do not have conceptual content. Still the curious being must be able to represent something unknown; without it there is no curiosity. That is why some animals or infants who exhibit novelty seeking behavior, and try to explore their environment are not necessarily curious beings on my account. Wandering is not wondering. To wonder, in the sense of being curious, one must have the ability to attempt to single out something unknown and seek it. This requires a higher order mental capability than simply having the instinct or drive to be attracted to novel things in the environment. So even if one may make a case that there can be curiosity with no conceptual content, there cannot be curiosity without the ability to represent the unknown. This kind of representation is what I take to be a form of purported reference, (which I call "inostensible reference".) Every curious being attempts to refer to an unknown; if there in fact is such a thing, then reference may succeed, if there is no such thing then it fails. Either way there is an attempt to refer to the object of curiosity. It is in this sense that curiosity is an intentional as well as intensional mental state. It is intentional in the sense that it is directed towards an object (though it may turn out that there is no such object), and it is intensional in the sense that it has representational content. That representational content when expressed in language is captured by an interrogative sentence. Given that sentences have conceptual content, it follows that human curiosity expressible in language has conceptual content. This is exactly why we cannot account for objectual curiosity by appealing to the notion of acquaintance alone. Acquaintance is an extensional notion, but what is needed is an intensional notion such as ostensibility. Objectual curiosity requires one to grasp a concept which determines the object of one's curiosity. The degree of curiosity is then a function of the degree of ostensibility of that concept. The notion of acquaintance is still relevant, but in an indirect way. We may define the ostensibility of a concept for a subject in terms of the degree of acquaintance of the object (determined by that concept) *under that concept*.

This account takes acquaintance as admitting of degrees. That was not the case for Russell who took "acquaintance" to be an absolute notion, an all or nothing affair. Russell did however distinguish between different stages of "removal from acquaintance":

> It will be seen that there are various stages in the removal from acquaintance with particulars: there is Bismarck to people who knew him, Bismarck to those who only know of him through history, the man with the iron mask, the longest-lived of men. These are progressively further removed from acquaintance with particulars...[15]

Once we lower Russell's standards of acquaintance, then it should follow that what he calls "the stages of removal from acquaintance" can simply be captured by the notion of "the degrees of acquaintance". Note that on Russell's account all these examples in the quote above are cases of what Russell called "knowledge of things". There were two forms of it; knowledge by acquaintance and knowledge by description: Bismarck knows himself by acquaintance, and we know Bismarck by description. In the terminology adopted here this would imply that Bismark's degree of acquaintance was maximum (say 1), but since our knowledge of him is indirect and mostly based on testimony of others our degree of acquaintance whatever it is, is less than 1. For a good historian who specializes in that period that degree may be quite high, for others it may be lower. The lower it gets the more room there will be for curiosity. If all I know about Bismarck is that he was a famous historical figure, I might then be curious about him. I could express this in a question form: Who is Bismarck? Now the problem with this classical piece by Russell is that it says a lot about knowledge but nothing about ignorance and obviously nothing about curiosity. There is, I believe, an intuitive cut off point between the four stages of removal from acquaintance that Russell talks about. Bismarck had knowledge of himself by acquaintance and we know him through history, but what about the man with the iron mask, and especially the longest-lived of men? These are supposed to be two cases of knowledge by description on Russell's account. It seems clear to me that they aren't. I do not know the longest lived of men. It is simply wrong to attribute to me knowledge of him. The degree of ostensibility reaches a very low level in such cases. That is why we easily become curious. That is I do not know of any individual as being the longest lived of men, which is exactly what makes the definite description "the longest lived of men" inostensible. If I have an interest in the topic, I could become curious about who in fact was the man with the iron mask or the longest lived of men. It will be more difficult to bring yourself to be curious about who Bismarck is, if you know a lot about him. The more you know the more difficult it will become. Just like a degree of belief that is short of complete certainty will allow for curiosity, a degree of ostensibility that is short of full acquaintance under a concept will leave some room for curiosity.[16]

---

[15] Russell (1910), p. 116.

[16] For a more detailed discussion of this see *Chapter 3, Knowledge by Acquaintance and Knowledge by Description,* in Inan (2012).

Finally let me note that in dealing with propositional curiosity the central notion that I have appealed to, namely, *degree of belief* can perhaps be cashed out in terms of the notion of *degree of ostensibility*. If one takes the object of propositional curiosity to be an unknown truth value, then we may translate every such case into an inostensible definite description that refers to one of the two truth values (or to some other value if one subscribes to many-valued system of logic.) If you are curious about whether there is life on other planets, then, under this account, what you wish to know is the referent of the definite description "the truth value of the proposition that there is life on other planets". This is the inostensible term that gives rise to your curiosity. If your degree of belief regarding this proposition is 0.5, then the degree of ostensibility of the definite description will be at its minimum, namely 0. And if you have a degree of belief that is higher, then the degree of ostensibility of the definite description "the truth value of the proposition that …" will also be higher. If so, then degree of belief will simply be a special case of degree of ostensibility. The degree of ostensibility of a whole declarative sentence is also a function of the degree of ostensibility of its constituent terms. Consider a simple sentence in the subject/predicate form, and suppose S knows with complete certainty the proposition expressed by it. In this case the degree of belief for S is 1. Now normally that would imply that the degree of ostensibility is also 1 for S. That however is not always the case. The degree of ostensibility of a sentence is a function of not only the degree of belief, but also the degrees of ostensibility of the constituent parts of the sentence. I know with complete certainty that the 12th perfect number is not a prime. That is because I know that no perfect number is a prime, not because I have calculated the 12th perfect number and discovered that it was not a prime. Though my degree of belief is 1, my degree of ostensibility is significantly lower. It cannot be higher than the degree of ostensibility of the subject term "the 12th perfect number", which is quite low because I do not know what number it refers to. All I can claim to know about this number is what I can immediately deduce from the description together with my background knowledge of perfect numbers. The degree of ostensibility also applies to the predicate term within that sentence. Now given that I know what a perfect number is, and that I know a few examples, my degree of ostensibility is quite high. That might not always be the case. There can be a predicate term that one grasps without knowing what property is denoted by it. If I ask you what color your lover's eyes are, and you answer by saying that they are your favorite color, I will have learnt something new, but I still may not know what color your lover's eyes are if I don't know your favorite color. So if you utter the sentence "my lover's eyes are my favorite color", the degree of ostensibility of the subject term would be reasonably high for me assuming that I know your lover to some extent, but the degree of ostensibility of the predicate term would be quite low for me given that I do not know your favorite color. We may also imagine that I have no clue as to who your lover is, and know nothing about her except that she is your lover and whatever else I can deduce from that. In such a case my degree of ostensibility of the subject term will also be very low. So it is possible to know that a proposition about a person is true even when you do not know who is being talked about and what is being said about her. Of course

I know that the proposition is about your lover, and if some third party were to ask me who you are talking about I could truthfully say "he is talking about his lover", but that does not change the fact that I do not know who your lover is, if we give the term "knowing who" what I have called its strict use.[17] The degrees of ostensibility for both the subject as well as the predicate term may be close to 0, and yet I may still know that the proposition is true. If I take your word for it, then I may come to know that your lover's eyes are your favorite color. My degree of belief might be very high, close to 1, but my degree of ostensibility is nonetheless very low. That is why I hold that there are two ways to satisfy propositional curiosity, de re and de dicto.[18] Merely coming to know that the proposition is true will give you de dicto satisfaction. You will come to know that there is a fact that makes the proposition true, but you are still in the dark as to what constituents that fact has. You know that a certain person has a certain property, but you neither know who that person is, nor what property is being predicated of her. That is why the degree of ostensibility of the whole sentence is very low. In order to satisfy your curiosity de re you need to raise it, and to do that you have to come to know that person and the property attributed to her. This will put you in touch with that fact. Merely de dicto satisfaction gives you what I have called inostensible knowledge, whereas de re satisfaction gives you ostensible knowledge. Propositional knowledge and belief attributions are not fine grained enough to distinguish between these two cases.

Acknowledgments I wish to thank Safiye Yiğit for comments on an earlier draft, and Abrol Fairweather, Alev Çınar, Ayça Boylu, and my students in my graduate seminar on *Truth and Reference* I gave in 2012 at Boğaziçi University for their support. This work has been funded by *Scientific Research Fund* of Boğaziçi University; Title: *The Epistemic, Semantic, and Ethical Dimensions of Curiosity*; Code Number: BAP 12B02P3.

References

Aristotle. 1924. *Metaphysics: A revised text with introduction and commentary*. Princeton: Princeton University Press.

Baehr, Jason. 2011. *The inquiring mind: On intellectual virtues and virtue epistemology*. Oxford: Oxford University Press.

Battaly, Heather. 2008. Virtue epistemology. *Philosophy Compass* 3(4): 639–663.

Baumgarten, E. 2001. Curiosity as a moral virtue. *International Journal of Applied Philosophy* 15(2): 169–184.

Brady, M. 2009. Curiosity and the value of truth. In *Epistemic value*, ed. A. Haddock, A. Millar, and D. Pritchard, 265–283. Oxford: Oxford University Press.

Braun, D. 2006. Now you know who Hong Oak Yun is. *Philosophical Issues* 16(1): 24–42.

Collingwood, R.G. 1940. Chapter 4: On Presupposing. In *An essay on metaphysics*, ed. R. Martin. Oxford: Clarendon Press.

Crisp, Roger. 2010. Virtue ethics and virtues epistemology. *Metaphilosophy* 41: 1–2.

[17] See *Chapter 2, Asking and Answering*, in Inan (2012).

[18] For a more detailed discussion of the distinction between de re and de dicto satisfaction of propositional curiosity see *Chapter 9, Conditions for the Satisfaction of Curiosity*, in Inan (2012).

Daston, L., and K. Park. 2001. *Wonders and the order of nature, 1150–1750*. Cambridge: MIT Press.
Descartes, R. 1989. *The Passions of the Soul*. Trans. and annotated by Stephen Voss. Indianapolis/Cambridge: Hackett Publishing Company.
Hookway, Christopher. 2003. How to be a virtue epistemologist. In *Intellectual virtue: Perspectives from ethics and epistemology*, ed. Michael DePaul and Linda Zagzebski. Oxford: Oxford University Press.
Huber, F., and C. Schmidt-Petri (eds.). 2009. *Degrees of belief*. Dordrecht: Springer.
Inan, I. 2010. Inostensible reference and conceptual curiosity. *Croatian Journal of Philosophy* X(28): 21–41.
Inan, I. 2012. *The philosophy of curiosity*. New York/London: Routledge.
Kvanvig, J. 2003. *The value of knowledge and the pursuit of understanding*. Cambridge: Cambridge University Press.
Leonard, H.S. 1957. *An introduction to the principles of right reason*. New York: Henry Holt.
Miscevic, N. 2007. Virtue -based epistemology and the centrality of truth: Towards a strong virtue-epistemology. *Acta Analytica* 22(3): 239–266.
Riggs, Wayne. 2010. Open-mindedness. *Metaphilosophy* 41: 1–2.
Roberts, R.C., and W.J. Wood. 2009. *Intellectual virtues: An essay in regulative epistemology*. Oxford: Oxford University Press.
Russell, B. 1910. Knowledge by acquaintance and knowledge by description. *Proceedings of the Aristotelian Society* 11: 108–128.
Schaffer, J. 2007. Knowing the answer. *Philosophy and Phenomenological Research* 75(2): 383–403.
Schmitt, F.F., and R. Lahroodi. 2008. The epistemic value of curiosity. *Educational Theory* 58: 125–148.
Subasi, A. 2009. *Dynamics of scientific curiosity*. Master's thesis in cognitive science. Bogazici University.
Whitcomb, D. 2010. Curiosity was framed. *Philosophy and Phenomenological Research* 81(3): 664–687.
Yigit, S. 2011. *Curiosity as an intellectual and ethical virtue*. Master's thesis. Bogazici University.
Zagzebski, L.T. 1996. *Virtues of the mind: An inquiry into the nature of virtue and the ethical foundations of knowledge*. New York: Cambridge University Press.

Epistemic Values and Disinformation

Don Fallis

1 Introduction

David Hume (1977 [1748], 77) famously said, "when anyone tells me, that he saw a dead man restored to life, I immediately consider with myself, whether it be more probable, that this person should either deceive or be deceived, or that the fact, which he relates, should really have happened." Of course, intentionally deceptive information on many topics (not just reports of miracles) can interfere with our ability to achieve our epistemic goals of acquiring true beliefs and avoiding false beliefs.[1] For instance, when a politician tells me that Saddam Hussein has *weapons of mass destruction* (WMDs), I have to consider "whether it be more probable, that this person should either deceive or be deceived" or that Hussein really does have WMDs. Also, when a used-car salesperson tells me that the little red Corvette runs like a dream, I have to consider "whether it be more probable, that this person should either deceive or be deceived" or that the Corvette really runs like a dream. And it is important to make the correct judgment because it can be extremely harmful (in terms of lives and lucre) to be misled by intentionally deceptive information.

Intentionally deceptive information, which I will refer to as *disinformation*, is not the only type of misleading information. Information may also be misleading as a result of an honest mistake, negligence, or unconscious bias. All of this misleading information can potentially be dangerous (even if it is not always intended to be). Nevertheless, there are reasons to focus specifically on information that is *intended* to mislead. Disinformation is likely to be especially dangerous. Someone who spreads disinformation goes out of his way to make sure that we end up with false beliefs, much like "some malicious demon of the utmost power and cunning [who]

[1] With the phrase "or be deceived," Hume might have wanted to include being "deceived" by one's senses. But I am reading deception as being intentional, as most philosophers do (see Carson 2010, 47).

D. Fallis (✉)
School of Information Resources, University of Arizona, Tucson, AZ, USA
e-mail: fallis@email.arizona.edu

A. Fairweather (ed.), *Virtue Epistemology Naturalized: Bridges Between Virtue Epistemology and Philosophy of Science*, Synthese Library 366, DOI 10.1007/978-3-319-04672-3_10,

has employed all his energies in order to deceive me" (Descartes 1996 [1641], 15). Furthermore, this person actively tries to avoid detection.

So, how can we deal with the dangerous problem of disinformation? The most obvious strategy is to get better at detecting disinformation. In addition to many methods, such as polygraphy, that look for signs of deception in the source of the information, there are also methods, such as linguistic analysis, that look for signs of deception in the information itself (see Newman et al. 2003; Farid 2009). Another strategy, however, is to design policies that will deter people from spreading disinformation. In order to implement either strategy, we need to understand what disinformation is. Moreover, in order to deter people from spreading disinformation, we also need to know *what* sorts of things affect the amount of disinformation and *how* they affect it.

In this paper, I begin by giving an analysis of what disinformation is. I then use that analysis to construct a simple game-theoretic model of the sending and receiving of disinformation. This model is based on a formal model of poker suggested by Reiley et al. (2008). It is also inspired by formal models of deceptive signaling in animals suggested by Elliott Sober (1994) and Brian Skyrms (2010, 72–82).[2] In principle, this model can be used to predict the amount of disinformation in various contexts. But more importantly, it allows us to identify what sorts of things affect the amount of disinformation and how they affect it.

In order to do this, this game-theoretic model appeals to philosophical work on *epistemic values* (e.g., Levi 1962; Goldman 1999; Riggs 2003; Fallis 2006, 2007). Thus, the present investigation falls within the scope of formal value-theoretic epistemology. It turns out that the amount of disinformation in a particular context depends largely on the epistemic value that receivers of information assign to acquiring true beliefs and to avoiding false beliefs. In particular, the amount of disinformation decreases as the value of avoiding false beliefs goes up relative to the value of acquiring true beliefs.

Most epistemologists, including René Descartes and Hume, think that the value of avoiding false beliefs is greater than the value of acquiring true beliefs. As Hume (1977 [1748], 111) put it, "there is a degree of doubt, and caution, and modesty, which, in all kinds of scrutiny and decision, ought for ever to accompany a just reasoner." By contrast, following Isaac Levi and William James, this game-theoretic model presented here makes no assumption about the relative value of acquiring true beliefs and of avoiding false beliefs. (It simply assumes that both are greater than, or at least equal to, zero.) But it is an interesting implication of this model that there is less disinformation whenever receivers of information *do* abide by the constraint that Descartes and Hume placed on epistemic values.

2 An Analysis of Disinformation

Like information in general, disinformation is "meaningful data" or "semantic content" that represents the world as being a certain way (see Floridi 2011, 260). For instance, a piece of disinformation might represent the world as being such that Hussein has

[2] Several economists (e.g., Tullock 1967; Schelling 1968; Davis and Ferrantino 1996) have also suggested formal models of lying and deception.

WMDs. But when people (e.g., Fetzer 2004; Jackson and Jamieson 2007) talk about disinformation, they are clearly interested specifically in meaningful data that is *intentionally misleading* (see Fallis 2011, 203–04). In fact, the *American Heritage Dictionary* defines disinformation as "deliberately misleading information." Thus, we might characterize *disinformation* as meaningful data that (a) is likely to mislead people (i.e., is epistemically dangerous) and (b) is intended to mislead people.[3]

Several philosophers (e.g., Fetzer 2004, 231; Floridi 2011, 260) have essentially equated disinformation with *lies* (see Fallis 2011, 206–07). In fact, according to James Fetzer (2004, 231), "disinformation ... should be viewed more or less on a par with acts of lying. Indeed, the parallel with lying appears to be fairly precise." However, as I argue in this section, lies and disinformation are not the same phenomenon. Moreover, drawing the distinction can help to clarify the concept of disinformation for purposes of this investigation.

For one thing, not all lies count as disinformation. In particular, there are lies that are not likely to mislead anyone. Lies and disinformation do share the property that, no matter how carefully constructed they are, they might not *actually succeed* in misleading the intended target. For instance, the potential customer might be a mechanic who can easily see that the Corvette is in bad shape. However, unlike disinformation, there are lies that do not even have a chance of misleading anyone. For instance, if there is not enough *admissible* evidence for a conviction unless he confesses, a guilty defendant might continue to assert his innocence even though he does not expect to convince anyone (cf. Fallis 2009, 42–43; Carson 2010, 20–22). As Roy Sorensen (2007, 252) points out, such "bald-faced lies do not fool anyone. They are no more a threat to truth telling than sarcastic remarks."

Of course, the traditional philosophical analysis of lying, going back to Augustine (1952 [395]), explicitly requires that the speaker intend to mislead (see Williams 2002, 96).[4] On this analysis of lying, bald-faced lies do not count as lies. However, even if we restrict our attention to lies that are intended to mislead, there are lies that are not likely to mislead anyone. For instance, while he is carousing in London, the protagonist of Oscar Wilde's *The Importance of Being Earnest* claims to be named Ernest, even though he believes that Ernest is not his name. While this character is clearly lying about his name, he is not likely to mislead anyone because what he intends people to believe is not actually false. When it is revealed in the final act that his name really is Ernest, Wilde's protagonist laments, "it is a terrible thing for a man to find out suddenly that all his life he has been speaking nothing but the truth."[5]

---

[3] I will not try to specify how likely it must be that the information will actually mislead people.

[4] Even on this analysis of lying, the speaker need not intend someone to believe a falsehood outright. She might simply intend to increase someone's degree of belief in a falsehood (see Fallis 2009, 45).

[5] A few philosophers (e.g., Carson 2010, 15) claim that, in addition to being believed to be false by the speaker, a lie must actually be false. Thus, they would say that Wilde's protagonist only *tried to lie* about his name. But in an earlier article (Fallis 2011, 207), I give an example of a lie that is false as well as intended to mislead, but that is not likely to mislead. A liar can be wrong about whether his claim is misleading even if he is right about its truth value.

But even if we could find an analysis of lying that was not too broad, it would still be too narrow to capture disinformation. That is, not all disinformation counts as a lie. First, in order to lie, you have to say (or write) something *to* the person that you intend to mislead (see Fallis 2009, 40). However, you can spread disinformation even if you only intend to mislead someone that you know to be eavesdropping on your conversation. For instance, in preparation for the D-Day invasions, the Allies sent a bunch of radio transmissions which they intended the Germans to intercept and which were meant to suggest that a huge force was preparing to attack Calais rather than Normandy (see Farquhar 2005, 71–72).

Second, in order to lie, you have to say (or write) something that you believe to be false (see Fallis 2009, 37–39). However, you can arguably spread disinformation even if you believe what you say. For instance, your neighbor, who has been opening your mail, might say, "Someone has been opening your mail" with the intention that you draw the false conclusion that it is not her (see Williams 2002, 96).[6] In fact, someone can spread disinformation without saying (or writing) anything at all. For instance, there are numerous examples of people attempting to mislead others by doctoring photographs (see Farid 2009, 95–99).

Finally, in order to lie, at least on the traditional philosophical analysis of lying, you have to intend to mislead. However, and despite the fact that most characterizations of disinformation require an intention to mislead, someone can spread disinformation without intending to mislead anyone. For instance, the politician's press secretary might believe what the politician says and innocently pass along the disinformation that Hussein has WMDs. In such cases, while the *original* source of the claim does intend to mislead people, the immediate source does not.

Moreover, disinformation can arguably be spread even if there is *no one* who intends to mislead anyone. For instance, there are numerous instances of "evolutionary lying" such as the case of the Monarch and Viceroy butterflies (see Sober 1994).[7] If a blue jay eats a Monarch butterfly, the bird will throw up. And Monarchs have developed gaudy coloration in order to warn blue jays that they are nasty to eat. By contrast, if a blue jay eats a Viceroy butterfly, the bird will get a nutritious meal. However, Viceroys have also developed gaudy coloration. Viceroys do not intend to mislead blue jays into believing that they are nasty to eat. (In fact, Viceroys do not form any intentions with respect to what blue jays believe.) But it is no accident that blue jays are misled. Viceroys who succeed in fooling blue jays with their gaudy coloration survive longer to pass on their genes to future generations.

While the coloration of a butterfly is not intended to represent the world as being a certain way, there are also instances of evolutionary lying that clearly do involve

[6] Several philosophers claim that meaningful data must be true in order to count as information (see Fallis 2011, 202–03). But even if one accepts this claim, this sort of case shows that there can be disinformation that counts as information.

[7] Despite being the standard example in the biological literature, Viceroys are not actually *Batesian mimics*. As David Ritland and Lincoln Brower (1991) discovered, Viceroys are actually as unpalatable to blue jays as Monarchs. But for purposes of this paper, I will follow Sober and treat this as an instance of Batesian mimicry of which there are many examples in nature.

meaningful data (see Fallis 2011, 211–12). For instance, many people have claimed that there is a "Jewish conspiracy to take over the globe" (see Farquhar 2005, 173–79). Even if everyone who ever made this claim believed it to be true (and, thus, did not intend to mislead anyone), many of them clearly benefited from spreading this falsehood. Thus, it is no accident this sort of falsehood continues to spread.[8] Given such cases, we should probably characterize *disinformation* as meaningful data that (a) is likely to mislead people and (b) has the *function* of misleading people.

If we were trying to develop techniques for detecting disinformation, it might be important to give an analysis that requires an intention to mislead. For instance, the immediate source of disinformation is probably not going to display the signs of stress that polygraphs detect unless she actually intends to mislead. However, the goal of this investigation is to figure out how we might deter people from spreading disinformation. Toward this end, in the following section, I construct a game-theoretic model that depends in large part on the costs and benefits to the senders and the receivers of disinformation. Even if only the original source (rather than the immediate source) intends to mislead, this sort of model still works. In that case, it is the costs and benefits to the original source that help to determine to whether or not disinformation is sent. Also, even though they do not consciously intend to mislead, the costs and benefits to "evolutionary liars" still help to determine to whether or not disinformation is sent.[9] Thus, it is not a problem that the foregoing analysis of disinformation does not explicitly require an intention to mislead.[10]

3 A Model of the Sending and Receiving of Disinformation

In this investigation, we are interested in what sorts of things affect the amount of disinformation. One specific version of this question is what sorts of things affect the amount of bluffing in a poker game. In order to win at poker, you have to deceive your opponents about what cards you hold. For instance, even though you have a fairly weak hand, you might make a big bet, thereby suggesting that

[8] In fact, reports of miracles are arguably another case where there is a systematic benefit to disseminating a falsehood. As Hume (1977 [1748], 78) pointed out, people tend to experience an "agreeable emotion," a sense of "surprise and wonder," when they hear that a miracle has occurred. As a result, the people describing the miracle can take "delight in exciting the admiration of others."

[9] We can essentially take the *intentional stance* toward "evolutionary liars," such as the Viceroy butterflies. That is, attributing beliefs and desires to such "liars" can allow us to predict how often they "lie" (cf. Sober 1994; Skyrms 2010, 72–82).

[10] In an earlier article (Fallis 2011, 210), I describe researchers who place false semantic content in *Wikipedia* in order to see how long it takes to be corrected. I also describe educators who place false semantic content on the Internet in order to teach students how to identify false semantic content. Even though they do not intend to mislead anyone, these researchers and educators seem to have created disinformation. They have certainly created something which has the *function* of misleading people. But since no one benefits from such disinformation being believed, it is not captured by the model that I construct in the following section.

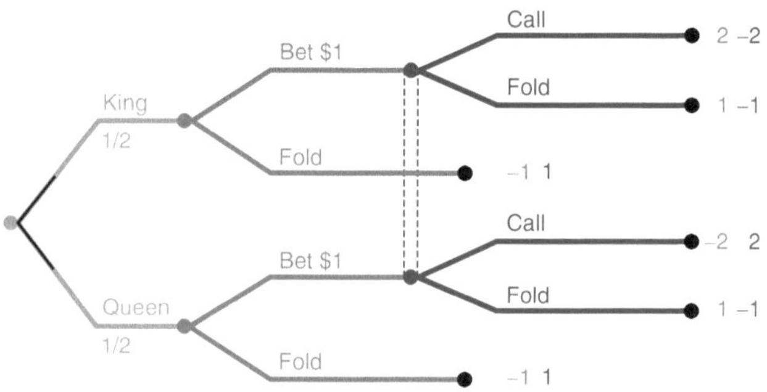

Fig. 1 The game tree for stripped-down poker

you have a strong hand, in order to induce your opponent to fold. How frequently do poker players bluff in this way?[11]

It is possible to answer this question using a formal model of "stripped-down poker" suggested by Reiley, Urbancic, and Walker.[12] After introducing their model, I show how it can be applied to disinformation more generally.

There are two players in stripped-down poker. Prior to every hand, each player antes $1 into the pot. Player A then randomly selects a card from a two-card deck containing a King and a Queen. (Player B does not get a card.) The King is the *winning hand* for player A and the Queen is the *losing hand*.

After looking at his hand, player A gets to decide between (a) betting $1 and (b) folding. If player A decides to fold, player B wins the pot (earning A's $1 ante). (Of course, whatever one player wins, the other player loses.) However, if player A decides to bet $1, player B gets to decide between (a) calling the bet and (b) folding. If player B calls the bet, player A turns over his card. Player A wins the pot (earning B's $1 ante and $1 bet) if it is a King. Player B wins the pot (earning A's $1 ante and $1 bet) if it is a Queen. If player B folds, player A wins (earning B's $1 ante).

Figure 1 shows the extensive form of this game. The first number at each terminal node is the payoff for player A. The dashed line indicates player B's *information set* (when player B has to decide between calling player A's bet or folding, he does not know whether player A has been dealt a King or a Queen).

The models of deceptive signaling in animals suggested by Sober and Skyrms are very close to stripped-down poker. For instance, consider the case of the Monarch and Viceroy butterflies. The butterflies are dealt a winning hand (i.e., they are born as an unpalatable Monarch) or a losing hand (i.e., they are born as a palatable Viceroy).

[11] It might be suggested that a big bet is simply an action rather than something that is clearly intended to represent the world as being a certain way. If so, not all bluffs may count as disinformation, strictly speaking. But we can easily imagine a poker game that requires you to say, "I have a winning hand" when you bet and to say, "I have a losing hand" when you fold.

[12] The poker games that people actually play, such as Seven-Card Stud or Texas Hold'em, are sufficiently complicated that they can be somewhat difficult to model. The model of stripped-down poker is sufficient for our purpose here of identifying what affects the amount of bluffing and disinformation.

The butterflies have a choice between betting (i.e., becoming gaudy) or folding (i.e., remaining plain). The birds then have a choice between calling (i.e., eating the butterfly) or folding (i.e., not eating the butterfly).

Viceroy butterflies do not individually decide whether to be gaudy or plain in the way that poker players individually decide whether to bet or fold. Instead, "evolution" decides whether the Viceroy population will be gaudy or plain (or, more precisely, it decides what percentage of the population will be gaudy and what percentage will be plain). But the game-theoretic structure of deceptive signaling is essentially the same as stripped-down poker. And the general structure of stripped-down poker is as follows:

There is a *sender* and a *receiver* of a piece of information. The sender is "dealt a winning hand" or she is "dealt a losing hand." The sender has a choice between telling the truth to the receiver or sending disinformation to the receiver. For instance, if the sender is dealt a losing hand, she can say, "I have a losing hand" or she can say, "I have a winning hand."[13]

In the case of stripped-down poker, a person is *literally* dealt a winning hand (or a losing hand). But in many other situations, a person is "dealt a winning hand" in a metaphorical sense. That is, the world has turned out well for this person in some respect, and she has no motivation to deceive anybody on the topic. For instance, some of the cars that the used-car salesperson wants to sell may actually run quite well. In such situations, telling the truth is the *dominant choice* for the sender. That is, it is the best choice for a player regardless of what the other player chooses to do (see Mansfield 1994, 370).

Payoffs to the Sender when she is dealt a winning hand[14]:

	Receiver does not believe	Receiver believes
Say "I have a winning hand."	2	1
Say "I have a losing hand."	−1	−1

In a similar vein, a person can be "dealt a losing hand" in a metaphorical sense. That is, the world might turn out badly for this person in some respect, and she would prefer that people believe otherwise. But while the sender would be happy to claim to have a winning hand if the receiver is going to believe her, she would prefer to just tell the truth if such disinformation will not be believed. In other words, the sender suffers a cost if she sends disinformation and the receiver does not believe it.

Payoffs to the Sender when she is dealt a losing hand:

	Receiver does not believe	Receiver believes
Say "I have a winning hand."	−2	1
Say "I have a losing hand."	−1	−1

[13] Since the sender knows whether or not the information is true, but the receiver does not, this is a *game of asymmetric information* (see Mansfield 1994, 47–48).

[14] The column label is short for "The Receiver does not believe that the Sender has a winning hand when she says that she has a winning hand."

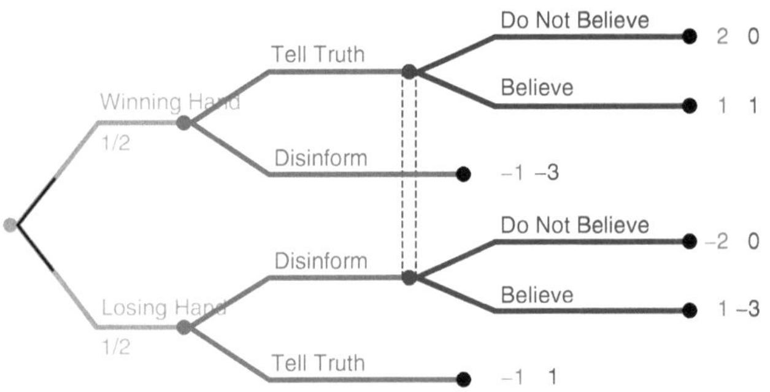

Fig. 2 The game tree for the disinformation game

Once the sender has sent her message, the receiver has to choose whether or not to believe the information. If the information is true, it is better for the receiver to believe it than to suspend judgment. But if the information is false, it is better for the receiver to suspend judgment than to believe it.

Payoffs to the Receiver when the Sender claims to have a winning hand:

	Sender has a losing hand	Sender has a winning hand
Do Not believe	0	0
Believe	−3	1

Note that the payoffs to the receiver here are not the same as in stripped-down poker. I have modified them to be more clearly in line with philosophical work on epistemic values. Not believing (or suspending judgment) gets a payoff of zero, believing truly gets a positive payoff, and believing falsely gets a negative payoff (see Fallis 2006, 188). But this modification does not affect the structure of the game.

Figure 2 shows the extensive form of this game (with the assumption that the probability of being dealt the winning hand is 1/2).

Sober and Skyrms have used this sort of model of deceptive signaling to refute Immanuel Kant's (1959 [1785], 19) claim that "universal lying" is not possible. But this sort of model can also be used to predict the amount of disinformation in various contexts (see Skyrms 2010, 79–80). In addition, it can be used (as I will do in this paper) to identify what sorts of things affect the amount of disinformation.

3.1 *Epistemic Values*

Sober and Skyrms give their models of deceptive signaling in terms of pragmatic utilities. But it is useful to focus specifically on *epistemic* utilities, such as the value of having true beliefs and the disvalue of having false beliefs. In a similar vein, philosophers (e.g., Good 1967) have often modeled the decision making of scientists (e.g., about whether or not to perform an experiment) using their utilities *all things considered*. But scientists primarily have epistemic goals. So, in order to understand their decision making, it can be helpful to focus specifically on epistemic utilities (see Fallis 2007). Thus, in this model of the sending and receiving of disinformation, we will characterize the receiver's payoffs in terms of epistemic values.

A few epistemologists (e.g., Goldman 1999, 89) have claimed that the value of a true belief has exactly the same magnitude as the disvalue of a false belief. However, it is more common for epistemologists (e.g., Descartes 1996 [1641], 12; Hume 1977 [1748], 111) to think that avoiding error is much more important than acquiring true beliefs (see Riggs 2003, 347; Fallis 2006, 183). For instance, in Fig. 2, the receiver's payoffs are such that the cost of having a false belief is three times greater than the benefit of having a true belief. In addition, there may also be circumstances where the benefit of having a true belief is greater than the cost of having a false belief (see James 1979 [1896], 31–32). Thus, in line with most formal models of epistemic values (e.g., Levi 1962, 56; Fallis 2007, 218), this model of the sending and receiving of disinformation puts no constraints on the relative magnitudes of these two epistemic values.

Furthermore, it should be noted that pragmatic considerations can influence these magnitudes. For instance, in stripped-down poker, the cost of having a false belief is greater than the benefit of having a true belief because of the monetary payoffs.[15] Such pragmatic influence does not mean that we are not still talking about epistemic values (see Fallis 2006, 183). The value of having a true belief is at least as great as the value of suspending judgment, and the value of suspending judgment is at least as great as the value of having a false belief.

While we can characterize the receiver's payoffs in terms of epistemic utilities, it is important to note that we cannot do so for the sender's payoffs. There are three issues that might seem to preclude characterizing the sender's payoffs in terms of epistemic utilities. But I argue that only the third issue is a real problem.

First, as with most deceivers, the ultimate goal of the sender of disinformation is clearly not epistemic. The goal is usually to manipulate the behavior of the receiver in some way (see Tullock 1967, 136).[16] For instance, in the poker case, she wants the other player to fold rather than to call her bet. In the butterfly case, she wants the blue jay not to eat her (cf. footnote 9). However, this does not preclude characterizing the sender's payoffs in terms of epistemic utilities. The sender wants to manipulate the behavior of the receiver *by manipulating his epistemic state*. For instance, in the poker case, she wants the receiver to falsely believe that she has a winning hand. In the butterfly case, she wants the blue jay to falsely believe that she is a Monarch butterfly (cf. footnote 9).

Second, the sender has a goal with respect to the epistemic state of the receiver rather than with respect to her own epistemic state. However, this does not preclude

[15] In fact, the benefit of having a true belief might even depend on what the belief is about (see Fallis 2006, 181–82). For instance, the benefit of truly believing that the sender has a winning hand might be greater than the benefit of truly believing that the sender has a losing hand.

[16] Some deceivers may simply value our having false beliefs for its own sake. As Augustine (1952 [395], 87) pointed out, some lies are "told solely for the pleasure of lying and deceiving."

characterizing the sender's payoffs in terms of epistemic utilities. While work in epistemology focuses primarily on how an epistemic agent can improve her own epistemic state, work in *social* epistemology often focuses on how we can improve the epistemic states of others. For instance, legislators are interested in which rules of evidence are most likely to lead to juries having true beliefs about the guilt of defendants (see Goldman 1999, 292–95).

Finally, the real problem is that the sender's goals with respect to the receiver's epistemic state are not *epistemic* goals. The sender would rather that the receiver suspend judgment than that the receiver truly believe that she has a winning hand. Also, she would rather that the receiver falsely believe that she has a winning hand than that the receiver suspend judgment. These are, at best, *doxastic* goals.[17]

But the fact that we cannot characterize the sender's payoffs in terms of epistemic utilities does not take this investigation of disinformation beyond the scope of epistemology. As I explain below, the receiver's payoffs are the main determinants of the amount of disinformation in a particular context. And that is the primary concern of this investigation.

3.2 Possible Extensions of the Model

This simple model of the sending and receiving of disinformation captures many situations in which disinformation is spread. Thus, it provides a useful tool for investigating *what* sorts of things affect the amount of disinformation and *how* they affect it. However, it must be noted that this model does not capture *all* situations in which disinformation is spread. Here are a few of the directions in which the model might be extended:

First, the particular sort of disinformation that will be the focus of this investigation is when players with a losing hand falsely claim to have a winning hand. However, in some situations, players with a winning hand might falsely claim to have a losing hand. Even if the sender is "dealt a winning hand," truth telling is not always the dominant choice. In fact, the sender might face the *same consequences* regardless of what hand she has been dealt. This is what happens in the case of the Monarch and Viceroy butterflies. If a blue jay eats a Monarch butterfly, the truth is immediately revealed (i.e., the bird gets sick). But this happens too late to do the butterfly any good. She has already suffered the cost of being eaten. Also, there are situations where the truth is not revealed at all.[18] For instance, even though there is a sense in which a defendant who is innocent has been dealt a winning hand, she still might get

[17] But Goldman (2002, 218–220) once tentatively suggested, to the contrary, that social epistemology should encompass attempts to bring about bad epistemic consequences.

[18] In fact, the truth not being revealed is quite common when it comes to models of epistemic utilities. For instance, scientists certainly have epistemic goals. But they never find out *for sure* whether or not their hypotheses are true. As a result, they always have to make do with *expected* epistemic utilities to guide their decision making (see Fallis 2007, 219).

convicted. Thus, innocent defendants sometimes confess to a crime that they did not commit in order to avoid the risk of a much longer sentence.[19]

Second, in this model, the receiver only has a choice about whether or not to believe what the sender says when she claims to have a winning hand. However, in most situations, the receiver will also have a choice about whether or not to believe what the sender says when she claims to have a losing hand. But if truth telling is the dominant choice when the sender is dealt a winning hand (as we are assuming here), the sender will only claim to have a losing hand when she really does.[20] So, we can simplify the model by assuming that the receiver always believes the sender when she claims to have a losing hand.

Third, in this model, the receiver only has a choice between believing what the sender says and not believing it. However, philosophers (e.g., Levi 1962, 55–56; Goldman 1999, 89) often model epistemic agents as having a choice between *three* doxastic states (see Fallis 2007, 222). In addition to believing what you are told or suspending judgment on it, you can also choose to believe the opposite of what you are told. But often, there is no difference in terms of payoffs between not believing what you are told and *disbelieving* what you are told. For instance, juries have to decide that the defendant is guilty or that he is not guilty. There is no third option.[21]

Fourth, this model focuses on the value of having a true *belief* and on the disvalue of having a false *belief*, which are the main values that are discussed in the philosophical literature on epistemic values (see Fallis 2006). However, there are many other epistemic values, such as *knowledge* and *understanding*, that might be incorporated. For instance, according to James (1979 [1896], 24), "we must *know* the truth; and we must avoid error—these are our first and great commandments as would-be *knowers*."

Fifth, this model does not distinguish between (a) not being believed and (b) being *caught* spreading disinformation. However, as Gordon Tullock (1967, 137) emphasizes, these are not the same thing. For instance, even if I do not believe what you say, I might be unwilling to call you a liar. So, it might be useful to have a model that distinguishes these two possible outcomes. But as in the butterfly case, these two outcomes have the very same consequences for the sender in many contexts.

Sixth, a receiver may often have several interactions with the same sender (see Tullock 1967, 138–39). So, it might be useful to study *repeated games* where the

---

[19] In addition, in a real poker game, a player who has been dealt a winning hand might want her opponent to think that she has been dealt a losing hand. That way, her opponent might continue to make bets that he will lose. But in this simplified model, there is no such motivation. Player B only gets to choose between calling player A's bet and folding. Unlike with Seven-Card Stud and Texas Hold'em, there are no additional betting rounds.

[20] In his model of the butterfly case, Sober (1994, 78) likewise assumes that plain butterflies are always palatable Viceroys.

[21] Regardless of whether the jury believes that he is innocent or simply suspends judgment on his guilt, the defendant will be found not guilty. In this case, while the value of truly believing that the defendant is innocent is at least as great as the value of suspending judgment, it is no greater. Also, while the value of suspending judgment is at least as great as the value of falsely believing that the defendant is innocent, it is no greater.

receiver has an opportunity to learn about the trustworthiness of a particular sender. But any single interaction can be captured with the present model. Everything that the receiver knows about the sender can be factored into the probability of the sender having a winning hand.

Seventh, in this model, there is a single receiver. However, senders of disinformation (such as advertisers and politicians) typically target *multiple receivers*. But in line with what Tullock (1967, 137) points out, "equations which dealt with many people would be much more elaborate in appearance ... but in principle they are merely summations of a set of individual equations. It seems unnecessary to confuse the issue by complicating the equations in this way."

Finally, this model assumes that truthful claims to have a winning hand are *indistinguishable* from false claims to have a winning hand. However, it is often possible to gather evidence to help determine whether or not a particular claim is disinformation.[22] Thus, a more realistic model would involve messages with varying degrees of convincingness. But the extreme case where the receiver has no way of distinguishing truths from falsehoods still bears analysis. As I describe in the following section, the present model provides a very instructive case for investigating the sending and receiving of disinformation.

4 Determining the Amount of Disinformation

We want to identify what sorts of things affect the amount of disinformation and how they affect it. Toward this end, the first step is to be able to determine, for any set of inputs to the model, how much disinformation there will be. In line with standard practice in game theory and economics, we do this by calculating how much disinformation there is when the game is at equilibrium. The second step is to determine how the equilibrium point and, thus, the amount of disinformation varies as the inputs to the model vary.

A game is at equilibrium when neither player has a motivation to unilaterally change her strategy. While there is no guarantee that, in actual practice, a game will reach the equilibrium point very quickly (or even that it will reach it at all), there is always pressure for the game to move toward it (see Mansfield 1994, 34–35; Sober 1994, 79–80).[23] After all, if the game is not at the equilibrium point, at least one of the players will have a motivation to unilaterally change her strategy. Thus, the amount of disinformation at the equilibrium point is likely to provide a reasonable approximation of the actual amount of disinformation. But more importantly, as

[22] Hume (1977 [1748], 75), for instance, recommends that "we entertain a suspicion concerning any matter of fact, when the witnesses contradict each other; when they are but few, or of a doubtful character; when they have an interest in what they affirm; when they deliver their testimony with hesitation, or on the contrary, with too violent asseverations."

[23] In general, a game may have more than one equilibrium point. But in this model of the sending and receiving of disinformation, there is always a unique equilibrium point.

long as there is such pressure, we can use this model to identify what sorts of things affect the amount of disinformation.

In order to find the equilibrium point, we make the standard assumption that both players are rational and fully informed about the structure of the game. In other words, we assume that both players are aware of the three inputs to the model:

1. the sender's payoffs for each possible outcome.
2. the receiver's payoffs for each possible outcome.
3. the probability that the sender is dealt a winning hand.

Notice that it is not a game of *perfect* information as the receiver does not know whether or not the sender has a winning hand.

4.1 Finding the Equilibrium

In order to find the equilibrium point, the first thing to check is whether there is an equilibrium in pure strategies. A *pure strategy* is a list of instructions that, for any situation that might arise in the game, tells a player to perform a definite action. For instance, an example of a pure strategy for the sender is to tell the truth whenever she is dealt a winning hand and to send disinformation whenever she is dealt a losing hand. However, with the inputs to the model given in Fig. 2, there is no equilibrium in pure strategies (see Reiley et al. 2008, 325–26). That is, if both players are playing pure strategies, then at least one of them will be motivated to unilaterally change her strategy.

If the sender is dealt the winning hand, then she should always tell the truth since that is the dominant choice. However, if the sender is dealt the losing hand, she should not always tell the truth and she should not always send disinformation. For instance, if the sender tells the truth whenever she has a losing hand, then it will be better for the receiver to always believe what the sender says. And if the receiver always believes, the sender will have a motivation to send disinformation at least some of the time. But if the sender sends disinformation whenever she has a losing hand, then it will be better for the receiver to never believe what the sender says. And if the receiver never believes, the sender will have a motivation to tell the truth at least some of the time.

There are some inputs to the model such that there is an equilibrium in pure strategies. In particular, the expected utility for believing what the sender says might be such that the receiver should believe no matter how often the sender sends disinformation. (In such cases, believing is not the dominant choice for the receiver; it just has a higher expected utility than not believing.) For instance, if the probability that the sender has a winning hand is sufficiently high, then it will always be best for the receiver to believe.[24] And if the receiver always believes, then the sender will

[24] If the payoffs are as given in Fig. 2, but the probability that the sender has a winning hand is greater than or equal to 3/4, then the receiver should always believe what the sender says.

send disinformation whenever she has a losing hand. In that case, the amount of disinformation is simply given by the probability of the sender being dealt a losing hand.

My primary concern in this investigation, however, is with the more interesting case where mixed strategies have to be played for the game to be at equilibrium. A *mixed strategy* is a list of instructions that, for at least one situation that might arise in the game, tells a player to perform one possible action with probability p and the other possible action with probability $1-p$.[25]

Players should only play such mixed strategies when they are indifferent between the two possible actions. In other words, both players should be perfectly happy to flip a coin to decide what to do. In the case of this model of the sending and receiving of disinformation, the sender must be indifferent between telling the truth and sending disinformation when she is dealt a losing hand. Also, the receiver must be indifferent between believing and not believing what the sender says when she claims to have a winning hand. If one of the possible actions in these situations were better for a player, then that player should always choose to perform that particular action. In other words, she should play a pure strategy. It would be crazy to play a mixed strategy.

Since we are interested in how much disinformation there is, we are interested in the mix that the sender plays when the game is at equilibrium. More explicitly, we are interested in the probability that the sender will say that she has a winning hand when she actually has a losing hand. The mix that the sender plays is going to be whatever makes the receiver indifferent between his two possible actions. Thus, in order to determine how much disinformation there will be, we need to determine when the expected utility to the receiver of believing what the sender says is the same as the expected utility of not believing.

Let q be the probability that the sender has a losing hand when she says that she has a winning hand. So, with the payoffs given in Fig. 2,

$$\mathrm{EU}_{\text{DO-NOT-BELIEVE}} = q \cdot 0 + (1-q) \cdot 0$$

$$\mathrm{EU}_{\text{BELIEVE}} = q \cdot -3 + (1-q) \cdot 1$$

Thus, the receiver will be indifferent between believing and not believing when q is 1/4.

But we are not interested in the value of q itself. We want to know the mix that the sender plays when she is dealt a losing hand. In other words, we need to determine the value of p, the probability that the sender says that she has a winning hand when she has a losing hand. Taking into account w, the probability that the sender is dealt a winning hand, we can do some simple algebra to determine the value of p.[26]

[25] This definition can easily be generalized to three or more possible actions.

[26] Since she will definitely say that she has a winning hand whenever she has a winning hand, w is also the probability that the sender has a winning hand and says that she has a winning hand. $p \cdot (1-w)$ is the probability that the sender has a losing hand and says that she has a winning

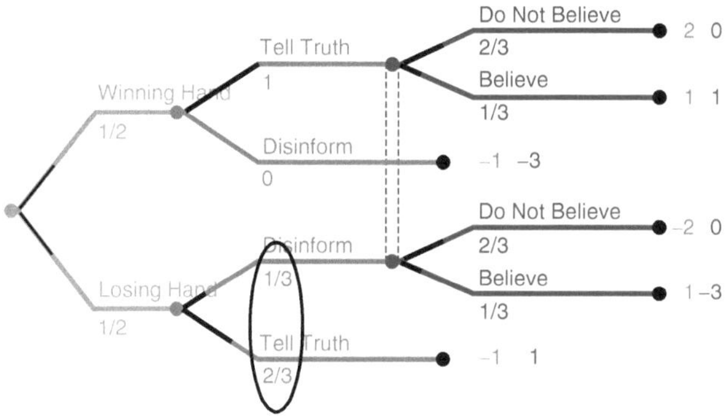

Fig. 3 The equilibrium point in the disinformation game

It turns out that, with the inputs to the model given in Fig. 2, the receiver will be indifferent between believing and not believing when p is 1/3. So, the sender sends disinformation **1/3 of the time** when she has a losing hand (see Fig. 3).[27, 28]

Actually, the amount of disinformation is given by the probability that the sender says that she has a winning hand *and* she actually has a losing hand.[29] We can determine this value just by multiplying the mix that the sender plays by the probability that she is dealt a losing hand. Since we are assuming here that she is dealt a losing hand 1/2 of the time, disinformation is sent **1/6 of the time**.

4.2 What Affects the Amount of Disinformation

Now that we are able to calculate how much disinformation there is when the game is at equilibrium, we are in a position to identify which things affect the amount of disinformation and how they affect it. As noted above, the game is at equilibrium when the expected utility to the receiver of believing what the sender says is the same as the expected utility of not believing. Also, it is clear from

hand. So, $w/(w+p\cdot(1-w))$ is the probability that the sender has a winning hand when she says that she has a winning hand. Thus, $1-q=w/(w+p\cdot(1-w))$. Given q and w, we can solve for p. $p=w/(1-w)\cdot q/(1-q)$.

[27] The equilibrium point was calculated (and the game tree was drawn) using *Gambit* open source software (see McKelvey et al. 2007). Murray Gell-Mann (2009, ix) has suggested that, in a wide variety of contexts, people and animals decide to send disinformation about 1/7 of the time. But the amount of disinformation can actually vary greatly depending, for example, on the costs and benefits to the receiver of the information.

[28] We can use an analogous technique to determine how often the receiver will believe what the sender says (see Sober 1994, 79). However, this arguably takes us beyond the scope of epistemology. The sender's payoffs are the main determinants of the receiver's level of credulity and, as noted above, we cannot characterize the sender's payoffs in terms of epistemic utilities.

[29] In other words, we are really interested in $p\cdot(1-w)$ rather than p itself.

the preceding discussion that only two of the inputs to the model have an effect on the expected utilities to the receiver:

1. the receiver's payoffs for each possible outcome.
2. the probability that the sender is dealt a winning hand.

Thus, these are the two inputs that affect the amount of disinformation. In other words, at equilibrium, the sender plays whatever mix is needed so that, given these two inputs, the receiver is indifferent between his two possible actions.

The receiver's payoffs can be cashed out as two distinct epistemic values. First, there is the benefit of believing what the sender says when it is true (compared to suspending judgment on it). Second, there is the cost of believing what the sender says when it is false (compared to suspending judgment on it). In addition, as noted above, the probability that the sender is dealt a winning hand is really the probability that the sender has no motivation to send disinformation. Thus, there are essentially three things that affect the amount of disinformation. But *how* do they affect the amount of disinformation?

First, the lower the benefit of truly believing that the sender has a winning hand, the less disinformation there will be. For instance, holding everything else fixed in Fig. 2, if the payoff for truly believing that the sender has a winning hand goes from 1 to 0.5 (i.e., if it is cut in half), then the sender's mix goes from 1/3 to 1/6. So, disinformation is sent 1/12 of the time rather than 1/6 of the time (given that the sender has a motivation to send disinformation 1/2 of the time).

The reverse tends to increase the amount of disinformation. That is, the greater the benefit of truly believing that the sender has a winning hand, the more disinformation there will be. For instance, holding everything else fixed, if the payoff for truly believing that the sender has a winning hand goes up from 1 to 2 (i.e., if it doubles), then the sender's mix goes from 1/3 to 2/3.

Second, the greater the cost of falsely believing that the sender has a winning hand, the less disinformation there will be. For instance, holding everything else fixed, if the payoff for falsely believing that the sender has a winning hand goes from −3 to −4, then the sender's mix goes from 1/3 to 1/4. So, disinformation is sent 1/8 of the time rather than 1/6 of the time (given that the sender has a motivation to send disinformation 1/2 of the time). And the reverse (i.e., decreasing the cost of falsely believing) tends to increase the amount of disinformation.

But it should be noted that it is actually the relative magnitude of these two epistemic values, rather than their absolute magnitudes, that matters. In other words, the amount of disinformation decreases as the ratio between (a) the benefit of truly believing that the sender has a winning hand and (b) the cost of falsely believing that she has a winning hand decreases.

Third, the lower the chances of the sender being dealt a winning hand (i.e., the greater the chances of the sender being dealt a losing hand), the less disinformation there will be. For instance, holding everything else fixed, if the chances of the sender being dealt a winning hand go from 1/2 to 1/3, then the sender's mix goes from 1/3 to 1/6. So, disinformation is sent 2/18 of the time rather than 3/18 of the time (given that the sender now has a motivation to send disinformation 2/3 of the time rather than 1/2 of the time). And the reverse (i.e., increasing the chances of the sender being dealt a winning hand) tends to increase the amount of disinformation.[30]

This is a fairly interesting result. If the chances of being dealt a winning hand go up, there will be fewer instances where the sender has a motivation to send disinformation. But the sender will send disinformation on more of those instances where she has a motivation to do so. And she will do so to such a degree that she will send disinformation more often overall.[31]

Finally, it should be noted that, if we consider a slightly different model where the sender faces the same consequences regardless of what hand she has been dealt (see Sect. 3 above), then there is another equilibrium point in addition to the equilibrium in mixed strategies that I have been discussing. Namely, the sender should claim to have a losing hand whether she has a losing hand or a winning hand. However, the existence of this additional equilibrium point does not undercut the results that I have discussed so far about what affects the amount of disinformation. First, everything that I have said so far holds when the game is at the equilibrium in mixed strategies. Second, when the game is at the other equilibrium, the probability of disinformation being sent is simply the probability of the sender being dealt a winning hand. So, the amount of disinformation only decreases when the chances of the sender being dealt a winning hand decrease.

4.3 Some Practical Implications

It is tempting to think that we could also decrease the amount of disinformation by imposing greater costs on people who spread disinformation. In line with this thought, Tullock's (1967) economic model of lying focuses only on the various costs (and benefits) *to the liar* of lying. But the first important practical implication of this investigation is that the sender's payoffs do not affect the amount of disinformation.

Admittedly, changing the sender's payoffs can have an effect. For instance, if the cost to the sender of not being believed goes up, then the receiver's level of credulity will increase. Holding everything else fixed in Fig. 2, if the payoff to the sender for not being believed when she sends disinformation goes down from −2 to −5, then the receiver believes 2/3 of the time rather than just 1/3 of the time. But just as much disinformation will be sent.

Also, changing the sender's payoffs can have an effect on what *type* of disinformation is sent. According to Frederick Schauer and Richard Zeckhauser (2009), the social cost of not being believed when you tell a lie is greater than the social cost of not being believed when you say something true but misleading. As a result, they

[30] It is a mixed bag if these three things change in different directions.

[31] Why does this happen? Basically, if more players are dealt a winning hand, players who are dealt a losing hand have more players that they can plausibly mimic. For instance, if there are more Monarchs out there, then blue jays have to be fairly credulous because most gaudy butterflies are telling the truth.

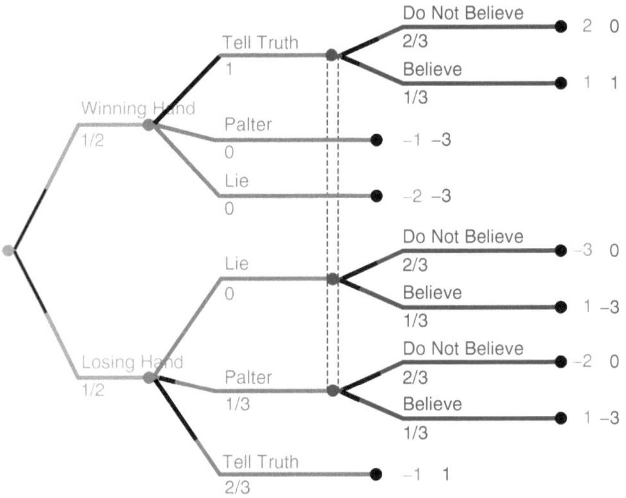

Fig. 4 The equilibrium point in an extended disinformation game

predict that such *paltering* will be a more common form of deception than lying.[32] Indeed, if we complicate our model to give the sender a choice between telling the truth, lying, and paltering, she will decide to palter rather than lie whenever she decides to send disinformation (see Fig. 4). But just as much disinformation will be sent.

Of course, the problem with most policies that punish senders of disinformation is that the senders only suffer the punishment *when they are not believed*, at least by the regulators (see Tullock 1967, 137–38). If senders suffered a cost for sending disinformation regardless of whether or not that disinformation was believed, and that cost was great enough, then telling the truth would be the dominant choice. In fact, there is empirical evidence that people have an "aversion to lying" that is independent of whether or not their lies are believed (see Serra-Garcia 2009, 6). Moreover, we could certainly implement educational policies "to try to reinforce this indoctrination … in various socially approved ethical principles" (Tullock 1967, 137). However, it seems unlikely that such policies could raise the "internal cost of lying" high enough to deter most senders of disinformation.

The second practical implication of this investigation is that it may depend on the details of the case whether or not there is anything that we can do to decrease the amount of disinformation. For instance, it may be feasible to change the receiver's

[32] For similar reasons, Michael Davis and Michael Ferrantino (1996) argue that we should expect to see more negative lies than positive lies in politics. During campaigns, politicians will be motivated to make false positive claims about themselves (and their policies) and to make false negative claims about their opponents (and their policies). However, if a politician is elected, voters will not have much opportunity to uncover any lies that she told about her opponents and their policies. So, the potential costs (e.g., in terms of gaining a reputation for insincerity) of such negative lies will be lower.

payoffs in some cases. In the butterfly case, the blue jays might discover a prophylactic that counteracts the toxicity of the Monarch butterfly. This would reduce the benefit of truly believing that a butterfly is a Monarch. So, a prophylactic would decrease the number of Viceroy butterflies who claim (with gaudy coloration) to taste nasty.

However, it will probably not be feasible to change the receiver's payoffs in all cases. For instance, consider defendants who go into court and claim to be innocent. According to an old saying, "it is better for ten guilty men to go free than for one innocent man to go to jail" (see Connolly 1987, 104). This saying arguably captures the relevant payoffs to juries and to society as a whole. It implies that the benefit of truly believing that a defendant is innocent is much larger, in terms of magnitude, than the cost of falsely believing that he is innocent. Thus, we should expect a lot of guilty defendants to claim to be innocent. Moreover, there is not much that we can do about it. We cannot easily change our moral convictions and just be happy with more innocent people locked up.

In this case, all that we can really do to reduce the number of guilty defendants who claim to be innocent is to decrease the chances of innocent defendants showing up in court.[33] In other words, we can reduce the amount of disinformation by decreasing the chances of the sender being dealt a winning hand. For instance, we might be more careful about who we charge with a crime (e.g., by improving our law enforcement techniques) so that fewer innocent people end up in the dock. Of course, this might reduce the amount of disinformation simply by reducing the number of defendants. But even if the number of defendants remained fixed, the amount of disinformation would go down under this policy.

5 Conclusion

We seem to be subjected to lies, spin, propaganda, and half-truths on an ever-increasing basis. One way of dealing with this problem is to get better at detecting such disinformation. However, another strategy is to design policies that will deter people from spreading disinformation in the first place. Toward this end, I have constructed a game-theoretic model of the sending and receiving of disinformation. Utilizing this model, we can see that the amount of disinformation decreases when (a) the benefits to the receiver of believing what the sender says when it is true go down, (b) the costs of believing what the sender says when it is false go up, or (c) the chances that the sender has a motivation to deceive go up.

Before we implement policies to deter people from spreading disinformation, however, we do need to investigate *all* of the consequences, epistemic and otherwise, of such policies. In some instances, we may actually be better off if there is more disinformation out there. For instance, Paul Rubin (1991, 681–84) has argued

[33] It is certainly possible to change the sender's payoffs in this case. For instance, we might increase the penalty for perjury. But as noted above, changing the sender's payoffs will not change the amount of disinformation.

that consumers may be less informed about products if laws against deceptive advertising are too strong. With weaker laws, suppliers may have more motivation to create deceptive ads about their own products, but competitors may also have more motivation to respond with ads that alert the public to any inaccuracies.

Acknowledgements I would like to thank Erika Au, Derek Ball, Tony Doyle, Abrol Fairweather, James Mahon, Kay Mathiesen, Andrew Peet, and Andreas Stokke for very helpful conversations and suggestions on this topic. I would also like to thank the Epistemology Research Group at the University of Edinburgh for their feedback. Much of this work was completed while I was a Visiting Fellow in the Centre for Ethics, Philosophy and Public Affairs at the University of St. Andrews.

References

Augustine. 1952 [395]. Lying. In *Treatises on various subjects*, ed. R.J. Deferrari, 53–120. New York: Catholic University of America.
Carson, Thomas L. 2010. *Lying and deception*. New York: Oxford University Press.
Connolly, Terry. 1987. Decision theory, reasonable doubt, and the utility of erroneous acquittals. *Law and Human Behavior* 11: 101–112.
Davis, Michael L., and Michael Ferrantino. 1996. Towards a positive theory of political rhetoric: Why do politicians lie? *Public Choice* 88: 1–13.
Descartes, René. 1996 [1641]. *Meditations on first philosophy*. Cambridge: Cambridge University Press.
Fallis, Don. 2006. Epistemic value theory and social epistemology. *Episteme* 2: 177–188.
Fallis, Don. 2007. Attitudes toward epistemic risk and the value of experiments. *Studia Logica* 86: 215–246.
Fallis, Don. 2009. What is lying? *Journal of Philosophy* 106: 29–56.
Fallis, Don. 2011. Floridi on disinformation. *Etica & Politica* 13: 201–214.
Farid, Hany. 2009. Digital doctoring: Can we trust photographs? In *Deception*, ed. Brooke Harrington, 95–108. Stanford: Stanford University Press.
Farquhar, Michael. 2005. *A treasury of deception*. New York: Penguin.
Fetzer, James H. 2004. Disinformation: The use of false information. *Minds and Machines* 14: 231–240.
Floridi, Luciano. 2011. *The philosophy of information*. Oxford: Oxford University Press.
Gell-Mann, Murray. 2009. Forward. In *Deception*, ed. Brooke Harrington, vii–xii. Stanford: Stanford University Press.
Goldman, Alvin I. 1999. *Knowledge in a social world*. New York: Oxford University Press.
Goldman, Alvin I. 2002. Reply to commentators. *Philosophy and Phenomenological Research* 64: 215–227.
Good, I.J. 1967. On the principle of total evidence. *British Journal for the Philosophy of Science* 17: 319–322.
Hume, David. 1977 [1748]. *An enquiry concerning human understanding*. Indianapolis: Hackett.
Jackson, Brooks, and Kathleen H. Jamieson. 2007. *Unspun: Finding facts in a world of disinformation*. New York: Random House.
James, William. 1979 [1896]. *The will to believe and other essays in popular philosophy*. Cambridge, MA: Harvard University Press.
Kant, Immanuel. 1959 [1785]. *Foundations of the Metaphysics of Morals*. Trans. Lewis W. Beck. New York: Macmillan.
Levi, Isaac. 1962. On the seriousness of mistakes. *Philosophy of Science* 29: 47–65.

Mansfield, Edwin. 1994. *Microeconomics*, 8th ed. New York: W. W. Norton & Company.
McKelvey, Richard D., Andrew M. McLennan, and Theodore L. Turocy. 2007. *Gambit: Software tools for game theory*. Version 0.2007.12.04. http://www.gambit-project.org.
Newman, Matthew L., James W. Pennebaker, Diane S. Berry, and Jane M. Richards. 2003. Lying words: Predicting deception from linguistic styles. *Personality and Social Psychology Bulletin* 29: 665–675.
Reiley, David H., Michael B. Urbancic, and Mark Walker. 2008. Stripped-down poker: A classroom game with signaling and bluffing. *Journal of Economic Education* 39: 323–341.
Riggs, Wayne D. 2003. Balancing our epistemic ends. *Noûs* 37: 342–352.
Ritland, David B., and Lincoln P. Brower. 1991. The viceroy butterfly is not a Batesian mimic. *Nature* 350: 497–498.
Rubin, Paul H. 1991. The economics of regulating deception. *Cato Journal* 10: 667–690.
Schauer, Frederick, and Richard Zeckhauser. 2009. Paltering. In *Deception*, ed. Brooke Harrington, 38–54. Stanford: Stanford University Press.
Schelling, Thomas C. 1968. Game theory and the study of ethical systems. *Journal of Conflict Resolution* 12: 34–44.
Serra-Garcia, Marta. 2009. Lying or truth-telling: Why does it matter in economics? *Aenorm* 62: 4–7.
Skyrms, Brian. 2010. *Signals*. New York: Oxford University Press.
Sober, Elliott. 1994. The primacy of truth-telling and the evolution of lying. In *From a biological point of view*, 71–92. Cambridge: Cambridge University Press.
Sorensen, Roy. 2007. Bald-faced lies! Lying without the intent to deceive. *Pacific Philosophical Quarterly* 88: 251–264.
Tullock, Gordon. 1967. The economics of lying. In *Toward a mathematics of politics*, 133–143. Ann Arbor: University of Michigan Press.
Williams, Bernard. 2002. *Truth and truthfulness*. Princeton: Princeton University Press.

Defeasibility Without Inductivism

Juan Comesaña

1 Four Principles

I will argue that four principles, all of which have to do with justification, give rise to a contradiction. I will be talking mostly about what has been called "doxastic" justification. Doxastic justification accrues to the mental state that a subject is in when she believes a proposition, and whether that state is doxastically justified depends on how the subject arrived at it. *Doxastic* justification is contrasted with *propositional* justification, which is a property that propositions can have irrespective of how the subject acquired a belief in the proposition, or indeed irrespective of whether the subject even believes the proposition at all.

First, the principle whose rejection I will be advocating:

> **Inductivism:** It is possible for S to be justified in believing **A** even when S's total evidence E does not entail **A**.

Second, a kind of closure principle for justification:

> **Closure:** If S is justified in believing that A and competently deduces B from A, then S is justified in believing that B.

I am following Williamson and Hawthorne in calling this a closure principle,[1] but it is worth noting that it shares aspects of what Davies and Wright would call a transmission principle.[2] The idea behind Closure is not just that if you are justified in believing A and know that A entails B then you are justified in believing B in addition. The idea is that if you perform the deduction competently, then what justifies you in believing B is your competent performance of the deduction itself.

[1] See Williamson (2000) and Hawthorne (2004).

[2] See, for instance, Wright (2004) and Davies (2004).

J. Comesaña (✉)
University of Arizona, AZ, USA
e-mail: comesana@arizona.edu

A. Fairweather (ed.), *Virtue Epistemology Naturalized: Bridges Between Virtue Epistemology and Philosophy of Science*, Synthese Library 366, DOI 10.1007/978-3-319-04672-3_11,

The third principle concerns evidential justification. We *may* be justified in believing a proposition ~~even~~ when we do not believe in that proposition on the basis of any evidence.

But when our justification does depend on our evidence, then part of our evidence must justify that proposition:

> **Evidential Justification:** If S is justified in believing A on the basis of some evidence, then at least some part of S's total evidence E justifies A.

Finally, a principle that sets a constraint on when a piece of evidence can justify a proposition:

> **Entailment:** If E is S's total evidence and A entails E (and it is not the case that E entails not-A), then no part of E justifies not-A.[3]

I take it that all four principles are, on reflection, highly plausible. The Entailment principle is perhaps the least obviously plausible of the four, but only a little reflection is needed in order to justify it. Let me first defend a more restricted versions of the Entailment Principle: if A entails E (and it is not the case that E entails not-A), then E itself does not justify not-A.

Notice, first, that if one accepts a probabilistic relevance constraint on justification, then one must accept this more restricted claim. A probabilistic relevance constraint says that E justifies a proposition A only if the conditional probability of A given E is higher than the unconditional probability of A. Now, it is a theorem of the probability calculus that if A entails E, then (provided that E and A have non-extreme probabilities) the probability of not-A given E is lower than the probability of not-A.[4] It follows from the probabilistic relevance constraint, then, that E does not justify not-A.

But one need not accept a probabilistic relevance constraint in order to find the restricted claim attractive. For if A entails E, then E only rules out one way for not-A to be true. Given that A entails E, the possibilities where E obtains are a subset of the possibilities where A obtains, and so one way for not-A to be true is for E to be false. The truth of E, then, rules out one way for not-A to be true–and, given that we are assuming that it is not the case that E entails not-A (otherwise, A would be guaranteed to be false), ruling out one way for it to be true is all that E does with respect to not-A. Therefore, whatever else we think about A (provided that we were not completely sure of its truth or falsehood to begin with), acquiring E as evidence should make us more confident of its truth, and therefore less confident of not-A. And if acquiring E should make us less confident of not-A, then it is hard to see how could E justify us in believing not-A.

---

[3] As explained later, if the parenthetical proviso A is not satisfied, then A is a contradiction (and, thus, not-A is a tautology).

[4] Given that A entails E, $Pr(E \mid A) = 1$, and so $Pr(E \mid A)\ Pr(A)/Pr(E) = Pr(A)/Pr(E)$. Given that the probabilities of A and E are both non-extreme, $Pr(A)/Pr(E) > Pr(A)$, and so (by the previous result) $Pr(E \mid A)\ Pr(A)/Pr(E) > Pr(A)$. By Bayes's theorem, $Pr(A \mid E) = Pr(E \mid A)\ Pr(A)/Pr(E)$, and so (by the previous result), $Pr(A \mid E) > Pr(A)$. Given that the probability of the negation of any proposition is (1 – the probability of that proposition), $Pr(\text{not-}A \mid E) < Pr(\text{not-}A)$.

Of course, it may well happen that some of our evidence does not justify not-A, but we are still nevertheless justified in believing not-A, because some other part of our evidence justifies not-A. Our evidence may be composed of considerations which partly support and partly undermine A in different ways, and if the overall balance comes out against A then we will be justified in believing not-A. In other words, our defense of the more restricted claim doesn't yet provide an adequate argument for entailment. But it does with only one modest additional assumption. The assumption is that one's total evidence E is a big conjunction, and that its parts are the individual conjuncts. Given this assumption, if A entails E then A entails every part of E. Therefore, for the special case when E is one's total evidence, the more restricted claim entails Entailment, and our argument for the more restricted claim therefore transfers over to Entailment.

2 The Problem

Plausible as they are, the four principles together lead to a contradiction, as shown by the following argument. We start by assuming that Inductivism is true:

1. S is justified in believing p on the basis of some evidence, and S's total evidence e doesn't entail p. (Inductivism)

Suppose then that S knows that p entails not-e or p and on that basis competently deduces not-e or p from p. We are therefore sidestepping issues having to do with our failure to be logically omniscient. Those issues are of course legitimate ones, but the answer to our problem could hardly be that we are ignorant of the relevant entailments. Given closure, then, we have:

2. S is justified in believing not-e or p. (1, Closure)

Now, given that e and not-p entails e and that not-e or p is (equivalent to) the negation of that proposition, Entailment gives us:

3. No part of S's total evidence e justifies not-e or p. (Entailment)

2 and 3 together entail that S is non-evidentially justified in believing not-e or p if justified at all. But S arrived at not-e or p via inference from p, and so at the very least S's justification for not-e or p depends on p, and very plausibly on e as well (given that S's justification for p itself derives from e). Therefore, S's justification for not-e or p can only be evidential.[5] So, by Evidential Justification:

4. S is not justified in believing not-e or p. (3, Evidential Justification.)

4 is the negation of 2, and so one of the four principles must be given up.

[5] I believe that this last comment undermines much of the appeal of neo-rationalist positions such as those of Wright (2004) and Cohen (2010) according to which we have a priori justification for believing propositions such as not-e or p. Whether or not we have propositional justification which is non-evidential for those propositions, the fact is that we seem to be able to acquire evidential doxastic justification, and that is all that is needed to generate the problem.

The problem can be illustrated as follows. Suppose that you are justified in believing that the wall in front of you is red on the basis of its looking red to you (as well as on the basis of normal background beliefs). You then notice that the proposition that the wall is red entails the proposition that either the wall is red or it doesn't look red,[6] and you come to believe this latter proposition by competently deducing it from the former. By Closure, then, you are also justified in believing that either the wall doesn't look red or it is red. But it cannot be your evidence that justifies you in believing this–if anything, your evidence should lower your confidence in this proposition, for it merely rules out one way for it to be true. However, your reason for believing the disjunction is that you competently deduced it from one of the disjuncts–so, either you are evidentially justified in believing it or you are not justified at all. Therefore, you are not justified at all in believing the disjunction, which contradicts our previous result.

I believe that this problem lies at the heart of many issues in contemporary epistemology, such as the bootstrapping problem for reliabilism urged by Vogel (2000 and 2008) and Fumerton (1995 and 2006), the easy knowledge problem presented by Cohen (2002) and the problems of defeasible justification discussed in different guises by Huemer (2001) and Vogel (forthcoming). Different solutions can be found in the literature to these related problems. But one solution which has elicited little if any commentary consists in rejecting Inductivism. In what follows I examine the prospects for such a solution.

3 Rejecting Inductivism

Let me first show how rejecting Inductivism gets us out of the problem. One might think that it doesn't, for even if e entails p, Closure guarantees that if we are justified in believing p we are also justified in believing not-e or p. And we can of course ask: what is it that justifies us in believing this proposition? But, if e entails p, we have an easy answer in this case. For, if e entails p, then not-e or p is a logical truth–and, supposing that the subject is sophisticated enough to notice that it is a logical truth, this gives her all the justification she needs to believe not-e or p.

Relatedly, when e entails p, we can no longer apply Entailment in step (3) of the argument. To see why, let us suppose that **A** is any proposition p and **E** is the conjunction p and q. In this case, **E** and not-A will be p and q and not-p. The corresponding instance of Entailment therefore is:

> **Instance of Entailment:** If p and q is S's total evidence and p and q and not-p entails p and q (and it is not the case that p and q entails Not-(p and q) or p), then no part of p and q justifies Not-(p and q) or p.

[6] For the sake of concreteness, I am treating the proposition the wall looks red as if it were your total evidence.

The first part of the antecedent is satisfied, but the parenthetical remark is not, because p and q does entail Not-(p and q) or p. Moreover, the parenthetical proviso is not there gratuitously. If **A** entails **E** and **E** entails not-**A**, that can only be because **A** is a contradiction and not-**A** a tautology (as in our example). In that case, it will not be the case that **E** lowers the probability of **A**, nor will it be the case that all **E** does is rule out one way for **A** to be false–it also rules out one way for it to be true.

So, if **E** entails **A**, then the argument is blocked because not-**E** or **A** will be a tautology, and so Entailment cannot be applied to step (3) of the argument. Therefore, giving up Inductivism solves the problem presented earlier.

The idea that the evidence a subject has must entail a proposition, for that proposition to be justified for the subject, is a traditional one. According to the textbooks, Descartes held it. According to the same textbooks, it is an ill-conceived idea that all right-thinking epistemologists ought to abandon. By far the most cited reason for thinking that rejecting Inductivism is a non-starter is that it will inevitably lead to skepticism. But there is also a different (though related) objection that one might have: rejecting Inductivism seems to entail the indefeasibility of justification. In the next two sections I pick up these objections in turn. My aim cannot possibly be to answer these objections in any detail. Rather, I will present a picture of the commitments that someone who rejects Inductivism, but doesn't accept skepticism or the indefeasibility of justification, needs to take on. To my mind, these commitments are nowhere near as disastrous as some philosophers fear.

4 Skepticism?

The objection that rejecting Inductivism leads to skepticism can be illustrated with the textbook presentation of Descartes' epistemology. According to this presentation, Descartes sought to justify all of his beliefs from the starting point of propositions about his own mental states. ~~Starting~~ from the cogito, the proposition which asserts his (Descartes') own existence, Descartes ~~then~~ derived the idea of God, from where he derived God's existence, and God's omnibenevolence assured Descartes that those ideas which where clear and distinct would be true. The problem is that many and perhaps all of the propositions derived from the starting points are not entailed by those starting points. Therefore, if one insists on allowing only entailment as the relationship that must hold between the starting points and further justified propositions, one is likely to end up as Descartes–with not much of interest legitimately entailed by those starting points.

But whether interesting propositions can legitimately be derived from the starting points depends not only on which relationships we allow to link starting points to further propositions, but also on the nature of the starting points themselves. In the Cartesian picture, those starting points were mental states of the subject in question. But if, for instance, we allow as starting points propositions that are already about the external world, the fact that we accept only entailment as the relationship between starting points and further justified propositions will not mean that we will not be able to get to interesting propositions. Thus, for instance, if we think that part of my evidence can be, not merely that it seems to me that there is a table in front of me, but that I see a table in front of me, then there will be no puzzle as to how I am justified in believing that it is not the case that I am hallucinating a table (and so that it is not the case that it looks as if there is a table in front of me but there isn't). Descartes himself, of course, would not have allowed that propositions about the external world can play that kind of evidential role. For him, only propositions about which we are infallible can play that role. Most contemporary epistemologists will not impose this requirement on evidence, and will indeed assert that we can be as fallible with respect to our basic evidence as we are with respect to anything else. If there is resistance about having propositions such as I see a table being part of our basic evidence, then, the reason for the resistance cannot be that we are not infallible about such propositions.

A better objection is that external world propositions just don't seem basic enough. There is, the objector may say, an obvious epistemic priority to propositions such as I seem to see a table, an epistemic priority which dictates that if I am ever justified in believing that I see a table, it must be partly in virtue of being justified in believing that I seem to see a table. Someone who rejects Inductivism, however, need not deny this latter claim, properly understood. That is to say, someone who rejects Inductivism can accept that the proposition that I see a table gets to be part of my evidence (and therefore justified) only in virtue of the fact that I seem to see a table (whether this latter proposition itself is part of the evidence or not is a different issue, one which the we need not legislate on). All we need to deny is that the proposition that I see a table (just as any other proposition) is justified by evidence which doesn't entail it. Thus, in admitting that the proposition that I see a table gets to be part of my evidence in virtue of the fact that I seem to see a table, we are not admitting that my basic evidence in this area consists only in the proposition that I seem to see a table, and the proposition that I see a table gets to be part of my evidence by courtesy, so to speak, by being inferred from the proposition that I seem to see a table (plus perhaps other propositions). For the objection to work, then, the intuitive idea that the seeming proposition has epistemic priority over the external world proposition needs to go beyond the claim that the external world proposition wouldn't be part of my evidence if the seeming proposition weren't true. But while there is admittedly something to the claim of epistemic priority, how best to capture that claim is open to argument.

A further problem with the idea that external world propositions such as that I see a table get to be part of my evidence arises when we consider victims of an evil demon or other similarly deceived creatures. They don't see a table, and yet many will like to say that they have the same evidence we do. There are two possibilities here. One is to go "disjunctivist" and deny the intuition that we have the same evidence as our massively deceived counterparts. Philosophers as different from each other as Williamson and McDowell have held precisely this position. The other one is to hold that we do have the same evidence as our massively deceived counterparts, not by shrinking my evidence to just the proposition that I seem to see a table, but rather by (what to some might look like) expanding our massively deceived counterparts' evidence so as to include the proposition that they themselves see a table. One could, for instance, hold that its seeming to a subject as if she sees a table is not only necessary for the proposition that they see a table to be part of their evidence, but it is also (ceteris paribus) sufficient. Taking this position involves accepting the claim that false propositions can be part of our evidence. Many philosophers see serious problems here–I do not. Although this is not the place to develop in detail a defense of the possibility of false evidence, I will say something more about it in connection with the issue of defeasibility.[7]

Even if I have convinced you that sometimes external world propositions (such as that I see a table) get to be part of our evidence, this is not enough to get out of the problem outlined earlier. For in order to get out of that problem, it has to be the case that every proposition we are justified in believing is entailed by our evidence. To take a particularly problematic case, think of the case of Alice. Alice has evidence bearing on whether the sun will come out tomorrow. If I am right, that evidence entails that the sun will come out tomorrow. Two pressing questions can be raised at this point. First, what is that evidence? Second, haven't I "solved" the problem of induction in a rather implausible way? I take those questions in turn.

What evidence does Alice have, such that it entails that the sun will come out tomorrow? One option here is to say that it is the proposition that the sun will come out tomorrow itself which is part of Alice's evidence. We need not think that it has always been the case that this proposition was part of Alice's evidence, but that is how it goes with all evidence anyhow. We can say that as evidence relevant to whether the sun will come out tomorrow accumulates, the proposition itself gets to be part of Alice's evidence. When exactly? This question is as difficult to answer as is the following one: when exactly does Alice become justified in thinking that the sun will come out tomorrow? Anyone who takes this first option will likely answer both questions the same way. As with the previous example about the table, however, we should be careful to distinguish between these two claims: on the one hand, the claim that the proposition 7 For more on false evidence see Comesaña and McGrath (forthcoming) that the sun will come out tomorrow becomes part of Alice's evidence when (and perhaps even in virtue of the fact that) Alice has accumulated enough evidence for its truth; on the other hand, the claim that the proposition that the sun will come out tomorrow gets to be part of Alice's evidence by being justified by non-entailing evidence that Alice possesses. The second claim conflicts with the rejection of Inductivism, but the first one doesn't.

Let us now go back to the second question: doesn't this view offer an implausibly easy solution to the problem of induction? I think that, once we are clear on what the problem of induction is, it doesn't. What bothered Hume about inductive arguments? The textbooks would have us believe that it was simply the fact that they were not deductive: that the truth of the premises of even the best inductive argument doesn't guarantee the truth of its conclusion. But notice that this characteristic of being non-deductive is a logical property of inductive arguments. Surely the problem of induction is not primarily a logical problem, but an epistemological

[7] For more on false evidence see Comesaña and McGrath (forthcoming).

one. What bothered Hume about inductive arguments (or what should have bothered him) was rather the fact that even being certain about the premises didn't offer one a good epistemic standing with respect to the conclusion. But maybe the textbook interpretation is right after all, for why is it that being certain about the premises of an inductive argument doesn't put one in a good epistemic position with respect to the conclusion? The obvious answer is: because it is possible for the premises to be true and the conclusion false. As I explain in the next section, I think that the obvious answer is wrong.

5 Defeasibility?

Let us say that your justification for a proposition *p* is *defeasible* if and only if it is possible for you to cease to be justified in believing *p* even while retaining your evidence for *p*. Most (if not all) of our justified beliefs are defeasibly justified. Moreover, it may be held with some plausibility that an acceptance of the fact that our beliefs are defeasibly justified if justified at all is a central virtue that an epistemic agent must possess. Many authors have claimed ("argued," as we shall see, it's not quite the right word) that only inductively justified beliefs are defeasible. If this is right, then abandoning Inductivism amounts to rejecting the idea that our beliefs are defeasibly justified at best. Am I, then, advocating a kind of dogmatism that is incompatible with the epistemic virtue of being open to rationally abandoning one's beliefs? I am not, for those authors are wrong in thinking that only inductively justified beliefs are defeasible.

Let me start by documenting the claims connecting defeasibility with Inductivism. John Pollock starts his article "Defeasible Reasoning" this way:

> There was a long tradition in philosophy according to which good reasoning had to be deductively valid. However, that tradition began to be questioned in the 1960's, and is now thoroughly discredited.

If Pollock is right that the idea that good reasoning must be deductive only began to be questioned in the 1960s, then it was a long tradition indeed. One wonders what Pollock thought of Mill and Bentham, not to speak of Aristotle. But I don't think that Pollock is ignoring these (and there are doubtless many more) inductivists of the past. Rather, Pollock is contrasting deductive reasoning with *defeasible* reasoning, the systematic study of which one perhaps can trace back only to the 1960s. But notice, then, that Pollock must be equating "deductive" with "indefeasible." This suspicion is confirmed by the following quote from *Contemporary Theories of Knowledge:*

> In chapter one (…) we encountered the assumption that a reason can be a good reason for believing its conclusion if it logically entails that conclusion (…). A frequently encountered variant of it is that reasons must be either entailments or inductive reasons. We feel that one of the most important advances of contemporary epistemology has been the rejection of both of these assumptions and the recognition of reasons that are neither inductive reasons nor logical entailments. (36).

Pollock and Cruz are obviously distinguishing between non-deductive reasons and inductive reasons. I do not. This is merely a terminological decision. But, independently of the terminology, they are here once again advancing the claim that deductive reasons are not defeasible reasons. In "The Skeptic and the Dogmatist," Pryor starts by giving a characterization of defeasible justification which is close to our own:

> Our perceptual justification for beliefs about our surroundings is always defeasible: there are always possible improvements in our epistemic state which would no longer support those beliefs. (517)

But a few sentences later, Pryor introduces the fallibilist in this way:

> A fallibilist is someone who believes that we can have knowledge on the basis of defeasible justification, justification that does not *guarantee* that our beliefs are correct. (518)

If we think (as it seems we should) that when our evidence *entails* a belief then our justification for that belief guarantees its truth, then Pryor's characterization of fallibilism indicates that he believes that only inductive justification can be fallible. This is confirmed by Pryor's remarks on Moore's unwillingness to offer any consideration in favor of his belief that he has hands–"even defeasible, ampliative considerations," (518) says Pryor–as well as by his claim that the fallibilist acknowledges that the support for our perceptual beliefs is "defeasible and ampliative" (520).

Another author who identifies defeasibility with inductive justification is Michael Huemer. In "The Problem of Defeasible Justification" Huemer gives the following explicit definition:

> I use the term "defeasible justification" to refer to the relation that obtains between a piece of evidence, *e*, and a conclusion, *h*, when
> (1) *e* provides sufficient support for *h* for one to be justified in believing *h* on the basis of *e*,
> but
> (2) *e* does not entail *h*. (375)

Of course, one is free to define "defeasible justification" as one wishes, and so, in particular, it may just be that Pollock and Pryor are thinking of defeasible justification just as Huemer does, as being *by definition* inductive justification (as I, but not Pollock, am using the term "inductive"). However, what one is *not* free to do is to assume that "defeasible justification," however defined, denotes an epistemically interesting notion. One undoubtedly epistemically interesting notion is that of justification that may be lost even while retaining one's evidence. My hunch is that Pollock and Pryor are thinking that defeasible justification just is justification that can be lost in this way, and that they advance the thesis that defeasible justification can only be inductive as a *substantive* (although perhaps obviously true) claim about defeasible justification. Even Huemer, who *defines* defeasible justification as inductive justification, may be just identifying justification that may be lost with inductive justification. Where does the need to make this identification come from?

I think that a strong motivation for this view stems from the idea that evidence can only grow: once a proposition is part of one's evidence, it stays as part of one's evidence forever. It is easy to see how this idea is compatible with the defeasibility of justification if we accept Inductivism. We start out having certain evidence which justifies us in believing a proposition which is not entailed by that evidence. Some time later we acquire new evidence. Our new evidence, together with our old one (which we retain) no longer justifies us in believing that proposition. Therefore, the justification that we had for believing the proposition is shown to be defeasible. But how can justification be defeasible if our evidence entails anything we are justified in believing? Not by the acquisition of new evidence, for if a set of propositions entails another one, then any super-set will still entail the same proposition. So it looks as if only by allowing non- deductive justification can we reconcile the fact that justification is defeasible with the idea that evidence can only grow.

Here's how people think that inductive justification can be defeated. Suppose that it is part of my evidence that the wall in front of me looks red, and that I believe that the wall in front of me is red on the basis of my evidence. Later on, I acquire an additional piece of evidence: that the wall is illuminated by red lights. The justification provided by the wall's looking red is defeated by the information that there are red lights shinning on the wall. This is what Pollock called an "undercutting defeater." Or suppose that John tells me that a certain wall is red, and on that basis I believe that it is. Later on, Mary tells me that the wall isn't red. Once again, the justification provided by John's testimony is defeated by Mary's. This what Pollock called a "rebutting defeater."

There can be rebutting and undercutting defeaters even for non-inductive evidence. Suppose that it is part of my evidence that the wall in front of me is scarlet. On the basis of my evidence, I believe that the wall in front of me is red. However, my usually reliable friend Mary tells me that the wall isn't red. If the details of the case are filled-in in the right way, I am no longer justified in believing that the wall is red. This is a case of a rebutting defeater for justification provided by entailing evidence.

Now, some may say that there is an important difference between this case and those of defeat of inductive evidence. In the previous cases, that the wall looks red is still part of my evidence, even after I acquire the information that it is illuminated by red lights, and that John said that the wall is red is still part of my evidence even after Mary tells me that it isn't. But in the present case, the objection goes, given that Mary's telling me that the wall isn't red defeats my justification for believing that the wall is red, it also defeats my justification for believing that it is scarlet, and so it is no longer part of my evidence that the wall is scarlet. The objection, then, is that the case of the previous paragraph is one where justification is lost, but not defeated.

It is not obvious that every way of filling in the details of the case will have as a consequence that I am no longer justified in believing that the wall is scarlet. But suppose it is. That was supposed to be a case of a rebutting defeater for non-inductive evidence. But there are also examples of undercutting defeaters for non-inductive evidence. Suppose, for instance, that it is part of my evidence that all swans are white, and on that basis I believe that the swans at the local zoo are white. However, a renowned logician tells me that universal instantiation is not valid. In that case, even though it is still part of my evidence that all swans are white my justification for believing that the swans at the local zoo are white is defeated. So even if the existence of rebutting defeaters for non-inductive evidence is problematic, that of undercutting defeaters is not.

There are two ways, then, in which one's justification for a proposition *p* can be lost even when one's evidence entails that *p*. It may happen that we receive evidence against *p*. In some cases, at least, this will have the effect that we are no longer justified in believing *p*. Perhaps it will also have the effect that we no longer have the same evidence as we did before receiving the evidence against *p*. It may also happen that we receive evidence which challenges the connection between our evidence and *p*. In that case, we will no longer be justified in believing *p* even if it is still the case that our evidence entails *p*. Rejecting Inductivism, then, does not have the unpalatable consequence that all justification is indefeasible, nor does it entail that there is no place for the virtue of open-mindedness.

6 Conclusion

In this paper I presented a problem about epistemic justification. The problem stems from the fact that four attractive principles about justification are incompatible with each other. I examined the prospects for solving the problem by holding that we are justified in believing only what our evidence entails. This solution must face a number of problems. Perhaps the most obvious one is that it seems to entail that justification is indefeasible. But I argued that this objection is misguided, for holding that we are justified in believing only what our evidence entails is compatible with our justification being defeasible, and with our having the epistemic virtue of open-mindedness.

References

Cohen, Stewart. 2002. Basic knowledge and the problem of easy knowledge. *Philosophy and Phenomenological Research* LXV: 309–329.
Cohen, Stewart. 2010. Bootstrapping, defeasible reasoning, and a priori justification. *Philosophical Perspectives* 24: 141–159.
Comesaña, Juan and Matt McGrath. forthcoming. False Evidence.
Davies, Martin. 2004. Epistemic entitlement, warrant transmission and easy knowledge. *Aristotelian Society* 78(1): 213–245.
Fumerton, Richard. 1995. *Metaepistemology and Skepticism*. Lanham,: Rowman and Littlefield.
Fumerton, Richard. 2006. Epistemic internalism, philosophical assurance, and the skeptical predicament. In *Knowledge and reality: Essays in honor of alvin plantinga,* eds. Thomas M. Crisp, Matthew Davidson, and David Vander Laan, 179–191. The Netherlands: Kluwer.
Hawthorne, John. 2004. *Knowledge and lotteries*. New York: Oxford University Press.
Huemer, Michael. 2001. The problem of defeasible justification. *Erkenntnis* 54(3): 375–397.
Pryor, James. 2000. The skeptic and the dogmatist. *Noûs* 34(4): 517–549.

Pollock, John. 2008. Defeasible reasoning. In *Reasoning: Studies of human inference and its foundations*, ed. Adler Jonathan and Rips Lance, 451–470. New York: Cambridge University Press.
Pollock, John, and Cruz Joe. 1999. *Contemporary theories of knowledge*. Lanham: Rowman and Littlefield.
Vogel, Jonathan. 2000. Reliabilism Leveled. *Journal of Philosophy* 97(11): 602–623.
Vogel, Jonathan. 2008. Epistemic bootstrapping. *Journal of Philosophy* 105(9): 518–539.
Vogel, Jonathan. forthcoming. E and not-H.
Williamson, Timothy. 2000. *Knowledge and its limits*. Oxford/New York: Oxford University Press.
Wright, Crispin. 2004. Warrant for nothing (and foundations for free)? *Aristotelian Society* 78(1): 167–212.

Part III
Virtues of Theories and Virtues of Theorists

Acting to Know: A Virtue of Experimentation

Adam Morton

1 Experiments Everywhere

Simple everyday experiments involve no special equipment, and ideas about experimental design are rarely consulted. Yet they fit the fundamental pattern that in order to learn something one does something, making information emerge which would not have otherwise. For example, you are on a committee interviewing candidates for a job which involves dealing with a range of people on a range of topics. A letter for one candidate says that he does not suffer fools gladly, that he is inclined to be brusque and visibly impatient with people who he takes to be confused or wasting his time. The letter may be exaggerating or malicious, and you would like some better evidence. So you ask a stupid question. You put a lot of thought into your stupidity, and at the interview you make an elaborate suggestion about his area of expertise that rests on a conflation of two similar-sounding words. The outcome is unpredictable. It may be that he seethes with contempt, that he patiently and tactfully unravels the confusion, that he deflects the question, or something in-between. Some of these outcomes will tell you more than others.

That experiments are causal interactions to epistemic ends was noted by Ian Hacking some time ago. See Chaps. 2 and 9 of Hacking (1983). The theme has been ignored in a lot of more recent work but see Radder (1996), and Woodward (2003). I do not find in any of this otherwise admirable work recognition of the continuity between the scientific and the everyday, of the kind that the interview example illustrates.

Several basic points are illustrated by informal experiments such as the interview case. Most basic of all, the experiment is an act – the realisation of an intention by causing some change in the world – which can be well-thought out or not, and can

A. Morton (✉)
Department of Philosophy, University of British Columbia, Vancouver, BC, Canada
e-mail: adam.morton@ubc.ca

A. Fairweather (ed.), *Virtue Epistemology Naturalized: Bridges Between Virtue Epistemology and Philosophy of Science*, Synthese Library 366, DOI 10.1007/978-3-319-04672-3_12,

be successful or not. It is an act whose purpose is epistemic, but the thinking behind it does not fall into a traditional category of belief-directed reasoning. One reason for this is that what belief it is that results depends on something unpredicted that happens outside the person's cognition. The opposite is also common, where you form a belief in order to achieve a practical aim, as when you look at the weather forecast in order to choose the best day for the picnic, but we are now concerned with walking to the hill where you can see the clouds in the west. Often, of course, we perform an action in order to gain knowledge in order to be able to do something: walking to the hill in order to predict the weather in order to time the picnic. In the interview example you do the experiment to learn if the candidate is tactful in order to appoint the best person. Epistemic and practical are usually entwined. Experimentation overlaps with thinking when a person wonders what she thinks on a topic ("would we be happier if nothing was secret", "are there really fundamental rights") by posing various hard questions to herself and seeing how she reacts. This is a kind of self-experiment similar to those one performs conversationally with other people. With oneself or with others, it produces information, material for thinking about, which one could not have got just by thinking or passively perceiving.

The interview experiment is also typical in that it has a cost. In asking the stupid question you make the candidate think less of you, and this may have repercussions. You use up time in the interview that could be used on other topics. You affect the atmosphere later in the interview. If you are thinking whether and how to perform the experiment you have to formulate these costs and risks, which have to be considered together with the benefits of the information you might gain.

Thirdly, this experiment like many others has an unpredicted outcome. The unpredictedness is hard to state carefully. It is reminiscent of epistemic paradoxes such as Kripke's observation that when one has good evidence for a belief one also has good evidence that evidence against it is likely to be misleading, and therefore to be ignored. (The idea comes from Saul Kripke, but its first appearance in print was p 148 of Harman 1973.) In the interview case you can expect several possible general types of response from the candidate. You may well consider some of these more likely than others. But you don't take yourself to know what the outcome of the experiment will be. Often one is surprised when an experiment turns out as it does, but in planning the experiment one does not take it for granted that it will not turn out this way. One does, though, make more elaborate contingency plans for following up the more expected outcomes than the less expected ones. In planning the interview you may think that it is pretty unlikely that the candidate will simply ignore the mistake in the question, but you still prepare a follow-up question to highlight it in case he does. You think it pretty likely that he will use some abusive language to you, and so on the one hand you prepare a pretence of injured pride in order to test his reaction to information that he is causing distress, and on the other hand you think how to get across to him later that no harm was done (except to his job prospects.) The situation also resembles the strategic interactions studied in game theory. There although one player may have expectations about what another is more or less likely to do, a prediction of the other's actions cannot be separated from a decision of what to do oneself (since the other is basing their action in part on a prediction of what the first player will do.) An experiment is a game against (or with) nature in this respect: your moves depend not on what you expect the other to do but on what values the possible outcomes have for you. (Considering experiments as games against nature opens up formal ideas, due to Abraham Wald. See Gigerenzer and others 1990. The similarity of experiment to strategic interaction described here is more basic than, and independent of, these ideas.)

These features are found in formal scientific experiments, too, and in innumerable everyday information-eliciting procedures. One finds out if someone is awake by whispering a message; one finds out if there is water in the well by dropping a pebble into it; one finds out if the enemy is still out there by sticking one's head above the parapet. It is important in distinguishing these from non-experimental inquiry to emphasise that the procedure has a causal effect that allows the information that would not otherwise be available to be produced. The nearest that simple perception comes to this is in some uses of one's tactile sense, as when one feels how many coins are in one's pocket by actively moving them around. (Fingers are special in that they both move and feel, in ways that are often inseparable.) Just opening one's eyes is an action, and can be intended to produce a situation in which information is available, as is flipping a light switch, but these should be seen as at most limiting cases of experimentation. I shall take it that in even very informal experimentation one performs an action, the action produces a situation that would not otherwise have existed, and the existence or features of this situation provide data one wants in order to form opinions. When a person opens her eyes she is producing a situation – her eyes being open and light striking her retinas – in which information is available, but it is information about the scene rather than about the state of her eyes or the effects of opening them. When the bandages are removed from someone recovering from an eye operation and she first opens her eyes, that *is* a real experiment.

2 Success

An experiment has gone well when the intended situation has been produced and it provides information that is relevant to the question at issue. (In science, experiments are usually directed at fairly specific questions; less so in everyday life, though in both there is a virtue of asking and probing at the right level of generality.) Then it has been successful, in more than the minimal sense of producing knowledge. (It is a frustrating success when you get the information you wanted, but in terms of the questions that matter to you, you are none the wiser.) Experiments often succeed inadvertently: a situation is produced which is not among those anticipated and one is not prepared for the information that results. When you ask your stupid question a feature of the wording may produce a reply that reveals a completely different flaw than the one you were probing for. The experiment has then failed in that you did not get an answer to a particular question, and has succeeded in that you did get an answer to a more general question that also interests you, such as "is he qualified?" Suppose that events take a completely unexpected turn and all that you learn is that the candidate has halitosis. Then, I would say, the experiment has failed *as an experiment* though it has provided information which other intellectual virtues can use. (As when you turn the switch on the accelerator and blow every circuit in Geneva, thus failing to learn anything about the Higgs but a lot about the Swiss power grid.) If an experiment has gone well then it is an accomplishment and results in knowledge. If it is conducted well then it exhibits virtues, of planning and anticipation and use of resources. Of course a well conducted experiment will often not go well. (Thinking of experiments as primarily sources of information is of course very common in the philosophy of science. For a discussion of evidence that makes a place for the results of experiment see Achinstein 2001.)

What is the relation between knowledge, accomplishment, and the virtues of experimentation? I am interested in intellectual virtues that are epistemic in the sense that they concern the conduct of inquiry but also practical in that they aim at an accomplishment, the production of specific information. One very basic such virtue is the capacity to devise the situation that will produce the information. Virtues are double-edged. In the external direction they are directed at the information-giving situations and the production of opinions from them. And in the internal direction they are directed at making use of the information, and thus at the cognitive economy of the agent, the use of her pattern of beliefs and desires and the shape of her reasoning. It is be a bad experiment if it produces loads of information which cannot be made sense of. And it is unreasonable to undertake an experiment if there is good reason to expect that instead of having the desired effect it will frustrate the experimenter's deeper aims. It is unreasonable for that agent at that moment even if it turns out to result in the perfect informative clue to the question.

Since experimentation is aimed at a result it is subject to a basic constraint of practical reason, the need to find a means to an end that accommodates other competing ends. You don't drive a Mercedes because although you would then be safe and elegant you would be hungry and indebted and harming the environment. Since experiments have costs, the experiment has to be designed so that it provides information without disrupting other projects. These other projects can themselves be epistemic, but there is not a lot to be gained by distinguishing between competition from epistemic and other aims, since there is no end of things it would be good to understand, most of which would gain from non-trivial experiments, and a very finite time for any single person to devote to them. Doing all but the simplest experiments means renouncing others. And accomplishing all but the simplest experimental or practical aim means renouncing others, of both kinds. So we might as well throw all a person's aims together into the same pan, all to be balanced against all.

One of the features of an experiment that is crucial to this balancing is the amount of light it might shed on questions the person has reason to be interested in. (Curiosity is a good reason, often.) The issue is impossibly complicated. An experiment – even a pretty trivial one as in the interview example – has many possible outcomes, and the facts any of these reveal can be inputs to many different lines of thought.

There are several kinds of linked imponderables. What will the physical outcome of the experiment be? How much relevant information will it provide? How successful will one be in exploiting the information, to refute a conjecture, formulate a new one, or adjudicate between existing hypotheses? Against these imponderables there is one manageable fact, the likely cost of the experiment. (It's certainly not a given, since the consequences of the experiment-as-act ramify into the future, but it is usually more nearly something one can get a comparative grasp on than the other questions.) The ability to handle situations of this shape, with these uncertainties deriving from these projects, is my main interest in this paper.

One important kind of experiment is a continuation or repeat of an experiment that has already been performed. After the candidate responds with only mild irritation to your question you ask him an even stupider one, in order to find his explosion point. In science it is important that experiments can be replicated, and in everyday life we sometimes fail to repeat them in part because of the familiar fact that we underestimate the importance of a person's situation on her actions. (We think that having behaved one way a few times tells us that this is her constant mode of operation.) A repeated or continued experiment produces more evidence to add to the evidence we already have, so in deciding whether to do it we have to decide whether the cost is justified given that we could instead do other experiments or throw a party to celebrate the results we already have. One particularly important case arises when the initial data in an experiment suggests that the experiment itself is doing harm. This can happen when a drug being tested is worsening the condition of subjects. Then the experiment is giving information about its own cost, and this is relevant to the question of continuing it. A pre-scientific analog is sticking your head above the parapet. If you immediately attract enemy fire you are reluctant to repeat the experiment in order to get more information about the number of enemy shooters.

So questions of cost are ubiquitous in experimentation. Not all experimenters have to face the most intractable forms of them. The budget for many scientific experiments is set in advance, or at any rate severely limited, by allocations in a department budget, a research grant, or other similar factors. So taking cost in this very narrow sense, there is often an upper bound to how much a proposed experiment can cost. Still, within a fixed budget, variant experiments are possible, and the experimenter has to decide which ones to run. That means comparing different possible experiments, and to do this one has to face the unpredictability of their results and the problems of anticipating what one will be able to make of these results.

3 Intractability

I am now in a position to describe the virtue that is the target of this paper. It is the capacity to evaluate possible experiments, in order to decide whether to do them. I do not mean simply the capacity to plan an experiment sensibly, maximizing the chances of getting desired results. I mean the externalist capacity actually to proceed when the objective situation will result in both the need for knowledge and the need for solvency (etc) being satisfied. There is obviously no such infallible capacity, and there are obviously many component skills of sensitivity to the environment and to one's own proclivities, different ones being relevant to different situations for different people. But this is a large part of what makes it a virtue and not a simple skill: its essence consists in getting a certain kind of result in a certain kind of situation.

One central consideration in the choice between experiments is that different experiments give different amounts of evidence. This can be a result of such familiar factors as sample sizes and the effort made to randomise within blocks. Generally speaking, the experiments that give more evidence cost more. Experiments that promise more significant evidence also tend to be more expensive. In one kind of experiment more varied samples are required, the randomization is more thorough, or the block structure allows protocols that might eliminate more alternative hypotheses. In another kind, more sensitive equipment is used, or it is applied to a richer variety of cases. The consequence is that one often has to decide how much and what quality evidence to try for. As a result, we do not, nor should we, always go for the most and the best. So how are these decisions made?

I do not think they can be made on the basis of a simple cost-benefit comparison. These are feasible – the project of making them makes sense – when there is a manageable variety of comparable values of specific outcomes and an intelligible probability distribution over them for every action under consideration. Under values of outcomes I am including gains of understanding and expenses of performance, and the actions in questions are ways of carrying out the experiment. In the simple ideal case there would be a series of ways of carrying out the experiment, graded in order of expense – you pay \$1 k and you get the basic experiment, you pay \$2 k and you get a more careful one, you pay \$100 k and you get a super one with many control groups and loads of randomization – and the likelihood of getting a given amount of information for a given expense could be assessed. But nothing like this is almost ever the case. There are problems of comparability and problems of prediction.

The major problems of comparability are between the costs of experimentation and the information gained. Suppose for the sake of argument that the costs can all be expressed in terms of money (though in many cases this does not seem plausible.) The benefit of an experiment is the light it throws on some uncertain question. The outcomes cannot normally be expressed in terms of units of information, as if outcome gives twice as much information as another. (Remember that in order to compare expected values we need cardinal comparisons of the values of outcomes, and not just an ordering of them.) Issues about comparability and the problem they make for cost-benefit or (equivalently) expected utility thinking are discussed in Morton (1990) and the essays in Chang (1996). Issues of incomparability have gone quiet lately, but they beg to be connected with questions about the value of knowledge raised in Kvanvig (2003).

Instead, the manageable way to think of the outcomes is as settling very simple questions, causing one to know their answers. Did the applicant lose his temper; did the subjects respond more quickly to the items they had been primed for? Then the benefits in question are the information these answers give to the questions of primary interest. Is the applicant likely to be a difficult colleague; is there an unconscious representation of some category of information, playing some given functional role? If we could measure the degrees of support that these possible simple answers give to the primary questions then we would have something to match against money. But the issue is notoriously hard and even with formalised simple hypotheses there is no consensus how to do it. The existing formal accounts of comparative strength of evidence will not apply, for example, when the hypotheses contain higher-order terms such as "there is some unknown factor which correlates phenomenon A with phenomenon B". And as noted above ordinal comparisons will not do: we would need numerical measures of evidential strength. All this is before we even try to introduce the different interests of the different hypotheses that might get the different degrees of support. Or factors other than support, such as understanding why a hypothesis might be true or how a causal mechanism might operate.

Some problems of comparability are mollified by the fact that an experiment often has a budget, with an upper limit, and we are often reluctant to leave any of it unspent. (We don't like returning any of the research grant, and we are not allowed to donate it to famine relief.) So some experiments are ruled out and a central question is simply "how can we best spend \$N?" Even then, less expensive ways of carrying out the experiment proper will have other benefits, some of them epistemic. We could do our consumer choice experiment with a large group of subjects, with payoffs in real money so that their motives are realistic, or we can save money by having a smaller sample and paying them with tokens for a lottery, and spend the rest of the grant on database software which will allow us to categorize the results of this and other studies. Comparability then re-enters the picture.

The problems of prediction are if anything greater than the problems of comparability. As noted above, the outcome of each proposed experimental avenue is open as a matter of principle: if we had much confidence how it would turn out we would have less reason to do the experiment. So it is hard to have more than the roughest assignment of probabilities of what I called the simple answers, just above, conditional on variant experimental procedures. And given these simple answers there is the problem of predicting the support they will give to ideas about the questions of interest. No doubt a competent experimentalist will have thought out the consequences of various anticipated outcomes, so that she can say that *if* one of them occurs then evidence of a given force for or against a given hypothesis will have been gained. But she will know that if one does occur then in thinking out its consequences, for example in preparing her results for publication, she will see more alternative possibilities, more complications. There is a kind of circular trap here: the more time she spends working out the likelihood that a hypothesis will have been confirmed to a given degree the less time she will have to do the same for other simple outcomes and other hypotheses, and the more indefinite her expectation of getting any particular degree of support for any hypothesis will be.

Consider a simple prediction-testing experimental situation. We know that general relativity predicts that the paths of particles will follow geodesics shaped by the presence of mass, and gives predictions about the exact paths involved. We are lucky enough to have a neutrino-measuring instrument on the moon and can measure the influence of the presence of the sun on neutrinos from a neutrino star. (This is evidently a science fictional experiment, so objections of unfeasibility or physical implausibility are to be put aside.) We can be pretty confident in advance that if the deviation of the paths of the neutrinos, for example in producing a double image, is exactly what general relativity predicts then we will have added confirmation for it, though it may be hard to assess how much. And we can be somewhat confident that if the deviation is extremely different then we will have significant disconfirmation for general relativity. Of course we would be surprised by either of these. The most likely outcome is something near to the prediction of accepted theory, with the difference ascribable to experimental error. But what will we conclude if the observed result is between these extremes? We will have to consider the possibility that we are wrong about the mass and shape of the sun, or the speed and mass of the kinds of neutrino, or the physics behind the neutrino detector. We may have made some relevant simplification in modeling the interaction of enormous and tiny objects. What will we say if two thirds of the particles are within the expected range but one third of them are weirdly deviant? Will that lead us, and others, to suspect that the theory is correct and some unknown factor is causing a random deviation, or that something physically mysterious is going on? It will obviously take a while for the physics community to digest such a result and predicting their verdict is not something you want to charge experimentalists with. (The capacities required to handle such situations are related to those discussed in Fairweather 2012.) The situation the experimentalist would prefer to be in is to be given a theory and a consequence of it that will appear in a novel situation, and a budget. Then the experimentalist doesn't question the budget but tries to produce the novel situation within its limits.

4 Experiment-Shopping

So, whether planning job interviews or testing relativity, we do not decide which experiments to run by doing a cost-benefit analysis. How we do it?

We do it by being good experiment-planners, knowing which and how much data we want to collect. There is an intellectual virtue here, a mixed epistemic-practical virtue. (For epistemic virtues see Zagzebski (1996) and Sosa (2007). My own approach is different, as suggested below.) It mixes the epistemic and the practical in that one's aims affect *how much* one knows, rather than the possibility debated in the 'pragmatic encroachment' literature (Fantl and McGrath 2010) of whether one's aims affect *whether* one knows. It is distinct from the experimentalist's virtue of ingenuity: being able to devise the setups that will force nature into the situations where unexpected things may happen. I have nothing to say about the psychology of the virtue in question, except that one place it is found is in the largely middle-aged experiment-managers who advise the ingenious ones on what they might try, approve and administer research grants, and generally shoulder the burden of deciding whether a data-producing project is worth the trouble. In saying that it is a virtue and that there are places to look for it I do not mean to claim that it is usually exhibited ideally, or even well.

The right way to approach intellectual virtues, I believe and have argued elsewhere (Chap. 2 of Morton 2012b), is in terms of their conditions for success. What situations are they applied to, and what outcomes do they aim at? The virtue we are discussing applies when there are several actions one can perform whose main benefit will be to provide evidence relevant to some questions of interest, and which have different costs one would like to minimize. The outcome it aims at has two sides, knowing the answer to the question and being satisfied with having paid what one did for it. Finally we know it well enough to name it: call this the virtue of experiment-shopping. It is a skill of buying a good enough experiment at a low enough price, of actually accomplishing these, not just worthily striving towards them or blindly fumbling in their direction. I have argued that we do not exercise this virtue by calculating and comparing costs and benefits. In fact we do not evaluate the desirability of outcomes and the likely results of courses of action independently at all. We consider whole situations, in which we or others face uncertainty about what to do in order to uncover uncertain information, and we assimilate new situations to them. At least that is what we do if the capacity here is a typical intellectual virtue of a bounded agent, and if I am right about how such virtues operate.

I have gestured at such an analysis in Morton (2004) and develop it at length in Morton (2012b). The essential elements are a database of past situations with the satisfactoriness of their solutions, and a similarity measure that can relate novel situations to stored ones. These will vary from one agent to another depending on their experience and how well they have assimilated it. Then given a new situation – a question needing information, a range of actions that might prompt it, background information – an agent can find solutions that are in a very general way like ones that have worked in the past – pushing out the boat for a grand and risky exploration, or a careful and tentative probe that might reveal whether the topic is fertile or recalcitrant. This may involve ingenuity and creative thinking, to see surprising similarities, or it may rely on rote learning of experimental paradigms in one's area of science. In either case it is likely to be very subject-specific: someone who makes the right probes when interviewing candidates may be disastrous in allocating money for DNA sequencing equipment.

Virtues understood in this way will be in a general way externalist, in that a capacity that is a virtue in one situation may not be a virtue in another, and the agent may not be able to tell one from the other. They could also be called reliabilist virtues, in that they are parts of reliable ways of getting true belief, and in fact reliable in ways that lead to knowledge, and more generally to accomplishment. For the purposes of this paper, these taxonomic issues are not important. What is important is the existence of the profitable species of thinking I have been describing, the necessity of using it throughout our activities, and the facts that it can be carried out more or less well.

The similarity of this virtue to others and my praise of my general analysis do not clinch the case. But look at the features of experimental choice that we can explain in this way. They can be gathered under three heads

- We make reasonable choices in situations whose complexity and incomparability prevent our thinking them out from first principles.
- We can train one another to make acceptable choices, even though in learning one acquires little information that one did not already have.
- We articulate many of the considerations we find relevant in threshold terms. Is the prospect of finding relevant enough to justify the expense? Does the design rule out enough alternatives? Have we collected enough evidence that we can now devote our resources to other tasks?

These have immediate explanations on the picture I am suggesting. But they become miraculous if we do not see them in terms of a specific acquired virtue. Acquiring the virtue is a central and indispensable part of any scientist's training. And acquiring the corresponding virtues in social interaction and in learning from others is essential to success in those areas. In social life one learns to probe, show emotions and provoke reactions, in ways that will lead others to reveal their emotions, intentions, and opinions. You frown when you want the other to explain more fully. In one's education one learns who to turn to for explanations, and how to do it effectively. Some people do some forms of each of these better than others, and everyone gets at least a little better at it with practice. These skills are part of a neglected but vital area of human capacity, the ability to do the right thing in order to know an interesting thing.

References

Achinstein, Peter. 2001. *The book of evidence*. New York: Oxford University Press.

Chang, Ruth. 1996. *Incommensurability, incomparability, and practical reason*. Cambridge, MA: Harvard University Press.

Fairweather, Abrol. 2012. Duhem-Quine virtue epistemology. *Synthese* 184: 1–20.

Fantl, Jeremy, and Matthew McGrath. 2010. Pragmatic encroachment. In *The Routledge companion to epistemology*, ed. D. Pritchard and S. Bernecker, 558–568. London: Routledge.

Gigerenzer, Gerd, Swijtink Zeno, and Daston Lorraine. 1990. *The empire of chance: How probability changed science and everyday life*. New York: Cambridge University Press.

Hacking, Ian. 1983. *Representing and intervening*. Cambridge: Cambridge University Press.

Harman, Gilbert. 1973. *Thought*. Princeton: Princeton University Press.

Kvanvig, Jonathan. 2003. *The value of knowledge and the pursuit of understanding*. Cambridge: Cambridge University Press.

Morton, Adam. 1990. *Disasters and Dilemmas: strategies for real-life decision-making*. Oxford, UK/Cambridge, MA: Blackwell.

Morton, Adam. 2004. Epistemic virtues, metavirtues, and computational complexity. *Nous* 38: 481–502.

Morton, Adam. 2012a. Accomplishing accomplishment. *Acta Analytica* 27(1): 1–8.

Morton, Adam. 2012b. *Bounded thinking: Intellectual virtues for limited agents*. Oxford, UK: Oxford University Press.

Radder, Hans. 1996. *In and about the world: philosophical studies of science and technology*. Albany: State University of New York Press.

Sosa, Ernest. 2007. *A virtue epistemology: Apt belief and reflective knowledge*, vol. I. Oxford: Oxford University Press.

Woodward, J. 2003. Experimentation, causal inference, and instrumental realism. In *The philosophy of scientific experimentation*, ed. H. Radder, 82–118. Pittsburgh: University of Pittsburgh Press.

Zagzebski, Linda. 1996. *Virtues of the mind*. New York: Cambridge University Press.

Is There a Place for Epistemic Virtues in Theory Choice?

Milena Ivanova

1 Introduction

The significance of the problem of underdetermination of theory by the data is a matter of dispute in the philosophy of science literature. While some believe that it is a purely philosophical problem not instantiated in the history of science, others believe that it undermines any form of realism about scientific theories. When providing a putative solution to this problem philosophers of science, independently of whether they are defending scientific realism, instrumentalism or constructive empiricism, often appeal to theory virtues. They believe theory virtues, such as simplicity and predictive power, determine why we should prefer one theory over its empirically equivalent rivals. I question this strategy and argue that theory virtues are inconclusive in cases of theory choice. I illustrate this point by discussing the case of underdetermination in quantum mechanics, where the choice of each rival can allegedly be justified because they exemplify different combinations of virtues, and consequently one's choice depends upon one's ranking of virtues and different preferences with regard to how they are measured. I furthermore look into recent discussions of Duhem's notion of 'good sense', which is supposed to offer a non rule-governed solution to theory choice where role is given to agents and their intellectual and moral virtues. I discuss the recent interpretation of this concept in terms of virtue epistemology and show its shortcomings. Furthermore, I argue that good sense is also inconclusive because we have no precise account of how it is compared in agents. I argue that the virtue epistemological interpretation does not show how good sense leads to conclusive choices and scientific progress. The problems with

M. Ivanova (✉)
Department of History and Philosophy of Science,
University of Sydney, Sydney, NSW 2006, Australia
e-mail: milena.ivanova@sydney.edu.eu

A. Fairweather (ed.), *Virtue Epistemology Naturalized: Bridges Between Virtue Epistemology and Philosophy of Science*, Synthese Library 366, DOI 10.1007/978-3-319-04672-3_13,

these strategies illustrates that they are often insufficient to provide a solution to the problem of theory choice and shift the underdetermination to another level.

The structure of this article is the following. In Sect. 2 I present the problem of underdetermination and the employment of theory virtues as its alleged solution. In Sect. 3 I argue that theory virtues are inconclusive in theory choice because they can justify a number of outcomes and do not single out a unique theory. I illustrate this point with an example from underdetermination in quantum mechanics. In Sect. 4 I present Duhem's employment of 'good sense' in theory choice. Section 5 presents David Stump's (2007) reading of good sense in terms of virtue epistemology. Section 6 highlights fundamental problems with the virtue epistemological reading of good sense. Section 7 is the conclusion.

2 Underdetermination and Theory Virtues

The problem of underdetermination is sometimes taken to originate in Duhem's holism about theory testing. According to Duhem, we test a particular hypothesis together with a set of auxiliary assumptions and initial conditions. If the experiment has negative outcome, it does not directly falsify the hypothesis we are testing; it condemns the whole body of the auxiliary assumptions together with the tested hypothesis. The experiment suggests that either the tested hypothesis or some of the auxiliary assumptions might be false but cannot guide which is the false one. As Duhem argues "the only thing the experiment teaches us is that among the propositions used to predict the phenomena and to establish whether it would be produced, there is at least one error; but where this error lies is just what it does not tell us" (Duhem 1954, 185). That means that neither deductive logic nor experience alone can dictate which is the false hypothesis or assumption. For Duhem, the holistic character of theory testing leads to the problem of underdetermination, since we can always modify the theory by changing the auxiliary assumptions and make it fit the evidence. This results, for Duhem, in underdetermination of the theory by the data. In the current literature, the underdetermination thesis is not necessarily linked to the holistic nature of confirmation. The problem of underdetermination simply states that at a given time more than one theories can equally well fit the empirical data.

We can identify several types of underdetermination. Internal underdetermination occurs within a particular theory that can be formulated using different values to a given parameter. Inter-theoretical underdetermination occurs when more than one theory is compatible with the same set of data. We can distinguish between weak and strong inter-theoretical underdetermination. Weak underdetermination is temporary and allows for future empirical evidence to discern between two rival equivalent hypotheses. Strong underdetermination occurs at the level of 'theories of everything' where no appeal to extra evidence can resolve the underdetermination. Last, metaphysical underdetermination occurs when a single theory is compatible with more than one metaphysical interpretations.

The problem of underdetermination of the theory by the data, in all its forms, challenges the scientific realist, who needs to defend why she believes that a particular theory is the approximately true theory and not its empirically equivalent rivals. It also calls for an explanation from the anti-realist as to why one theory is employed as empirically adequate rather than another. Such defense is often offered by appealing to the virtues of a theory. Theory virtues, such as simplicity, unification, fertility, explanatory power, etc., are employed in cases of underdetermination, to justify why we privilege one out of a set of empirically equivalent theories. Scientific realists believe that these virtues have epistemic significance and are evidence of the approximate truth of the theory (Psillos 1999). Constructive empiricists and conventionalists regard these virtues as mere pragmatic devices for choice, which leads to the employment of a more convenient theory (van Fraassen 1980 and Ben-Menahem 2006). However, both views assume that theory virtues lead to a conclusive choice between underdetermined theories.

Philosophers often talk about the virtues of a 'perfect theory' and believe that we can choose a unique theory from a set of empirically equivalent rivals by simply pointing to the amount of virtues the chosen theory possess. In *The virtues of a good theory* (2009), Ernan McMullin classifies the theory virtues into three categories: internal, contextual and diachronic. The 'internal' virtues include simplicity, internal consistency (lack of contradiction between the hypotheses) and internal coherence (absence of ad hoc elements). The 'contextual' virtues include consonance (the theory being consistent with other theories), optimality (the theory being the best available explanation of a set of data). Some 'diachronic' virtues mentioned by McMullin include fertility, consilience and durability. Fertility is understood as novel predictive power – the ability of the theory to predict phenomena which it was not designed to account for. Consilience, a term introduced by William Whewell (1989), means the power of the theory to unify distinct sets of phenomena into the same 'scheme'.[1] Another diachronic theory virtue, mentioned by McMullin, is that of durability which is understood as the ability of the theory to account for the phenomena over a long period of time, after the discovery of new phenomena.

Despite their disagreement on the epistemic significance of theory virtues, both realists and anti-realists (constructive empiricists, instrumentalist or conventionalists) often assume that theory virtues can determine the choice of one theory from a set of empirically equivalent rivals.[2] This presupposition is questioned in the next section where I argue that theory virtues do not determine a unique outcome of choice.

[1] Whewell takes consilience to be a criterion of confirmation of a theory because "the evidence in favour of our induction is of a much higher and more forcible character when it enables us to explain and determine cases of a *kind different* from those which were contemplated in the formation of our hypothesis. The instances in which this have occurred, indeed, impress us with a conviction that the truth of our hypothesis is certain" (Whewell 1989, p. 87–8).

[2] It must be stressed that scientific realists often employ inference to the best explanation and argue that in case of underdetermination, we should choose the best available explanation. In light of this argument, they can argue that there is only one theory virtue – explanatory power – thus they are immune to the argument for the inconclusiveness of theory virtues. However, the problem of inconclusiveness appears when one tries to account for which explanation is the best. Since theory virtues, such as simplicity, unity, fertility, are used to justify a particular theory as the best explanation, it is presupposed that they can conclusively do so.

3 Are Theory Virtues Conclusive?

The aim of this section is to illustrate the claim that appeal to theory virtues in theory choice is inconclusive. This idea is first suggested by Duhem and is further discussed by Kuhn (1977). As Duhem claims:

> No doubt the physicist will choose between these logically equivalent theories, but the motives which will dictate his choice will be considerations of elegance, simplicity, and convenience, and grounds of suitability which are essentially subjective, contingent, and variable with time, with schools, and with persons; as serious as these motives may be in certain cases, they will never be of a nature that necessitates adhering to one of the two theories and rejecting the other, for only the discovery of a fact that would be represented by one of the theories, and not by the other, would result in a forced opinion. (Duhem 1954, 288)

Kuhn further stresses the fact that theory virtues are subjective and vary with time and from one scientific community to another.

There are two problems with theory virtues that makes their employment insufficient as a solution to theory choice. First, in order to compare two (or more) theories by a given virtue, we need to provide specific criteria of how this virtue is to be measured. These criteria, even if specified, do not guarantee a unique outcome of choice because individuals can still prioritise a particular measurement criterion of this virtue. As a consequence, the comparison between the theories, even if the criterion of measurement is specified, becomes incommensurable since there is nothing in the virtues themselves to justify the uniqueness of a specific measurement criterion.

Second, genuine empirically equivalent rivals usually exemplify a number of important theory virtues. In order to make a choice, a ranking of these virtues needs to be provided. However, there is no unique ranking of theory virtues. Even if we grant that we can globally agree on a fixed list of theory virtues, we have the freedom to rank these virtues differently. The lack of justification of a unique ranking of virtues makes comparing theories by their virtues an inconclusive procedure; unless one can provide a justification of a particular ranking, a number of alternative theories can be chosen by employing the same list of criteria. This simply shifts the underdetermination from theory choice to choice between theory virtues.

The fact that each theory virtue can be understood and measured in a number of ways is evident when we consider the virtues of simplicity and fertility. Even though many regard simplicity as the primary virtue of a theory,[3] there is no consensus in the literature as to how we determine whether simplicity is exemplified by a theory and how it is to be measured. When is a theory simple? When it posits the most minimal ontology? When it relies on less free variables? When the number of equations is minimal? When the equations have only first order derivatives? And respectively, when is a theory fertile? Scientific realists often take the novel

[3] Swinburne argues that "the simplest hypothesis proposed as an explanation of phenomena is more likely to be the true one than is any other available hypothesis, that its predictions are more likely to be true than those of any other available hypothesis, and that it is an ultimate *a priori* epistemic principle that simplicity is evidence for truth". (Swinburne 1997, 1).

predictive power of a theory as the ultimate test of its truthlikeliness.[4] But how do we understand novel predictive power? Is it when a theory has actually predicted some previously unobserved phenomenon? Or do we take into account the fact that the theory could have predicted this phenomenon despite the contingent circumstances of whether such a phenomenon was empirically detected? In the current literature there are several different accounts of novelty, all disagreeing with how novelty should be understood and whether specific examples, such as the phenomenon of white spot predicted by Fresnel's theory, should be taken as novel predictions.[5]

The two problems with theory virtues seem to challenge the appeal to theory virtues as conclusive criteria for choice. Despite the fact that theory virtues can justify a choice, since we can always explain how we measured and ranked the virtues of a theory, we cannot in principle rule out other choices based on other ranking and measurement of virtues. I want to illustrate that theory virtues do not guarantee conclusive choices by discussing the case of underdetermination in quantum mechanics.[6]

3.1 Underdetermination in Quantum Mechanics

There are three main rival theories of quantum mechanics in the modern literature: the Everett, or 'many worlds' formulation; 'hidden-variables' theories (e.g. Bohmian

---

[4] For Duhem, novel predictive power is the only test of theory being a 'natural classification'. Psillos (1999) also takes the novel predictive power of a theory to be indicative of its truth.

[5] One unpopular account of novelty is 'temporal novelty', which treats any prediction of a phenomenon, which was entailed by the theory prior to the observation of that phenomenon, as novel. Many oppose this definition because it gives too much role to the time in which observations are made and seem to make the fact whether a theory indeed was fertile arbitrary. For example, the phenomenon of perihelion of Mercury was observed before the formulation of the general theory of relativity but why should this historically contingent fact determine whether the prediction was novel? In trying to avoid this objection, accounts of 'epistemic novelty' suggest that we determine a novel prediction when the scientist constructing the theory is not aware of the phenomena prior to constructing the theory. Worrall (1994) suggests that knowledge of a phenomena is not indicative of novelty and what we should focus on is whether the phenomenon was considered when the theory was constructed. He argues that the perihelion of Mercury, predicted by general relativity, was known to Einstein, but he did not construct the theory in order to account for this phenomenon. It was simply entailed by the theory. Leplin (1997), however, argues that epistemic novelty relativises the novelty to the intentions of the scientist – whether Einstein really intended to save this phenomenon or it was simply entailed by the theory. He puts forward an account of 'use novelty' according to which a phenomena is novel if it was predicted by the theory but was not used in the derivations of the theory. Ladyman and Ross (2007) defend a 'modal' account of novelty according to which what should be taken in consideration is whether a theory could have predicted some unknown phenomenon despite the historically contingent facts.

[6] The same point can be made by considering several cases of underdetermination from the history of science. For a discussion of the underdetermination between Lorentz' ether theory and Einstein's special theory of relativity, see Friedman (2001).

mechanics); and 'dynamical collapse' theories (e.g. Ghirardi-Rimini-Weber (GRW) theory). Each of these provides a distinct solution to the 'measurement problem' of quantum mechanics. Consider the x-spin of an electron. According to quantum mechanics, the x-spin state of the electron can be either spin up, spin down, or a 'superposition' of both options, such that the electron is not in a definite state with respect to the two observable states. If the electron interacts with a reliable measurement device, then according to quantum mechanics, the measurement device reads either "spin up", "spin down" or goes into a superposition of both readings. The problem concerns how to make sense of this third option, given the belief that measurements produce unique outcomes.

The Everett interpretation makes sense of this by denying that measurements do have unique outcomes. On this interpretation, it is tenable that the entire system, indeed the entire world, can be in a superposition with respect to some set of observables, and that there are branches within such a world in which there is the *appearance* of unique outcomes to experiments. Bohmian mechanics makes sense of the measurement problem by denying that standard quantum mechanics (i.e. the Schrödinger equation and the quantum mechanical wavefunction) is a *complete* description of a quantum system. Finally GRW theory offers a dynamical mechanism by which the quantum mechanical wavefunction of particles spontaneously localises, such that sufficiently large systems do not stay in superpositions, but repeatedly collapse to definite states with respect to observables.

According to GRW (Ghirardi et al. 1986), every fundamental particle at any point in time has a small probability of collapse. If the system is composed of a very few particles, the probability of collapse in some finite period of time is small but it becomes significantly great for large systems. This theory holds that collapse occurs randomly and is not caused by measurement.

Bohmian mechanics (1952) is a 'hidden variables' theory insofar as it does not take the quantum state to be a complete description of the system. This theory denies the occurrence of collapse and introduces the 'guiding force' which is responsible for the motion of all particles and their definite, well defined trajectories at all times. According to Bohmian mechanics the evolution of all quantum systems is causal and deterministic and subject to the non-local potential.

The Everett interpretation of quantum mechanics (1957) does not add new dynamics to the formalism, it does not have a collapse postulate nor does it add trajectories to the particles. The wavefunction is a complete description of the system. When an x-spin measurement is performed on the electron, two branches of the wave function evolve independently of each other, one of which is associated with the device reading 'spin up' for the electron's spin in the x direction while the other branch is associated with the device reading 'spin down'. Both branches are considered equally real and causally independent. Since we ourselves are also branch-bound, this theory explains why we always observe determinately either spin up or spin down, never a superposition of both states, and as a consequence, it provides a solution to the measurement problem.

The above three theories of quantum mechanics are regarded as empirically equivalent.[7] When choosing between these theories, a case can be made in favour and against each rival by prioritising a particular theory virtue over another or by employing different assessment criteria to the very same virtue. As a consequence, the virtues of each rival are insufficient for a conclusive choice.

Here it should be acknowledged that one can immediately object against these being alternative theories rather than different formulations of the same theory. There is no agreement amongst philosophers whether these count as separate theories or simply different mathematical formulations of quantum mechanics. The problem of giving precise conditions to identify a theory is an open and debatable philosophical problem and many believe that the problem of underdetermination is meaningless unless the problem of identifying a theory is settled (Ben-Menahem 2006). For my purposes here, I adopt the recipe provided by Cushing (1994) that a theory is just its mathematical formalism and its interpretation.

3.2 *In Search of a Unique Outcome*

How is one to choose between the different theories in quantum mechanics? The obvious response is to point to the theory virtues they possess. Other things being equal, one might argue that one of these alternative theories might exemplify more virtues or exemplify a really important virtue which should render it the overall winner. But is such a ranking of theories possible and how would one justify preference towards one of the competing rivals in quantum mechanics?

It can be argued that one of the strongest virtues of Everettian quantum mechanics is its simplicity or mathematical elegance. Preservation of the standard formalism makes this theory much more mathematically simple than its rivals since no extra equations are added to the standard formalism. Hence, proponents of Everettian quantum mechanics believe their theory is superior to formulations that introduce new mathematical structure to the standard formalism. If simplicity is to be regarded as the most important virtue, then the Everettian quantum mechanics could be considered as the higher scoring theory on our ranking list. But this is not straightforward. The problem of how we understand and measure simplicity becomes relevant when we consider the objections to this theory. Everettian quantum mechanics is attacked on the grounds of *not* being simple. That is, it is argued that it does not satisfy the criterion of ontological economy, since according to the theory the world

[7] Note that this claim is not uncontroversial. Albert (1993) holds that ordinary quantum mechanics has no empirical content, since it is a solution to the measurement problem but what counts as measurement is something quantum mechanics does not answer. Also, Bohmian mechanics is empirically equivalent to the rivals formulations of quantum mechanics only given boundary conditions.

splits into many equally real copies when a measurement is performed.[8] If ontological economy is taken as the most fundamental virtue, then it seems that Everettian quantum mechanics should not be ranked high on our list. But does this theory really postulate a superfluous ontology? Proponents of the Everettian formulation could argue that their theory does not multiply the ontology of the world beyond necessity, and thus scores higher on the virtue of simplicity. By appealing to the distinction between qualitative and quantitative parsimony, introduced by Lewis (1973, 87), we can argue that the branching ontology does not postulate a multitude of different entities but postulates multiple copies of the same ontological kind. Nolan (1997), for example, argues that "not only ought we not multiply types of entities beyond necessity, but that we should also be concerned not to multiply the entities *of each type* more than is necessary." (Nolan 1997, 330)

Using the distinction between qualitative and quantitative parsimony, proponents of Everettian quantum mechanics can argue that what the theory implies is that the world splits into equally real copies of the same kind and therefore the objection that the theory is ontologically extravagant looses its grip.[9] Barrett (1999) stresses the "trade-off" between different understandings of simplicity and argues that while the ontology of the theory might be regarded to violate Ockham's razor,[10] the linear dynamics is simple and can "be expressed in a covariant form, which is something that is typically difficult to do with the auxiliary dynamical laws that have been proposed for quantum mechanics (like the standard collapse dynamics or Bohm's law for particle motions)." (Barrett 1999, 156) But in order to resolve the problem of whether the Everettian quantum mechanics really exemplifies the virtue of simplicity and should be ranked higher than its empirically equivalent rivals, we need to have a consensus as to how we should 'measure' and understand simplicity and how we should rank it with respect to other virtues. Since we lack such a consensus, the issue of whether the Everettian quantum mechanics exemplifies the virtue of simplicity and should be ranked higher than its rivals remains unresolved.

If one wanted to defend the alternative theories, one could attack the Everettian formulation of quantum mechanics on the grounds of it being unintuitive. But this is also controversial. The theory postulates that after measurement the observer has split into many equally real copies of themselves and in each of the branches they have obtained a determinate measurement result, something we are not aware of in our experience. As DeWitt puts it "the idea of 10^{100} + slightly imperfect copies of oneself all constantly splitting into further copies, which ultimately becomes unrecognisable, is not easy to reconcile with common sense. Here is schizophrenia with

[8] This objection applies to the 'many worlds' interpretation of quantum mechanics. Another understanding of the Everettian formulation of quantum mechanics is given by the 'many minds' interpretation, according to which mental states are discontinuous and probabilistic while physical states are deterministic and causal. This interpretation presupposes a strong mind-body dualism (Albert and Loewer 1988).

[9] Note here the trade-off between ontology and ideology. Postulating a richer ontology can be preferred for the sake of a simpler theory, predictive accuracy or unification (i.e. postulating an extra planet – Neptune – contributed to Newtonian mechanics' predictive accuracy).

[10] According to Ockham's razor, theoretical entities should not be employed beyond necessity.

a vengeance." (DeWitt 1971, 161) However, as Everett himself argues "[a]rguments that the world picture presented by this theory is contradicted by experience because we are unaware of any branching process, are like the criticism of the Copernican theory that the mobility of the earth as a real physical fact is incompatible with the common sense interpretation of nature because we feel no such motion. In both cases the argument fails when it is shown that the theory itself predicts that our experience will be what it in fact is." (Everett 1957, 321)

Adherents of Bohmian mechanics can argue that their theory is superior because it provides a deterministic solution of the measurement problem. The guiding equation describes particles moving under the influence of the forces, one of which is due to the 'quantum potential', and they have continuous trajectories in space-time. Everettians can also claim that their theory is deterministic. Despite the fact that there is no unique future branch according to the multiple future 'branching' introduced by the Everettian formulation, the Schrödinder dynamics is deterministic. Dynamical collapse theories, however, offer an indeterministic solution to the measurement problem. The stochastic parameter introduced by dynamical collapse theories is indeterministic, which can make these theories appear inferior to their alternatives if determinism is to be prioritised over other virtues.

Proponents of Bohmian mechanics can further claim that their theory is continuous with classical mechanics. For example, Bohmian mechanics represents particles in a similar way to classical mechanics – as points in configuration space – and not as vectors and operators on a Hilbert space. One can nevertheless question whether Bohmian mechanics is really 'classical', since it is not equivalent to classical mechanics with an additional force. The velocities of particles according to Bohmian mechanics are constrained by the guiding equation, and are not independent of their positions. But even if one could make a strong case that Bohmian mechanics is indeed deterministic and 'classical' in some sense and its rivals are not, determinism and conceptual continuity with an established framework does not entail this theory is superior.

There is also significant disagreement when it comes to determining whether Bohmian mechanics is simple or not. Contrary to the Everett formulation, which preserves the standard formalism of quantum mechanics, Bohmian mechanics introduces the 'guiding' equation.[11] This, intuitively at least, makes its mathematical structure less elegant and simple in comparison to the Everett formulation. The Schrödinger equation is linear. The modified Hamilton-Jacobi equation figuring in Bohmian mechanics is non-linear. Also, the introduction of 'the quantum potential' makes the theory less ontologically simple. However, not everyone is willing to agree that Bohmian mechanics does not exemplify the virtue of simplicity. In his (2005) Putnam defends the simplicity of Bohmian mechanics and argues that "[t]he formula for the velocity field is extremely simple: you have the probability current in the theory anyway, and you take the velocity vector to be proportional to the current.

[11] Consideration of elegance and simplicity enter also in the derivation of the guiding equation as well. For example, in their derivation, Durr et al. (1993) choose to use only first and not higher order derivatives. Thanks to Bryan Roberts for pointing this example to me.

There is nothing particularly inelegant about that; if anything, it is remarkably elegant!" (Putnam 2005, 622) This is another example of a theory virtue being understood and evaluated differently, leaving the issue of ranking of the theory unresolved.

Another point on which there can be a significant disagreement regards whether Bohmian mechanics and dynamical collapse theories are ad hoc. Let us take the former first. It is often argued that the additional force, the 'quantum potential', is introduced in a rather ad hoc way and is regarded as 'unnecessary', 'redundant' and 'unnatural'.[12] But proponents of Bohmian mechanics disagree that the guiding equation is introduced in an ad hoc manner (Cushing 1994). They argue that Bohmian mechanics was formulated prior to the Copenhagen (standard) quantum mechanics. What is regarded as an 'additional mathematical structure' added to the standard formulation of quantum mechanics has actually been, according to this argument, part of the original theory, developed prior to what is now regarded as 'the standard formulation of quantum mechanics'. In his (1994), Cushing provides an argument along these lines and draws our attention to the historically contingent factors responsible for regarding the Copenhagen formulation as prior to Bohmian mechanics. He presents a counterfactual history in which the objection against Bohmian mechanics being an ad hoc modification of the standard formalism no longer holds.

Dynamical collapse theories are attacked on the grounds of not being simple, given the introduction of the collapse postulate to the theory, and are also regarded as ad hoc and physically unmotivated.[13,14]

When it comes to the virtue of unification, it is again difficult to determine whether any of the rival theories exemplify this virtue and, if they do, how much better in comparison to their alternatives. Everettians point out that their theory exemplifies the virtue of unification, while Bohmian mechanics and dynamical collapse theories do not. Since the latter are non-local, that is, they allow for action at a distance, they are not obviously consistent with relativity. This apparent inconsistency with the special theory of relativity makes it difficult to see how Bohmian mechanics and collapse theories of quantum mechanics can be unified in a relativistic quantum field theory. There has nevertheless been attempts at showing how hidden variable theories can be made compatible with the general theory of relativity (Struyve and Westman 2006) as well as how dynamical collapse theories can be made relativistic (Tumulka 2006). However, there is no consensus on this issue, making the evaluation of these theories with respect to the virtue of unification inconclusive. Moreover, it is not clear whether any of the current competing theories,

[12] This objection is discussed in detail in Barrett (1999).

[13] Ladyman and Ross (2007) argue that it is an open question whether collapse is a genuine physical process and this question is going to be answered by future physics.

[14] Both GRW and Bohmian mechanics could be confirmed, in principle, by further evidence. Both dynamical collapse, introduced by GRW, and the additional equation of motion for the particles' trajectories, introduced by Bohmian mechanics, could eventually produce new empirical consequences which could serve as a confirmation boost. However, this fact by no means resolves the issue of underdetermination, it simply shows that these theories are not strongly underdetermined.

apart from the Everettian quantum mechanics, will be compatible with relativistic quantum field theory.

The case of underdetermination in quantum mechanics is interesting not only because it is a current philosophical problem but also because it exemplifies the general point that, because there is freedom in the ranking and measuring of virtues, theory virtues are insufficient to guarantee a unique choice.

4 Employing 'Good Sense' in Theory Choice

The inconclusiveness of theory virtues implies the involvement of judgement in theory choice. The outcome of choice depends on the ranking preferences and particular measurement of theory virtues that can differ significantly in individuals or scientific communities. The employment of judgement has driven attention to the role played by agents in choosing among empirically equivalent theories. Pierre Duhem (1954) famously employs the concept of 'good sense' to stress the role of the scientist's intuition when faced with the problem of underdetermination. Duhem suggests that good sense is the judge when experience cannot lead to a decisive choice between theories. The aim of this section is to provide a brief description of the properties and function of 'good sense'.[15]

The starting point for Duhem is the holistic nature of confirmation which allows, in light of negative outcome of an experiment, to modify either the auxiliary assumptions, used to derive the predictions and perform the experiment, or the testing hypothesis itself. The scientists are faced with two paths: a 'timid' path, which calls for modifications of the auxiliary assumptions to accommodate the new evidence, or a 'bold' one, which forces the formulation of a new hypothesis. The holistic nature of theory testing, for Duhem, always results in underdetermined theories. Which path the scientist should follow is not dictated by some "absolute principle". When neither logic nor experiment can guide a scientist to a decision, they are guided by certain 'reasons' or 'considerations' that Duhem calls 'good sense'.[16] One of the crucial features of good sense is that it cannot be reduced to an algorithm:

> [T]he rules of syllogistic logic are not adequate. They must be assisted by a certain sense of soundness that is one of the forms of good sense [...] good sense will intervene at the moment at which one realizes that the consequences of a preconceived idea are either contradicted or confirmed by the experiment. (Duhem 1991, 23)

[...]

> What a delicate task, concerning which no precise rule can guide the mind! It is essentially a matter of insight and ingenuity! Truly, in order to perform this well, it is necessary that good sense should transcend itself, that is, push its strengths and its suppleness to their very limits, that it become what Pascal called the intuitive mind. (ibid., 24–25)

---

[15] For a systematic presentation of the properties of good sense see Ivanova and Paternotte (2013).

[16] For Duhem 'good sense' is greatly captured by Pascal's claim that 'the heart has reasons which reasons does not know'.

In situations of underdetermination no strict rules can be employed: "no absolute principle dictates this inquiry, which different physicists may conduct in very different ways without having the right to accuse one another of illogicality." (Duhem 1954, 216)

Duhem often describes good sense as certain 'reasons' or 'intuitions' that guide scientists when faced with theory choice.[17] The notion of good sense is far from sharp and Duhem devoted very few pages to it. However, despite the difficulty in describing exactly what good sense amounts to, for Duhem it is this quality of agents that enables them to make decisions in absence of empirical (and deductive) means. Duhem believes there must be an explanation of the fact that underdetermination tends to be resolved prior to the availability of new evidence so he believes there must be such a faculty of good sense in scientists that enables them to make the correct judgments.[18] He is also convinced that exactly because scientists have good sense, which leads them to make the correct choices, they should aim to cultivate it and develop it further. In that sense, good sense is not only a descriptive but also a normative concept. Duhem argues that cultivating good sense can lead to the acceleration of scientific progress: "physicists may hasten this judgment and increase the rapidity of scientific progress by trying consciously to make good sense within themselves more lucid and more vigilant." (Duhem 1954, 218).[19]

Apart from being irreducible to an algorithm and accelerating scientific progress, good sense is correlated with the employment of theory virtues. Duhem believes theory virtues figure in theory choice but do not always resolve it. They are necessary but not sufficient because they are "subjective, contingent, and variable with time, with schools, and with persons." (Duhem 1954, 288) As a consequence, part of the judgement will be based on the theory virtues.

[17] As noted in Ivanova and Paternotte (2013), experimental science is not the only domain of application of good sense. In his (1991) Duhem argues that good sense also figures in mathematics and history. However, the properties of good sense as well as its role are different depending on the context in which it is employed. Good sense in mathematics is equated with common sense and is the ability to 'feel' self-evident mathematical truths and anticipate the results of mathematical deductions. (1991, 6–11) In history, good sense is necessary for the acquisition of truth and its main characteristic is that of impartiality, "detachment from all interests and all passions." (1991, 42–44)

[18] There are plenty of examples in the history of science where an individual scientist or a research group managed to choose the theory, which would eventually become a fruitful research programme and lead to scientific progress. For example, we would like to be able to claim that in the dispute between atomists and energeticists in the beginning of the twentieth century, atomists had good sense while defenders of energetics (amongst which was Duhem himself) lacked it. Moreover, in the debate between Lorentz' ether theory and the special theory of relativity, we would like to be able to claim that Einstein had good sense to choose the latter, which let to the general theory of relativity. Last, in the debate over the completeness of quantum mechanics between Bohr and Einstein, we would like to claim that Einstein lacked good sense since he promoted the search of a hidden variables theory and claimed that quantum mechanics, which has been a highly successful research programme, was incomplete.

[19] The importance of the acceleration property for Duhem's notion of good sense is first discussed in Ivanova and Paternotte (2013).

When trying to articulate what good sense is, Duhem also draws our attention to the role of moral virtues in scientific judgment. He states that "in order to estimate correctly the agreement of a physical theory with the facts, it is not enough to be a good mathematician and skillful experimenter; one must also be an impartial and faithful judge." (Duhem 1991, 218) He condemns personal interest, stresses the importance of impartiality in science and suggests that developing one's impartiality would lead to scientific progress "nothing will delay the decision which should determine a fortunate reform in a physical theory more than the vanity which makes a physicist too indulgent towards his own system and too severe towards the system of another." (Duhem 1954, 218)

Another property of good sense is that it is not equally instantiated in all scientists. Some scientists have cultivated their good sense more and thus can make better judgments than others.

> [T]hese reasons of good sense do not impose themselves with the same implacable rigour that the prescriptions of logic do. There is something vague and uncertain about them; they do not reveal themselves at the same time with the same degree of clarity to all minds. (Duhem 1954, 217)

When in disagreement, two scientists can both insist they have good sense but Duhem says there is nothing that can resolve their disagreement. This is because there is no way for good sense to be evaluated and measured in situations of underdetermination. We can decisively claim that one of the scientists has good sense only retrospectively, after new evidence has become available and was accommodated by one of the theories:

> [T]he day arrives when good sense comes out so clearly in favour of one of the sides that the other gives up the struggle even thought pure logic would not forbid its continuation. (ibid.)

However, this kind of response is unconvincing since Duhem does not believe in crucial experiments that decisively confirm one theory and not another. Given his claim that any evidence can be always made to fit the data and his rejection of crucial experiments, how can we explain why Duhem appeals to new evidence as a decisive condition of good sense? Duhem discusses the example of Biot, who abandoned the emission hypothesis after Foucault's experiments showed that light travels faster in air than in water. He is careful to claim that this was not a crucial experiment that supported wave optics, however, Duhem argues, it would have been lack of good sense if Biot continued to resist wave optics. But we are left in the dark as to what exactly resolves theory choice – is it good sense or new evidence? What made Biot realise he would have lacked good sense were he to try and accommodate another experimental result in his theory? We could understand Duhem's claim by having a broader notion of confirmational support that considers which theory was potentially a successful research project.

Despite its 'vague' nature, we have managed to identify the following characteristics of Duhemian good sense in the context of theory choice, mentioned in both his (1954) and his (1991). Good sense cannot be reduced to an algorithm; it enables scientists to make decisions when the evidence is insufficient; it is not equally instantiated in all agents; scientists should cultivate it in order to fasten scientific progress;

it is cultivated with experience; it takes into consideration theory virtues; it always resolves the underdetermination; it is connected to the virtue of impartiality.

The most problematic thing about good sense is how it is to be identified in agents. Duhem dedicated very few pages on good sense and the question of how we can identify good sense is left unexplained despite it being the central problem with good sense raised by Duhem. But if good sense is to be of any help to resolutions of theory choice, as he seems to aspire, we need to understand how good sense is identified and compared. The fact Duhem claims that good sense leads to theories that are later confirmed by further evidence, makes good sense appear reliable, since it always leads to more confirmed theories.[20] But this claim is trivial since good sense is only judged post hoc. Prior to the availability of further evidence Duhem does not admit that there can be any way of distinguishing which scientist has good sense.[21] I discuss this problem in Sect. 6, but before that, let us examine what good sense shares with virtue epistemology.

5 Good Sense and Virtue Epistemology

As mentioned in the previous section, properties of good sense include moral qualities of scientists, in particular, impartiality. It is this emphasis on moral virtues that makes Stump (2007) see an important link between current virtue epistemology and the concept of good sense and argues seeing good sense in light of virtue epistemology can be useful in providing a coherent reconstruction of good sense. Stump sees a strong analogy between Duhem's emphasis on the virtues of scientists and the role of intellectual and moral virtues of agents in current virtue epistemology, where knowledge depends on the intellectual and moral virtues of the agent.[22,23] He suggests that "there

[20] Note that for Duhem good sense cannot lead to true theories, because a true theory would reveal not only the real relations between appearances, but also the nature of the unobservable reality. His skepticism of science discovering the nature of the unobservable reality results in his defence of structural realism, where knowledge only of the unobservable relations is allowed (Duhem's structuralism is discussed in Worrall (1989)).

[21] I develop the objection that good sense is judged post hoc in Ivanova (2010).

[22] Stump appeals in particular to the work of Linda Zagzebski (2003).

[23] In my Ivanova (2010) I raise two concerns with Stump's interpretation of Duhem as virtue epistemologist. First, Duhem has a different epistemic aim from virtue epistemologists. Given his structural realist commitments and the pessimism expressed towards our ever being acquainted with the nature of the unobservable reality, the concept of 'good sense' is not employed in the same sense as within the virtue epistemologist account, according to which the virtues of the agents ultimately justify their true beliefs. Duhem does not believe a true theory is possible and good sense can at best lead to theories that are 'natural classifications'. Second, while virtue epistemology is concerned with the virtues of the agent in order to explain knowledge, Duhem is not concerned with justifying scientific knowledge when employing 'good sense'. He is simply describing a situation of theoretical underdetermination, where two scientists can disagree which theory should be employed and does not argue that good sense has any role to play in the construction of a scientific theory. (see Ivanova (2010, 2011))

is an overlap between Duhem and the virtue epistemologists—making values primary and rejecting rule-based decision procedures" (Stump 2011, 16) and argues that this concept can provide important insight for solutions to the problem of theory choice. Since no algorithmic rules can be offered as a solution to theory choice, the weight of decision is on the agent. The virtuous agents can choose the right theory because they exemplify important intellectual and moral virtues.

Stump suggests that Duhem's argument for good sense illustrates the importance of intellectual and moral virtues of agents for the acquisition of knowledge and promotes a non rule-governed approach to theory choice. As he suggests: "[i]n Duhem's account of scientific theory choice, there is openness, since strict rules do not apply, but also objectivity. The source of this objectivity is the epistemic agent—the scientist who acts as an impartial judge and makes a final decision" (Stump 2007, 155).

Stump's reading of good sense has attracted attention to Duhem's concept and its relation to virtue epistemology. In Ivanova (2010), I offer a possible interpretation of good sense, while Abrol Fairweather (2011) has supported Stump's virtue theoretic reading by providing a hybrid account of good sense.[24] According to him good sense is developed in non-underdetermination situations with scientific practice, which sharpens scientists intellectual character and employed in underdetermination situations. In a recent article (Ivanova and Paternotte 2013) we have offered another possible reconstruction of Duhem's concept and have outlined the recent literature as well as the shortcomings of each account. There is no doubt that there are similarities between good sense and virtue epistemology and it is important that the moral dimension of good sense has been brought to our attention. But how deep are these similarities and what can we learn about good sense from them?

6 Analyzing the Analogy

An important characteristic of virtue epistemology is that it reverses the traditional order of analysis. Usually we focus on the beliefs and not the agents; virtue epistemology focuses on the agents possessing the beliefs and analyses their character in terms of which they achieve epistemic success. It is the epistemic virtues of agents (moral or intellectual) that have normative weight. Epistemic virtues of agents are valued not for the products they achieve but in themselves. As we argue in Ivanova and Paternotte (2013), good sense departs from this dimension of virtue epistemology since it is not valued in itself but only for its products – theories that are chosen

[24] This hybrid reading takes elements from both Stump's and my accounts – that good sense can choose a unique theory (Stump) and that good sense is only judged post hoc and as a consequence cannot give epistemic significance to the theory it has chosen, only the availability of new evidence that confirms it can do so (Ivanova). According to Fairweather, good sense can confer uniqueness in situations of underdetermination but cannot provide epistemic standing, such standing is provided by new supporting evidence. The shortcomings of this reconstruction are discussed in Ivanova and Paternotte (2013).

in situations of underdetermination and are later evidentially supported and incorporated into what Duhem calls a 'natural classification'.[25]

An interesting question that needs to be addressed is how the properties of good sense are interrelated and whether Duhem attributes any importance or priority to some of them. Here Stump's reading can be suggestive; on his reading moral qualities would bear some priority over other characteristics of good sense. We saw that Duhem takes impartiality to be part of good sense, and believes that judgment results from the exercise of some 'moral conditions'. The question is whether moral virtues are prior to other epistemic virtues and any further considerations that constitute good sense. This is of crucial importance when it comes to comparing properties of good sense in agents. Virtue epistemologists are in disagreement as to whether it is moral virtues (character traits) that are responsible for truth acquisition, or it is the intellectual virtues, or it is both. If one adopts Stump's reading, moral qualities would be regarded prior to other characteristics of good sense. Even though this reading gives us an answer to this question about good sense, it is not clear that Duhem actually addresses this issue of priority. We have seen that judgment in theory choice is a product of several considerations, moral virtues being one of them. On several occasions Duhem states that impartiality is important in theory choice. He appeals to Claude Bernard to state that "the sound experimental criticism of a hypothesis is subordinated to certain moral conditions; in order to estimate correctly the agreement of a physical theory with the facts, it is not enough to be a good mathematician and skillful experimenter; one must also be an impartial and faithful judge". (Duhem 1954, 218) It is this claim that has driven Stump (2007, 2011) to argue that good sense is equated with moral qualities (that moral qualities have some priority). However, it is not clear whether moral qualities receive any priority over other virtues and considerations in theory choice. Duhem states that personal interests and passions hinder good sense (Duhem 1954, 217). He also argues that scientific progress can be accelerated if scientists cultivate their good sense by omitting their interests and passions. But he does not explicitly argue that these 'reasons' of good sense that guide scientists' choices boil down to moral qualities or are driven exclusively by moral qualities. Duhem is explicit that there are other 'considerations' that figure in theory choice apart from moral qualities, such as theory virtues and intellectual virtues, which suggests that he did not take moral qualities to be primary.

This takes us to another crucial issue in understanding Duhemian good sense. As mentioned earlier, according to Stump good sense amounts to the exemplification of moral qualities and there is textual evidence to support this claim. However, there is textual evidence that undermines the claim that choice from good sense is driven by

[25] For Duhem "the aim of physical theory is to become a natural classification, to establish among diverse experimental laws a logical coordination serving as a sort of image and reflection of the true order according to which the realities escaping us are organised" (Duhem 1954, 31). This 'true order' is not fully epistemically accessible. Duhem believes that we cannot know the nature of the unobservable entities our theories postulate, but our natural classifications capture increasingly better their structure.

moral qualities. Recall the above discussion of how we judge who has good sense. Duhem claims that because good sense varies in scientists, they can rarely end up defending the same theory and thus in situations of underdetermination it is always possible to face disagreement. How do we decide who has good sense? Duhem says that "only the discovery of a fact that would be represented by one of the theories, and not by the other, would result in a forced opinion" (ibid., 288). If good sense is judged by its products, then good sense should amount to whatever 'reasons' or 'considerations' led a scientist to choose a theory that turned out to be a successful research programme. This immediately presents a problem for the virtue epistemological reading of good sense not only because good sense is judged by its products but also because judgment need not be driven by moral virtue but could be driven by vice.

There is nothing in Duhem's claims about good sense that implies that good sense amounts only to the exemplification of moral virtues and absence of vice. The appeal to impartiality Duhem proposes should not be taken as a universal rule; good sense is not reducible to rules and impartiality is not the only consideration that figures in scientific judgment. The virtue epistemological reading of good sense implies that moral (and/or intellectual) qualities are all that matters in theory choice. But it is difficult to understand how theory choice can be described uniquely by the employment of moral qualities of agents because more considerations are involved. One can be impartial but fail to exemplify other essential qualities to make the right decision. But more importantly, good sense seems perfectly compatible with the omission of virtues and the instantiation of epistemic or moral vice. Recall Millikan's famous oil drop experiments performed to measure the charge of the electron. Millikan did not instantiate the moral qualities the virtue epistemologists are after, that is, he was not intellectually honest and impartial, when he erased the results from the second set of experiments he performed in his calculations. Instead, he took into consideration only the results of the first set of experiments in his calculation of the charge of the electron. Were he to stick to intellectual honesty and impartiality, he would have produced results much further from the ones he indeed produced and which we now regard as highly accurate. His bias guided him to ignore the significance of the newly acquainted data and thus to epistemic success.[26] Interests and passions are vices which Duhem advises us to avoid. However, avoiding passions and interests should not be seen as a rule that should be followed unconditionally. Straying away from this 'recipe' is not incompatible with good sense. Insofar as bias figures in theory choice and leads to epistemic success, it is part of good sense. This implies that good sense is not only driven by virtues but also vices.

[26] One can suggest that epistemic success is not necessary for good sense. An internalist reading of good sense would focus on the internal coherence of the agent's beliefs, their attitude towards new evidence, etc. so that the agent's beliefs are rendered rational despite of whether the agent achieved epistemic success (in unfavourable conditions agents can be rational but fail to achieve epistemic success). However, I am sceptical that such an internalist reading captures Duhem's understanding of good sense, since it is evaluated in terms of its performance and thus is an externalist notion.

Here, one could argue that what appears to be a vice is actually a virtue. A lot of recent philosophical work points at the positive effects bias and partiality can have on scientific progress.[27] Private motives, bias and dogmatism are important for the success of science because they promote diversity in the scientific community and the pursuit of different research programmes. This shows that good sense has a much broader scope than the virtue epistemological reading allows and that virtues can be understood in multiple and conflicting ways.

Last, the main question that concerns Duhem regards whether good sense can lead to consensus in theory choice. He turns to the problem of how we compare good sense in agents, how we establish who has good sense and who does not. But this is exactly where we stumble into difficulties. According to Duhem, good sense is judged only retrospectively, and its choice is justified only after a theory has received further empirical support.[28] Good sense includes moral virtues but could also allow for moral or intellectual vice. The worry is that we cannot choose the scientist with good sense because we have no unique and precise conditions for this choice. Moral virtues are problematic because: (1) good sense is not exhausted by them; (2) they can be omitted for other considerations; (3) they are not clearly defined (as mentioned above, what counts as an exemplification of a virtue could also be seen as a vice depending on the context). This makes it difficult to establish how good sense can be compared in agents and thus how it can lead to a resolution of theory choice. There are a number of non-exhaustive considerations mentioned by Duhem that constitute good sense and a multiple ways of understanding them. Even if the notion of good sense is to be sharpened and such conditions were made precise, good sense still faces the objection of being inconclusive. This is because depending on a particular understanding of which considerations are more important (e.g. the instantiation of intellectual virtues might be prioritised to moral virtues, or a specific moral virtue can be prioritised over another moral virtues, etc.) and on how we choose to understand a particular virtue, we could argue that conflicting agents both exemplify good sense.[29]

The above arguments show that despite the fact that the virtue epistemological reading of good sense has brought to our attention that good sense is partly governed by moral qualities, this reading leaves unresolved important questions about good sense that Duhem also left unexplained. We have seen that good sense does not promote the reverse analysis advocated by virtue epistemologists, it is not exhausted by moral (or even epistemic) virtues, and since it can only be judged post hoc, it could allow for moral or epistemic vice. Apart from these difficulties, it remains unclear how, even if good sense were to be understood in terms of moral

[27] See in particular Hull (1988), Kitcher (1993) and Strevens (2003).

[28] As noted in Sect. 4, this support need not necessarily be provided by further confirmation in light of new evidence which is accommodated into one of the theories. We can have a broader notion of confirmational support that evaluates which theory was potentially a successful research project. This is in accordance with Duhem's notion of 'natural classification' which unifies distant set of theories into the same mathematical framework.

[29] Note also the problem of the temporal dimension of good sense. As noted in footnote 19, we could say that Einstein had good sense when he defended special relativity but lacked good sense when he defended the incompleteness of quantum mechanics.

qualities of agents, it would lead to the resolution of theory choice and the acceleration of scientific progress.

7 Conclusion

I have considered how theory virtues can be appealed to in cases of underdetermination. Theory virtues can justify a number of rivals depending on one's ranking preferences of virtues and measurement criteria. As a consequence, theory virtues lead to inconclusiveness in theory choice and shift the problem of underdetermination to another level. The notion of good sense, which takes theory virtues to be only one of a number of consideration that need to be taken into account in theory choice and regards the scientist's intuition as part of the decision making process, still remains puzzling. The virtue epistemological reading of good sense has not resolved problems with this notion that Duhem himself left unresolved. In particular, the virtue epistemological reading does not explain how we judge which agent exemplifies good sense. Moreover, this reading does not explain why moral virtues should be taken to have more epistemic importance in comparison to other considerations that comprise good sense, why moral qualities would lead to the resolution of theory choice and the acceleration of science and how this reading fits with Duhem's overall epistemology of science.

Acknowledgments I would like to thank Abrol Fairweather for inviting me to be part of this volume. I am grateful to Matt Farr, James Ladyman, David Stump and Bryan Roberts for their helpful comments on an earlier draft of this paper. This work was funded by the British Society for Philosophy of Science and The Royal Institute of Philosophy while the author was at the University of Bristol.

References

Albert, D. 1993. *Quantum mechanics and experience*. Cambridge, MA: Harvard University Press.

Albert, D., and B. Loewer. 1988. Interpreting the many worlds interpretation. *Synthese* 77: 195–213.

Barrett, J. 1999. *The quantum mechanics of minds and worlds*. Oxford: Oxford University Press.

Ben-Menahem, Y. 2006. *Conventionalism: From Poincaré to Quine*. Cambridge: Cambridge University Press.

Bohm, D. 1952. A suggested interpretation of the quantum theory in terms of "hidden" variables, I and II. *Physical Review* 85: 166–193.

Cushing, J. 1994. *Quantum mechanics: Historical contingency and the Copenhagen hegemony*. Chicago: The University of Chicago Press.

DeWitt, B. 1971. The many-universes interpretation of quantum mechanics. In *Foundations of quantum mechanics*, ed. B. d'Espagnat. New York: Academic Press. Reprint in *The many worlds of quantum mechanics*, ed. B. DeWitt and N. Graham. Princeton University Press (1973).

Duhem, P. 1954[1906]. *The aim and structure of physical theory*. Princeton: Princeton University Press.

Duhem, P. 1991[1915]. *German science: Some reflections on German science: German science and German virtues*. Trans. John Lyon. La Salle: Open Court.

Durr, D., S. Goldstein, and N. Zanghi. 1993. A global equilibrium as the foundation for quantum randomness. *Foundations of Physics* 23: 721–738.
Everett, H. 1957. Relative state formulation of quantum mechanics. *Reviews of Modern Physics* 29: 454–462.
Fairweather, A. 2011. The epistemic value of good sense. *Studies in the History and Philosophy of Science Part A* 43(1): 139–146.
Friedman, M. 2001. *Dynamics of reason*. Stanford: CSLI Publications.
Ghirardi, G.C., A. Rimini, and T. Weber. 1986. Unified dynamics for microscopic and macroscopic systems. *Physical Review* 34: 470–491.
Hull, D.L. 1988. *Science as a process: An evolutionary account of the social and conceptual development of science*. Chicago: University of Chicago Press.
Ivanova, M. 2010. Pierre Duhem's good sense as a guide to theory choice. *Studies in History and Philosophy of Science* 41: 58–64.
Ivanova, M. 2011. 'Good Sense' in context: A response to Kidd. *Studies in History and Philosophy of Science* 42: 610–612.
Ivanova, M., and C. Paternotte. 2013. Theory choice, good sense and social consensus. *Erkenntnis* 78(5): 1109–1132.
Kitcher, P. 1993. *The advancement of science: Science without legend, objectivity without illusions*. Oxford: Oxford University Press.
Kuhn, T. 1977. Objectivity, value judgment, and theory choice. In *The essential tension*, ed. T. Kuhn, 320–353. Chicago: The University of Chicago Press.
Ladyman, J., and D. Ross (with Spurrett, D., and J. Collier). 2007. *Every thing must go: Metaphysics naturalised.* Oxford: Oxford University Press.
Leplin, J. 1997. *A novel defense of scientific realism*. Oxford: Oxford University Press.
Lewis, D. 1973. *Counterfactuals*. Oxford: Basil Blackwell.
McMullin, E. 2009. The virtue of a perfect theory. In *The Routledge companion to philosophy of science*, ed. Martin Curd and Stathis Psillos. London: Routledge.
Nolan, D. 1997. Quantitative parsimony. *British Journal for the Philosophy of Science* 48: 329–343.
Psillos, S. 1999. *Scientific realism – How science tracks truth*. London: Routledge.
Putnam, H. 2005. A philosopher looks at quantum mechanics (again). *British Journal for the Philosophy of Science* 56: 615–634.
Strevens, M. 2003. The role of the priority rule in science. *The Journal of Philosophy* 100(2): 55–79.
Struyve, W., and H. Westman. 2006. A new Pilot-Wave model for quantum field theory. In *Quantum mechanics: Are there quantum jumps? And on the present status of quantum mechanics, AIP conference proceedings, 844*, eds. A. Bassi, D. Dürr, T. Weber, and N. Zanghì, 321–339. New York: American Institute of Physics.
Stump, D. 2007. Pierre Duhem's virtue epistemology. *Studies in History and Philosophy of Science* 38: 149–159.
Stump, D. 2011. The scientist as impartial judge: Moral values in Duhem's philosophy of science. New perspectives on Pierre Duhem's *The aim and structure of physical theory* (book symposium), *Metascience*, vol. 20, 1–25. London: Routledge.
Swinburne, R. 1997. *Simplicity as evidence of truth*. Milwaukee: Marquette University Press.
Tumulka, R. 2006. A relativistic version of the Ghirardi–Rimini–Weber model. *Journal of Statistical Physics* 125: 821–840.
Van Fraassen, B.C. 1980. *The scientific image*. Oxford: Oxford University Press.
Whewell, W. 1989. *Theory of scientific method*. Indianapolis: Hackett.
Worrall, J. 1989. Structural realism: The best of both worlds? *Dialectica* 43(1–2): 99–124.
Worrall, J. 1994. How to remain (reasonably) optimistic: Scientific realism and the "Luminiferous Ether". In *PSA 1994*, vol. 1, ed. M. Forbes and D. Hull. East Lansing: Philosophy of Science Association.
Zagzebski, L. 2003. The search for the source of epistemic good. In *Moral and epistemic virtues*, ed. M. Brady and D. Pritchard, 13–27. Malden: Blackwell.

Bridging a Fault Line: On Underdetermination and the *Ampliative Adequacy* of Competing Theories

Guy Axtell

> *Discussion of theory virtues exposes a fault-line in philosophy of science that [separates] very different visions of what the natural sciences are all about.*
>
> – *Ernan McMullin*[1]

1 Introduction

Logical empiricists posited a singular logic of inquiry across all sciences, and pied-pipered social scientists to follow them by emulating the methods of the "hard" sciences. Post-positivist thinkers denounced the pied-piper, but themselves often imposed an equally rigid egalitarianism among academic fields and cognitive styles as sources of knowledge.[2] More careful attention to problems related to underdetermination, however, indicates a relationship between the sciences at once both less neat but far more intriguing than either of the foregoing views: A more dappled relationship where some bits of history or sociology may be more reliable than some bits of physics, and where worries about underdetermination are felt as regularly in such areas of natural science as contemporary theoretical physics and cosmology, as in the social sciences like economics where we usually locate them.

[1] "The Virtues of a Good Theory" (2009), 506.

[2] Connected with this is Sankey's point that "while empiricists explain consensus but have a hard time with disagreement, post-empiricists emphasize dissensus at the cost of being unable to explain how agreement is arrived at. But [any] adequate philosophical model of scientific rationality must explain both consensus-formation and the existence of widespread disagreement" (1996, 1).

G. Axtell (✉)
Philosophy and Religious Studies, Radford University, Radford, USA
e-mail: gsaxtell@radford.edu

A. Fairweather (ed.), *Virtue Epistemology Naturalized: Bridges Between Virtue Epistemology and Philosophy of Science*, Synthese Library 366, DOI 10.1007/978-3-319-04672-3_14,

Attention to the underdetermination problem in the sciences is also one among a number of bridges between philosophies of science and virtue epistemologies (hereafter VE), or so I will argue. As Ernan McMullin writes in "The Virtues of a Good Theory" (2009), "The assessment of theory is a form of inference quite different from induction over a set of observation reports resulting in a law-like generalization" (501). The verificationist conception of objectivity he points out moved under criticism (around mid-century) to a fall-back position closer to the hypothetico-deductive account; yet the assessment of theory as the critics of positivism pointed out is more often a comparison of extant rivals, and less often a sheer encounter between a stand-alone theory and an experimental test, as in the Popperian notions of a "crucial experiment" implying a quick kill of a theory that faces recalcitrant evidence as a result of disappointing test data.

The deeper reasons for these criticisms of the received view of theory choice involve a long discussion of methodological holism in response to recognition of certain kinds of worries about the underdetermination of theory by data, and again about the underdetermination of theory choice by methodological rules and standards. The former kind, sometimes called *logical underdetermination* or *Humean underdetermination*, I think of as a 'global' but at the same time a rather weak thesis that no theory is strictly-speaking proven or entailed by its confirming instances/predictions. The latter is a much stronger thesis, but I will argue as the same time a *localized* concern or problem rather than a global one. This is how Larry Laudan formalizes the first ("Humean") kind in "Underdetermination Demystified" (1990, 323):

> (HUD) For any finite body of evidence, there are indefinitely many mutually contrary theories, each of which logically entails that evidence.

Famously, "Quinean holism" as presented by the early Quine was tied into his support of much stronger claims than (HUD) about the underdetermination of theory by data; but he later repudiated his early views, saying that his statements about holism were both stronger than was needed to challenge the dogmas of empiricism, and stronger than he wished he would have made.[3]

The so-called Duhem-Quine Thesis is now widely regarded as a mistaken designation, since Duhem's views were considerably milder than Quine's; but for present purposes we needn't go into these matters.[4] (HUD) already shows us a quite substantial sense in which theory-choice turns upon values at work in science, which is to say, upon non-deductive or ampliative desiderata, which primarily include the virtues of a good theories (theory virtues). Sometimes in

---

[3] As Laudan points out, in "Two Dogmas of Empiricism" Quine "propounded [but did not give any good reasons for believing] a thesis of normative, ampliative, egalitarian underdetermination" (334). Laudan defines and then argues against Quinean underdetermination:

> (QUD) Any theory can be reconciled with any recalcitrant evidence by making suitable adjustments in our other assumptions about nature.

[4] For an admirably clear explanation of the differences between Duhem and Quine, and the Quinean history of retracting (QUD), see Massey (2011).

the literature these are called cognitive values. McMullin writes that "Calling them 'virtues' rather than 'values' draws attention to their status as attributes at once objective and desirable".[5] As an aside to be developed later, virtue theorists of all kinds present a Janus-faced (or compatibilist) conception of the relationship between naturalism and normativity. If reasoning about the *ampliative adequacy* of theories works through thick concepts, as I think it plainly does, this arguably strongly supports such a compatibilist view, whether one prefers to characterize these concepts as cognitive values or theory virtues.[6]

The second kind of underdetermination problem that should especially concern us is what we'll call *ampliative underdetermination,* or the underdetermination of theory *choice* by methodological standards. It is a worry that ampliative desiderata in the form of theory virtues and the good sense of the researchers themselves fail in providing a unique preference weighting in a particular situation. Alex Rosenberg (2012) argues, and I think rightly, that "The problem of empirically equivalent but logically incompatible theories becomes especially serious as science becomes more theoretical."[7] But some thinkers do not want to treat this as the "local" issue I have presented it as being, and that Rosenberg's point would seem to support. They rather want to claim that *ampliative underdetermination* is global, and associate it with a thesis of "non-uniqueness" that is supposed to impugn the legitimate functions of the theory virtues in theory-choice. I think these claims are over-wrought as much discussion of underdetermination and holism is over-wrought, but these issues will have to concern us more directly later.

It may well be that the kind of intractable, irresolvable theoretical disputes that underdetermination seems to make possible are almost never actual. Still, the logicist conception of objectivity associated with logical empiricism is already shattered by recognition that "[B]esides the test of observation, theories are also judged on other criteria: simplicity, economy, explanatory unification, precision in prediction...consistency with other already adopted theories...amount of allowable experimental error, etc.," and that while there are disagreements and sometimes very great disagreements among theorists, "yet over time these disagreements are settled, to almost universal satisfaction," by reasoning about *ampliative adequacy* (Rosenberg, 212; 214). For those who acknowledge the genuineness of the problem

---

[5] E. McMullin, 501.

[6] On virtue epistemology's role in respect to recent calls for the "thickening" of epistemology, see the papers in the 2008 *Philosophical Papers* edition, *Epistemology Through Thick and Thin*, 37(3). These include Guy Axtell and Adam Carter (2008), "Just the Right Thickness," which identifies and challenges an epistemological analogue of the (ethical) "centralist" thesis (of the primacy of thin concepts over thick) that Bernard Williams criticized.

[7] Rosenberg, 212. I lean on Rosenberg especially here because he seems to acknowledge that even strong empiricists see a substantial role for theory virtues and for ampliative reasoning more generally in theory choice. Any attempt at a more directly empiricist justification for the methodological rules we employ in theory choice, he concedes, "is circular as an argument against the threat of underdetermination," (214) and appeals to them as *a priori* are unavailable to empiricists. Thus neither the rationalist nor the empiricist account of an algorithm of theory choice is at all satisfactory (214).

of underdetermination for the sciences—without going so far as to draw relativist implications from it—theory-confirmation and disconfirmation needs to be conceived of as utilizing a "toolbox" of theoretical virtues: "Theory choice is a continual process of iterative applications of this same toolbox of considerations in order to assess the implications of empirical observation in making theory choices" (214).

The connections we are beginning to see here between theory virtues and virtue epistemology are not merely ornamental. Both directly concern ampliative (non-deductive) reasoning in the sciences. Moreover, application of theory virtues to choice among competitors involves weighing these theory virtues against one another, and thereby calls upon what Pierre Duhem would call the *bon sens* of the scientist. Hence a condition of character is implied even where the explicit appeal is only to impersonally-framed theory virtues. Indeed both the appeal to theory virtues and the appeal to *bon sens* remind us that the scientist qua scientist makes value judgments. This is all the more so when theory assessment is a comparative choice among extant rivals. So what we will here term the *ampliative adequacy* of a theory is conceived largely as a matter of comparison of extant rival theories, and not as a question only of a single theory and its relationship to observational data.

Moreover, virtue epistemology as will be argued further is what helps us to find continuity between contexts of scientific assessment that are not beset by localized underdetermination worries, and those that are. This continuity and the attendant sense of how theory virtues and the intellectual virtues of researchers themselves aid objectivity in science, shows clearly how tying virtue epistemology into our philosophy of science allows us to accommodate 'the turn to practice,' while also avoiding relativism. Indeed the present view like McMullin's directly responds to the shared assumption of logicists and their radical historicist critics, that *if* theory choice isn't fixed by observations or some kind of observation-linked algorithm, then it is still fixed, but by non-epistemic factors, like personal bias, desire for authority, fame and fortune, etc. Virtue theory shows us how to say 'neither, nor' to that bogus kind of 'either, or.'[8]

In this paper I urge a virtue epistemology extricated from any over-strong interpretation of holism or of underdetermination, and one that isn't strongly committed to a particular position on the realism/anti-realism debate. We will address different concerns about underdetermination as well as different versions of virtue epistemology. I will have one (relatively independent) thesis for each of the four sections. In Sect. 2 we examine the relationship between theory virtues and personal

[8]Like Daston and Galison in their book *Objectivity* (2007), I would argue that the concept of scientific objectivity has a history, but that the epistemic norms that have informed scientific practice can be historicized without leading to relativism. I have elsewhere argued that considerations stemming from underdetermination problems motivate the claim that historicism *requires* agent-focused rather than merely belief-focused epistemology, and that this is partly what makes it possible to distinguish weak or moderate historicism from radical historicism about the epistemic values recognized in science. See Axtell, "The Dialectics of Objectivity," (2012) in a special topical issue of *Journal of the Philosophy of History*, on intersections of historicism, naturalism, and virtue epistemology.

intellectual character traits and introduce a taxonomy of theory virtues that addresses both prescriptive guidance and normative assessment. Section 3 discusses a thesis that Richard Dawid argues for, which describes a "substantial shift" he sees occurring in contemporary fundamental physics: "the increasing importance of assessments of scientific underdetermination" (2011, 2).

Dawid's thesis is a reasonable one, nicely descriptive of problems of theory choice in high energy physics and the localized reliance there on standards of *ampliative adequacy*. I argue that it thereby also indicates a need in philosophy of science to utilize a kind of virtue epistemology. But what kind of virtue epistemology, specifically? One that essentially stops with theory virtues shared by a community of inquiry, or one that appeals as well to the virtues, or *bon sens*, of good researchers themselves? That raises the issues of what we'll call *ampliative underdetermination*. Section 4 develops these further connections between ampliative reasoning and theory choice by joining a recent debate among philosophers of science over Pierre Duhem's account of the *bon sens* or good sense of scientific practitioners. I work out my differences from Abrol Fairweather, David Stump, and Milena Ivanova in their respective interpretations of Duhem. Section 5 develops what we termed a Janus-faced conception of the descriptive and normative aspects of theory virtues, and argues that inquiry-focused virtue epistemology coheres with and adds substantial support to meta-scientific pluralism, and to normative naturalism.

2 The Virtues of Empirical and Ampliative Adequacy

McMullin provides a useful taxonomy of theory virtues, a taxonomy that as one commenter puts it, "preserves the epistemic character of scientific theory without confining the epistemic values merely to first-order 'empirical adequacy' as van Fraassen understands it."[9] Is empirical adequacy always the 'thin' notion associated with a test, or does it in function in scientific debates function more like set of virtue concepts? McMullin explains that the association of theory confirmation with deductive implications of observations and tests should be restricted to the synchronic and retrospective virtue of *empirical fit*. *Empirical Adequacy* actually refers to a more over-arching class of cognitive virtues than does *empirical fit*, and on close inspection contains some forward-looking sub-virtues. "Empirical fit should be distinguished from empirical adequacy, as this is defined in van Fraassen's constructive empiricism. Empirical adequacy refers to all of the consequences of the theory, regardless of whether they have ever actually been drawn or checked against observation" (502).

McMullin sub-divides what I call the theoretical virtues of *Ampliative Adequacy* into internal, contextual and diachronic virtues. These virtues he presents as *complementary* to the central theoretical virtue of epistemic fit. Resolving a localized

---

[9] Allan (2006), 81.

situation of underdetermination through a new test providing decisive empirical advantage of one theory is the best outcome.[10] But where it is not to be had, then with McMullin we must "argue for the relevance of a whole series of confirmatory virtues that complement the central virtue of epistemic fit, transforming natural science from a mere saving of the phenomena to a genuinely explanatory and ontologically expansive enterprise" (2009, 502).

"Internal" virtues like *internal consistency* are basic requirements while others limit the degree of allowable ad hocness. "Contextual" virtues include *external consistency* and *consonance,* which address consistency with background knowledge, and *optimality,* which involves not only retroduction but also comparative merit of a theory. But the most unusual and useful feature of McMullin's taxonomy is his close attention to "Diachronic" theory virtues, including especially *fertility,* along with *consilience* and *durability. Fertility*, almost an executive virtue for McMullin, is Janus-faced, looking backwards to novel facts predicted and confirmed, as well as forwards to potential for the hypothesis to issue new, bold predictions. Unlike logicist metascience, defenders of the relevance of diachronic virtues like *fertility*, *consilience,* and *durability* to theory confirmation can easily maintain a lively understanding of the importance of history of science to philosophy of science. McMullin's emphasis on fertility as epistemic desiderata opens up the *diachronic* aspects of theory assessment lost with a logicist account of scientific objectivity. But we earlier claimed that the more serious underdetermination problems are localized ones, and that these do not track the conventional distinction between soft and hard sciences. Problems of local underdetermination arise everywhere that fields of study become more theoretical and less directly experimental. In the next section we consider an example of this from contemporary theoretical physics and cosmology, and at what implications situations of local underdetermination may have for the centrality of ampliative reasoning in our conception of scientific objectivity.[11]

3 Underdetermination and Theory Choice: The Case of String Theory

The logicist's notion that theory confirmation should be strictly rule-governed, and that accordance with this logic constitutes the objectivity of science or the rationality of particular scientists, is challenged by underdetermination problems. Scientific language and a practice based on the use of instruments makes such problems unavoidable. In actual practice, underdetermination problems are not 'solved,' but

[10] "In scientific research one always hopes for determination: that the world should *determine* the observations we make of it; that evidence should *determine* the theories we adopt; that the practice of science should *determine* results independent of the sort of society in which that practice takes place" (McMullin 1995, 233).

[11] Rosenberg 2012, 212.

they typically are *resolved* after a period of time. They often need to be if researches are to continue to identify and pursue successful research strategies. It is left to philosophers to sort out the epistemic status of the chosen theory, and to debate what philosophers' distinctions should be drawn, for instance, between theory 'pursuit' and theory 'acceptance,' and between narrowly epistemic more broadly cognitive values, etc. With these self-appointed tasks philosophers have not always done very well. As a case study of this, let us look at contemporary String Theory, and the quandary in which standard accounts of theory confirmation leave it.

Richard Dawid writes that,

> The canonical understanding of scientific progress...strictly distinguishes assessments of scientific underdetermination from the core elements of scientific progress, which are (1) the development of a scientific hypothesis and (2) the empirical testing of that hypothesis... Assessments of scientific underdetermination, to the contrary, are taken to constitute mere instances of auxiliary reasoning that may be of some relevance by channeling scientific activity towards more promising investigations but do not directly contribute to the generation of scientific knowledge. Put in terms of the old conceptual dichotomy between context of discovery and context of justification, one may say that assessments of scientific underdetermination were always acknowledged as playing some role in the context of discovery but were denied any role in the context of justification.[12]

While philosophers and non-practioners tend to accept the canonical view, among high energy physicists themselves Dawid claims there is a substantial "shift" taking place as they increasingly question whether the canonical understanding of theory assessment is adequate for grasping String Theory's merits. He argues that theoretical virtues must supersede strict dependence on empirical fit, and that appeals to desiderata of ampliative adequacy "amount to assertions of limitations to scientific underdetermination." This means that dependence on ampliative desiderata should moderate the demand that a scientist be agnostic about the parts of her theory not in practice open to direct testing. Dawid of course is not suggesting that assessments of scientific underdetermination can ever replace empirical confirmation, but rather that we need an epistemology for the sciences that "can establish an intermediate epistemic status for theories that lie between 'empirically confirmed' and 'pure hypothesis.'"[13] Dawid more constructively sees the old dichotomy between empirical confirmation and mere speculation replaced "by a continuum of degrees of credibility, where the available elements of empirical corroboration and non-empirical

[12] Dawid 2011, 4. "Assessments as to how likely it is that no or few alternative theories can be fit to the available data thus lie at the root of all considerations regarding the prospective viability of a so far empirically unconfirmed or insufficiently confirmed theory. We want to call such assessments 'assessments of scientific underdetermination' (2011, 3).

[13] "The emerging new paradigm moves away from an understanding...that attributes the status of mere hypotheses to scientific theories which have found no empirical confirmation. But Dawid also qualifies his claim in certain ways: "Non-empirical theory assessment thus crucially relies on empirical testing and can never fully replace it. Nor does non-empirical theory assessment award the same status to a theory as strong empirical confirmation. It is vaguer and less conclusive than the testing of theories by empirical data. Its vagueness induces the risk that its deployment might be overstretched...." (2011, 18–19).

theory assessment jointly contribute to an overall evaluation of theory's chances of being viable" (2011, 19).

Rejecting the canonical view associated with empiricism and the hypothetico-deductive model means turning in certain fields of research from objectivity through direct testing to objectivity through ampliative reasoning. Dawid's proposal helps make sense out of problems of theory choice in contemporary theoretical physics, and I hold that this shift is one that virtue epistemologies help us to articulate and implement. If we do need an altered conception of theory assessment in high energy physics and scientific cosmology, it is a conception in which reasoning through theoretical virtues plays a more central role in theory assessment.

4 Duhem and the Role of *Bon Sens* in Scientific Practice

Thus far we have associated a virtue epistemology for the philosophy of science with the study of ampliative reasoning utilizing *impersonal* theory virtues. But to what extent will a virtue epistemology draw us also into study of the *personal* intellectual virtues of scientists themselves—the good sense or *bon sens* of the inquirer? I want to argue that there will always be an interesting research program with respect to the personal virtues of scientific practitioners for the reason that personal habits are always active in inquiry. Study of the personal traits and the scientists' 'doings' are always relevant when we take a practice-focused approach to scientific reasoning.

But one might push the question: Could the personal *bon sens* of the scientist ever directly contribute to the epistemic status of the theory which that scientist chooses? By-and-large the role of what Abrol Fairweather terms the "methodological cognitive character" of the scientist—"the set of abilities, skills and dispositions a scientist acquires and expresses through the structured forms of inquiry involved in applying scientific methods" (141)—simply plays a role supportive of their prowess or reliability in deductive and ampliative reasoning. To this extent I would think their study might be of interest more in a sociology than in an epistemology of the sciences. But there may be exception cases, and if there are then these cases can also be delineated by the types of underdetermination worries that dog inquiry. When localized problems become severe about what cognitive values to accept, or how to weigh them against one another (see Kuhn 1977) then inquiry is taking place under another level or type of underdetermination problem. Our primary focus becomes underdetermination of theory *choice* by methodological standards, including the theory virtues. Let us call this type *ampliative underdetermination,* and consider now the philosophical concerns it raises and the resources that virtue epistemologies have for responding to them.

Paralleling our treatment of *logical underdetermination*, I want to say with respect to *ampliative underdetermination* that the sheer possibility that ampliative criteria will not result in a "unique" choice from one scientist to the next is a global, but also only a weak claim. It is not much of a worry since it really only restates how we got to this point: ampliative reasoning by definition does not meet deductive standards of entailment; if we cannot read theories off of their empirical consequences, the notion of an algorithm for theory choice is off the table and so to re-impose a logicist conception of rationality or objectivity is simply inappropriate for beings such as we are. *Logical underdetermination* forcefully shows us both that "falsifications do not undermine one particular statement and [that] confirmations do not uniquely support one particular set of statements" (Rosenberg, 287). Situations of local underdetermination, where multiple theories compete in some area of science, heighten the need for more holistic evaluation of evidence and of the scientific merits of the competing theories. Further, just as it is always desirable but not always possible that theory choice be based on experimental findings that confer empirical adequacy uniquely upon one theory, so I would hold that it is always desirable but perhaps not always possible to distinguish sharply between the desiderata of *impersonally-framed* 'theory virtues' and *personal* intellectual traits of good researchers themselves. This is what we can call a 'tiered' account of the epistemic relevance of *impersonal* standards and *personal* expertise or judgment.

With McMullan and against van Fraassen (1980) we have held that ampliative reasoning clearly contributes to epistemic status. And we have held that personal virtues and vices (probably both intellectual and ethical) are active and implicated in ampliative reasoning—most obviously in the kind of weighing that inference to the best explanation demands. But also on the present view, we must remain wary of "collapsing" the theory virtues into a set of personal virtues of scientists themselves. My view isn't shared by all self-described virtue epistemologist, however, and this is why I bring it up. Since there are a number of different extant versions of virtue epistemology it is unsurprising to find them running the full gamut of views in relationship to the underdetermination problem. Some authors neglect the theory virtue/personal virtue distinction by not recognizing the importance of the researcher's character and "doings," while others collapse the theory virtues into personal virtues, attaching no importance to impersonally-framed theory virtues.

Some accounts, though not present one, lean upon a strong interpretation of underdetermination and/or holism, and this is one primary way to collapse the distinction. Perhaps the clearest example of this is Lynn Holt's book, *Apprehension: Reason in the Absence of Rules* (2002). Holt develops an "Apprehensionist" virtue epistemology, and one that takes "Methodism" as it's opposite and as its stalking horse. He contrasts the apprehensive virtues of understanding (*nous*) and of practical wisdom (*phronesis*) with "the non-apprehensive elements of expertise—calculative reasoning, technical skills" (44). But theory virtues don't fit well with this dichotomy and seem almost entirely left out of his account of theory choice. Holt's view of theory assessment and the epistemic status of theories is basically that it is whatever reflects the judgments of the experts, those who possess the phronesis relevant to their field.[14] This might then be the form of virtue epistemology most appealing if

[14] One of the most common objections is that it is circular; another is that it is simply intuitionism in new garb—apprehension or *bon sens* as 'the Emperor's new intuitions.' Holt acknowledges keen

one took a strong stance on holism or on *ampliative underdetermination* and saw it as motivating an either-or choice between Apprehensionism and Methodism. It results in the very strong claims that "the way to adjudicate between rival traditions is to ask the wise" (72) and that "a genuine epistemology ought properly to be regarded as *virtuoso* epistemology: an account of who is best able to judge truth from falsity in virtue of his or her possession of wisdom" (73).

While he neglects to examine Duhem's account of theory choice, I take it that Holt's central distinction between apprehensive and non-apprehensive expertise and the personal traits that constitute the former strongly overlaps with Duhem's distinction between "intuitive" and "mathematical" reasoning. Not surprisingly, some neo-Aristotelian virtue epistemologists find substantial interest in Duhem's account of *bon sens.* David Stump's paper "Pierre Duhem's Virtue Epistemology" offers a virtue-theoretic account of the role of *bon sens* in Duhem's philosophy of science. Stump argues that despite the fact that Duhem is sometimes read as a conventionalist arguing that there is simply no cognitive way to decide between empirically equivalent theories, closer examination reveals that through good sense of practitioners consensus typically does emerge, and not for purely conventional or epistemically irrelevant reasons.[15]

Whether rightly or not, Milena Ivanova associates Stump's reading of Duhem with the apprehensionist variety of virtue epistemology. She associates it with a very strong "change in the direction of analysis" thesis she takes all form of virtue epistemology to be committed to, where the merits of the agent's character are determinate of the epistemic standing of particular beliefs.[16] Although I don't think that thesis is held by reliabilist or 'mixed' forms of virtue epistemology, and don't see Stump himself as making all the strong claims she seeks to refute, given what we have said above I do agree with much of her criticisms of Apprehensionist virtue epistemologies. She objects that it seems to negate the need for scientists to look for future evidence to evidentially distinguish the theory chosen by good sense. If this were correct it is easy to see why calling Duhemian good sense a virtue theoretic solution to underdetermination will arouse suspicions. As Abrol Fairweather puts it, "it will be controversial to locate some share of the epistemic value of our currently accepted scientific theories in properties of the scientist, rather in properties of the

critics of his view, including Hintikka (2002), who argues that intuitionism is a failed view in the philosophy of science, and "apprehension" is nothing but a re-working of intuitionism. Holt's Chapter 3, "Apprehension and the Apprehensive Virtues", offers a direct reply.

[15] Stump (2007), 149–159, 149–150. The personal habits that comprise the scientist's methodological cognitive character describe real or ideal excellences of inquirers, not of theories or hypotheses *per se.* Perhaps for this very reason, they are less purely intellectual, and indeed those who emphasize their role in inquiry, from Pierre Duhem to Daston and Galison, often want to *insist* that they are or include character traits in the full Aristotelian sense, engaging motivations and sometimes crossing boundaries between the epistemic and the ethical. See Stump (2011) for a further development of this view.

[16] In fact it represents only one strong form of virtue epistemology that would define justified true belief in terms of what an intellectually virtuous person would believe. Stump (2011) does come close to endorsing such a view.

science itself" (2012, 140). Even independently of whether Stump is right to see Duhem's account as proto-virtue epistemological, Ivanova finds Stump's own views about the epistemic value of good sense unsatisfactory. She also argues for a different reading of Duhem's account of good sense. I will not have space to go very far into Ivanova's or Fairweather's interesting responses to Stump, but would like to draw out a few general points and to try to straighten out what I see as misconceptions that hinder progress in this debate over the role of impersonal and personal virtues in scientific reasoning.

Ivanova writes that "[Duhemian] good sense does not determine the construction of a theory and it is not what justifies a belief in the truth of a theory. It determines the scientist's choice, but not uniquely. It restricts the scientist's choice by excluding some of the possibilities with which he is faced in theory choice. It does not lead to justified true belief, but simply to a temporary acceptance of a theory" (2010, 62). While I can largely agree, I have certain quibbles. Developing a distinction Fairweather suggests, I would say that a virtue epistemology helps us recognize the contribution to epistemic *value* of the scientist's methodological cognitive character, but that it leaves open a range of views about the epistemic *status* of the theory that good sense selects. But there is also something important in Fairweather's point that "Method and evidence reign when they can, but epistemic normativity becomes aretaic in UD inquiry with the express purpose of resolving underdetermination" (141). In my own terms, the question of how to parse the differences between empirical testing, theory virtues, and the personal *bon sens* of researchers themselves admits of no *general* answer, but depends crucially upon *local* issues about the relative normality of inquiry being pursued under conditions of underdetermination.

The success condition of Duhemian good sense seems merely to be its breaking of the empirical stalemate in an appropriate way, not in a way that confers uniqueness (across competent scientists or the epistemic community as a whole) on the choice made. But while respecting the distinction between theory pursuit and acceptance or belief, I also think that distinction should not be made too rigid, like the psychology/logic or discovery/justification dichotomies on which logicist metascience depended. What Fairweather argues is missed when such dichotomies are assumed, is the "continuity" a virtue theoretic account provides to and from the movements between contexts of UD (underdetermination) and non-UD inquiry: "In *UD inquiry* we are trying to resolve the problem of theory choice, whereas in *non-UD inquiry* we either have not yet faced the problem, or have resolved it for the time being. The virtues of good sense do not have a constitutive role in generating the epistemic standing of theories in non-UD inquiry...The virtue theoretic reading exhibits axiological continuity between the two contexts of inquiry and thus provides a constraint on admissible resolutions to underdetermination by precluding the introduction of radically new epistemic values" (141).

Another problem I find with Ivanova's account is with her own claim about non-uniqueness. In her discussions of theory virtues, she jumps too quickly from the *possibility* of ampliative underdetermination or non-uniqueness to the generalization that a unique choice is *never* indicted by the criteria of ampliative adequacy.

"Even though criteria to describe theory choice can be found, they cannot determine the choice uniquely" (60); "These criteria can help us to describe, explain and justify the scientist's decision, but they do not do so uniquely" (63). Cannot? Do not? This seems to me too great a generalization, which our own account of the *local* nature of ampliative underdetermination worries should serve to undo. The *possibility* of differential weighting applied to the theory virtues, etc., does not mean they *must* differ so significantly; it would be wrong to presuppose that every time two theories are empirically equivalent (i.e., logical underdetermination prevails) there must also be underdetermination of theory *choice* by methodological standards (i.e., ampliative underdetermination) (Laudan 1990). If we are not implicitly identifying a "unique choice with a logically or evidentially *forced* choice, the claims Ivanova repeatedly makes that virtue epistemological treatments must be "unsatisfactory" because they fail of "solve" the underdetermination problem but only move it to a new level, are simply un-motivated. "Solving" underdetermination problems was never in the cards, and one who bases satisfactoriness of *resolutions* on that measure will always be disappointed. The sense of uniqueness that should be in play is that of consensus within a scientific community, not that of "conclusiveness" as she uses it. And just as serious underdetermination worries are local, it is quite possible that a consensus emerges that one of a pair of competing empirically equivalent theories is uniquely selected on the basis of ampliative adequacy. Nor ought we to take forced beliefs as paradigmatic of rationality or objectivity.

This leads to my final point of criticism.[17] Ivanova arguably goes too far in the direction of reading Duhem as holding definitively like van Fraassen that only a later experiment that gives advantage in empirical adequacy provides grounds for justified belief. McMullin held that debate over the deductive and ampliative reasoning in the sciences "usually masks a deeper difference about the epistemic function of theory itself" (507). This is certainly evident in the present discussion over how to interpret Duhem's notion of good sense, which by all accounts wasn't very well developed by Duhem anyway. Duhem was always walking a thin blue line between empiricist conventionalism and realism, and my reading of him has him walking a similar line here. It is true that Duhem did in one passage characterize ampliative criteria as "essentially subjective, contingent, and variable with time, with schools, and with persons" (1954, 288). But this seems to be an obvious over-generalization on his part and anyway, by way of counterpoint, he also claimed that "Pure logic is not the only rule for our judgments: certain opinions which do not fall under the hammer of the principle of contradiction are in any case perfectly unreasonable"

[17] I tend to agree with Laudan that we should distinguish the (true but weak) general claim that theories are underdetermined by data from the (false because over-generalized) claim that theory-choice is always underdetermined by methodological standards. On my view Ivanova presents a straw-man version of McMullin in numbering him among those who supposedly "believe that we can always choose a unique theory from a set of empirically equivalent rivals by simply pointing to the amount of virtues the chosen theory possesses". Nor is it true that "both views [realism and empiricism] assume that theory virtues lead to a conclusive choice between underdetermined theories" (this volume).

(1954, 217). Arguably also, identifying Duhem with van Fraassen's constructive empiricist account of theory confirmation would challenge rather than support the axiology of "natural classification" Ivanova (2011) highlights in Duhem's philosophy of science. McMullin's emphasis on the substantial role of theory virtues and ampliative reasoning in theory confirmation arguably *better* serves the axiology that Duhem's "natural classification" evokes than does saddling him with the stance that only differences in empirical adequacy contribute to epistemic value or provide rational grounds for belief. For as McMullin writes, "Those who deny the ability of theory to reveal underlying structure will also tend to see empirical fit as the only feature of theory worth worrying about, with possible pragmatic concession for such features as lend themselves to convenience of use or utility of application...[whereas] those who saw in theory the way to discover real underlying causes of macroscopic regularities are likely to stress a variety of epistemic virtues and to insist that saving the phenomena is not enough."[18] It is important that a virtue epistemology in philosophy of science refute the view that reduces the epistemic to the purely synchronic notion of evidential fit. That view militates against recognition of diachronic and "explanatory" virtues that scientific realists have drawn attention to. But establishing the relevance of a whole series of confirmatory theory virtues should not be thought to entail any grand conclusions about realism or anti-realism. The claim McMullin makes about how ontologically expansive the recognition of these theory virtues in natural science is, reflects back upon the 'very different visions' of scientific reasoning he finds in philosophy of science. But our point has been that a 'tiered' account of the role of theory virtues and expert judgment allows us to largely set these axiological differences between the realists and empiricists aside in favor of recognizing the richness of the ampliative desiderata available to scientists to resolve problems of local underdetermination. Theory virtues of both the synchronic and diachronic sort are enablers of rational preference.

5 Rational Reconstructionism Meets Normative Naturalism

So what implications for the epistemology and methodology of the sciences might follow from acceptance of the usefulness of virtue theory for philosophy of science? We haven't the space to develop detailed answers to this question, but I make three suggestions here from the perspective of "inquiry focused" VE, recognizing that proponents of other versions might draw somewhat different implications. The first

[18] McMullin 1996, 17. Indeed McMullin sees this as extending to Duhem himself: "Theory assessment involves the faculty of good judgment (*bon* sens) which permits disagreement between competent scientists....What tends to decide the issue between competing theories is how they develop over time, to what extent their response to anomaly appears ad hoc, and so forth" (17; see Duhem 1954, 216–218, and compare Duhem 1991).

is the one we began with, the idea of a more varied or "dappled" conception of the relationship between the sciences, which it can be argued is an implication of our thesis of the *localized* nature of the most worrisome kinds of underdetermination. I purposely alluded to a term from Nancy Cartwright's *A Dappled World* (1999) to describe this thesis, in part because I suspect there are substantial lines of support that could be developed between virtue epistemology and the thesis of *metascientific pluralism* she argues for. Pluralism as a metascientific level thesis presents an alternative to both the "unity of method" that Hempelian logical empiricists demanded, and to the epistemological relativism of some of the post-positivists.[19] Indeed the connection with a theory of epistemic virtues has already been made from the other direction by the editors of a notable collection, *Scientific Pluralism* (Kellert et al. 2006), when they explain,

> Philosophers of science have begun to advance pluralism at the metascientific level, most notably with respect to epistemic virtues. A variety of views regarding the role, status, and identity of scientific or epistemic virtues has been advanced in the philosophical literature.... [Some pluralists claim] that which virtues should hold what degree of regulative status in any given research project is a function of features specific to the problem and of the particular aims of the research (2006, x).

My second implication is a quite different way of approaching questions of demarcation and of the relationship between disciplines or fields of research. This is the view of John Dupre, who more explicitly than other pluralists has suggested that "we try to replace the kind of epistemology that unites pure descriptivism and scientistic apologetics with something more like a virtue epistemology." If "no strong version of scientific unity of the kind advocated by classical reductionists can be sustained" (1993, 242), then "the successor to the quest for demarcation criteria between science and non-science may be an account of theory virtues that characterize scientific reasoning…we are much better off to think in terms of epistemic virtues, features of an investigative practice that confer credibility. No doubt the cardinal empirical virtue is a proper connection with empirical evidence, which is the large grain of truth in the criterion of falsificationism" (2010). The implication is an abandonment of the hierarchy of the sciences ideal, together with its presumption of clear demarcation criteria: "Many plausible epistemic virtues will be exemplified as much by practices not traditionally included within science as by paradigmatic scientific disciplines…No sharp distinction between science and lesser forms of knowledge production can survive this re-conception of epistemic merit. It might fairly be said, if paradoxically, that with the disunity of science comes a kind of unity of knowledge" (1993, 243).

These two purported bridges between philosophy of science and virtue epistemology are not very original, I am afraid, having been drawn by others. But I would

[19] Moderate historicism, according to which the 'units of selection' in theoretical enterprises of all types are historical research programs, and a moderate confirmation holism, according to which a test of one theory always depends on other theories and/or auxiliary and background assumptions, are well-suited to provide this kind of pluralism, but more radical versions of historicism and holism are not.

like to end by developing something that I think isn't already found in the literature, which is a relationship of mutual support between virtue theory, often described as "Janus-faced," and normative naturalism. This also ties back to the debate over Duhem's *bon sens,* since both Stump and his critic Ivanova seem to think that if Duhem's account of good sense is of value, it is due to its usefulness in some kind of project of rational reconstruction, whereas normative naturalism quite explicitly rejects the project of rational reconstructionism as a central task of the epistemology of the sciences.

I suppose there is a sense in which the person of good sense is best judged retrospectively, as the person who chose wisely from the perspective of future science. But this certainly isn't the Duhemian meaning of good sense, nor is it the import of a study of a scientist's methodological cognitive character. These by contrast are clearly *prospective* posits, concerned with guidance and best bets for a path of research to pursue. Ivanova seems at odds with herself in attributing to Duhem (and to virtue epistemologists) a project of using good sense to retrospectively "explain" the fact that underdetermination tends to be resolved prior to the availability of new evidence. To do this is to steep him in the assumptions of later rational reconstructionists, and to import along with this certain dichotomies that arguably make that project impossible to fulfill. My skepticism about this is not because retrospective explanations aren't ever possible or helpful. Surely at times they are. It is because like Laudan and other normative naturalists I would repudiate the ideal of rational reconstructionism as central to the normative tasks of philosophy of science: "The requirement of rational reconstructibility is neither wanted nor needed" (1987, 21).

Like Duhem, Laudan rejects the notion that justification may proceed algorithmically, while also insisting that non-empirical conceptual considerations are crucial to scientific practice and meta-methodology. Positivists and many post-positivists alike misconstrue the underdetermination problem because they either mistakenly assume that theories possessing the same positive instances must be regarded as equally-well confirmed, or because "they assume that the only rational basis for rejecting [dismissing] a theory or hypothesis is if it has been definitely refuted." Epistemic values and virtues (scientific axiology) may change somewhat as science develops, but we are still able to view rules possessing normative force as grounded in factual means-end relations.[20] Methodological rules are fixed by means-end relations, but our conception of ends—scientific axiology—is neither given nor timeless. We need also an axiology of inquiry whose function is to certify or decertify certain aims

[20] Methodology so conceived is basically "restricted to the study of means and ends,"; they are "best understood as relativized to a particular aim" and judged by whether they guide inquiry to its achievement. But far from the Quinean version of epistemology naturalized qua replacement thesis for normative epistemology, Laudan holds that "methodology gets nowhere without axiology," and that "We thus need to supplement methodology" with an investigation into an *axiology of inquiry* (1987, 29). Axiology in turn is multi-faceted, and while generally naturalistic also "preserves an important critical and prescriptive role for the philosopher of science" (29).

as legitimate, for "methodology gets nowhere without axiology." Against rational reconstructionism Laudan proposes normative naturalism:

> [E]pistemology can both discharge its traditional normative role and nonetheless claim a sensitivity to empirical evidence ... normative naturalists hold that the best methods for inquiry are those which produce the most impressive results ... the naturalist uses the simple method of induction to 'bootstrap' his way to more subtle and demanding rules of evaluation which, in their turn, become the license for subsequent and yet more highly refined rules and standards .... (1987, 44, 58).

For the normative naturalist, as another of its proponents puts it, "there have got to be other criteria, coherence, simplicity, predictive fertility, explanatory power, that an epistemology, like a scientific theory, must meet, and it must meet them, not because they are intrinsic goals of science, but because they are instrumental ones, instrumental to the goal of attaining knowledge."[21] Thus responsibilist and reliabilist concerns combine in the present view, which fits better the Janus-faced understanding of virtue theory as both a descriptive account and one aiming to provide prescriptions for the improvement of practice.[22]

In this more naturalistic alternative to the overt or 'closet' intuitionism of the rational reconstructionists, we need not preoccupy ourselves with the question of whether we can always replicate the choices of past scientists as rational. This is no grand mark of the adequacy of a methodology of science anyway. Instead we simply "inquire about which methods have promoted, or failed to promote, which sorts of cognitive ends." History of science still plays a key role here, and indeed may be center stage in the evaluation of proposed methodological standards. But history of science is not in the mission Imre Lakatos (and even the early Laudan!) gave it of insulating an "internal" (rational) from an "external" (social) explanation for past successes. Agreeing with Longino and other social and feminist philosophers who call for deconstructing the rational/social divide constitutes my biggest divergence from Laudan normative naturalism. While sounding simple, I think of normative naturalism as actually demanding that we disassemble (along with the myth of universally valid methodological and epistemological standards), the rational-social dichotomy, the logic/psychology dichotomy, and the dichotomy between internal and external history of science—each of which rational reconstructionists have appealed to in a misguided attempt to maintain the independence or scientific 'reason' from the vagaries and messiness of scientific 'practice.'

[21] Rosenberg 1990, 42–43.

[22] "In light of the requirement that the means *reliably* conduce to the desired end, normative naturalism might appear to be a form of reliabilist epistemology. There do, however appear to be a number of salient differences between normative naturalism and reliabilism, at least as it is classically understood (e.g., Goldman (1979)). First, for Goldman a reliable method is one which leads reliably to truth, whereas for Laudan the cognitive ends in question are typically something other than truth. Second, reliabilism is a theory of the justification of an agent's epistemic states, whereas normative naturalism is a theory of the justification of method. Thus, rather than take a reliabilist view of individual epistemic rationality, Laudan operates with an instrumental account of rationality on which an agent's belief that an action will lead to their aim is required for the act to be rational" (Sankey).

What would seem to be obviously false dichotomies between the rational and the social have much staying power, however. One main reason for this is that because underdetermination problems leave us with questions about how scientific theories are chosen when empirical evidence fails to determine one theory as uniquely choice worthy, they also *appear* to present us with a referendum on the rationality of science. They do not. Only a philosophy of science in the service of rational reconstructionism, or a radical historicism that uncritically assumes the same dichotomies in order to take the opposite, relativistic side of the issue, pushes us towards any such referendum. So for instance we hear that to ground theory choice in anything else but hard data impugns the objectivity of the theory chosen, and of science itself; we must perforce seek sociological explanations of scientists' cognitive choices. Or we hear that if assumptions of some sort are required to mediate the relation between data and hypothesis, these assumptions "can be the vehicles on which cultural ideology or social values ride 'right into' the rest of science".[23] Or again we hear that because appeal to theory virtues are a means of "persuasion" between advocates of different empirically equivalent systems, these concepts introduce a rhetorical dimension that should be foreign to science were it really objective.

But the correct response to each of these mistaken claims, it seems to me, is to reject from the outset the notion that scientific reasoning should be characterized by a sort of epistemic purity that social practices of other sorts lack. Again, rhetorical strategies of persuasion, like underdetermination worries, are ever-present in some areas of research yet rare in others, and their absence or presence never did and never will neatly divide scientific from other forms of inquiry. To the extent that we take the advice of Longino and others to "disassemble the rational-social divide" the global referendum notion simply fades away, while local problems where guidance is actually needed come more clearly into focus. So rather than setting the rational and the social, or again epistemology and history, against one another as the project of rational reconstructionism has done, a virtue epistemology tied to normative naturalism would take another path. "We may still be able to construct a philosophy of science that derives *both* from the learning that has gone on in history *and* from a more general logical and epistemological framework" (McMullin 1984, 57). If this is indeed a viable goal, then virtue epistemologies I suggest will need to be disassociated from rational reconstructionism and developed in tandum with a more naturalistic project of scientific meta-methodology, yet one that preserves important "critical roles" for the philosopher of science (Laudan 1987, 29).

Acknowledgements Special thanks go out to Abrol Fairweather for comments on an earlier draft, as well as to Lynn Holt, Milena Ivanova, David Stump and James Kidd for comments and discussion on related 2010–2011 posts at *JanusBlog: The Virtue Theory Discussion Forum,* http://janusblog.squarespace.com.

[23] Longino 1990.

References

Allan, P.L. 2006. *Ernan McMullin and critical realism in the science-theology dialogue*. Aldershot, UK: Ashgate.

Axtell, G. 2012. The dialectics of objectivity. *Journal of the Philosophy of History* 6: 339–368.

Axtell, G., and J.A. Carter. 2008. Just the right thickness: A defense of second-wave virtue epistemologies, *Philosophical papers* (topical edition on *Epistemology through thick and thin*) 37(3): 413–434.

Daston, L., and P. Galison. 2007. *Objectivity*. New York: Zone Books.

Dawid, R. 2011. Theory assessment and final theory claim in string theory. *Foundations of Physics* (Online First). doi:10.1007/s10701-011-9592-x.

Duhem, P. 1954[1906]. *The Aim and Structure of Physical Theory*. Trans. P.P. Wiener. Princeton: Princeton University Press.

Duhem, P. 1991[1915]. *German Science: Some Reflections on German Science: German Science and German Virtues*. Trans. John Lyon. La Salle: Open Court.

Dupre, J. 1993. *The disunity of things*. Cambridge, MA: Harvard University Press.

Dupre, J. 2010. The disunity of science. Interview with P. Newall at http://www.galilean-library.org/site/index.php/topic/3302-john-dupre-the-disunity-of-science/

Fairweather, A. 2012. The epistemic value of good sense. *Studies in History and Philosophy of Science* 43: 139–146.

Goldman, A. 1979. What is justified belief? In *Justification and knowledge*, ed. G. Pappas, 1–23. Dordrecht: D. Reidel.

Hintikka, J. 2002. The emperor's new intuitions. *Journal of Philosophy* 96(3): 124–47.

Holt, L. 2002. *Apprehension: Reason in the absence of rules*. Farnham: Ashgate.

Ivanova, M. 2010. Pierre Duhem's good sense as a guide to theory choice. *Studies in History and Philosophy of Science* 41: 58–64.

Ivanova, M. 2011. Good sense in context: A response to Kidd. *Studies in History and Philosophy of Science* 42: 610–612.

Kellert, H., H. Longino, and C.K. Waters. 2006. *Scientific pluralism*. Minneapolis: University of Minnesota Press.

Kuhn, T. 1977. Objectivity, value judgment, and theory choice. In *The essential tension*, ed. T. Kuhn, 320–353. Chicago: University of Chicago Press.

Laudan, L. 1987. Progress or rationality? The prospects for normative naturalism. *American Philosophical Quarterly* 24(1): 19–31.

Laudan, L. 1990. Demystifying underdetermination. In *Scientific theories*, Minnesota studies in the philosophy of science, vol. 14, ed. C.W. Savage, 267–297. Minneapolis: University of Minnesota Press.

Longino, H. 1990. *The fate of knowledge*. Princeton: Princeton University Press.

Massey, J. 2011. Quine and Duhem on holistic hypothesis testing. *American Philosophical Quarterly* 48(3): 239–265.

McMullin, E. 1984. The goals of natural science. *Proceedings and Addresses of the American Philosophical Association* 58(1): 37–64.

McMullin, E. 1995. Underdetermination. *Journal of Medicine and Philosophy* 20(3): 233–252.

McMullin, E. 1996. Epistemic virtue and theory appraisal. In *Realism in the sciences*, ed. I. Douven and L. Horsten, 13–34. Leuven: Leuven University Press.

McMullin, E. 2009. The virtues of a good theory. In *The Routledge companion to philosophy of science*, ed. M. Curd and S. Psillos. New York: Routledge.

Rosenberg, A. 1990. Normative naturalism and the role of philosophy. *Philosophy of Science* 57(1): 34–43.

Rosenberg, A. 2012. *Philosophy of science*, 3rd ed. New York: Routledge.

Sankey, H. 1996. Normative naturalism and the challenge of relativism. *International Studies in the Philosophy of Science* 10(1): 37–51.

Stump, D. 2007. Pierre Duhem's virtue epistemology. *Studies in History and Philosophy of Science* 38: 149–159.
Stump, D. 2011. The scientist as impartial judge: Moral values in Duhem's philosophy of science. *Metascience* 20: 1–25.
Van Fraassen, B.C. 1980. *The scientific image*. Oxford: Oxford University Press.

Epistemic Virtues and the Success of Science

D. Tulodziecki

1 Introduction

Scientific knowledge, that is, knowledge about the entities and structures that populate the physical world, is often taken to be one of our best and most firmly grounded kinds of empirical knowledge. Much of this knowledge involves reference to entities, processes, mechanisms, and events that go beyond what we can observe in straightforward ways. However, while talk about and reference to such 'unobservables' is part and parcel of what constitutes our scientific theories, claims of this nature are not uncontroversial: there are those who believe that we can never be justified in believing any claims of this sort. The main argument that proponents of this rather dim view appeal to is the argument from underdetermination of theories by evidence. Informally, the argument goes like this:

1. The available observational evidence (including all *possible* future evidence) always supports two (or more) theories that cannot both be true.
2. Our only reason for believing our scientific theories to be true is the observable evidence on which they are based.
3. Hence, we have no reason to prefer any one of these theories to any other.

Since scientific knowledge typically comes in the form of scientific theories, and, since, according to this argument, we are not justified in believing in *any* theory, it is a consequence of this argument that scientific knowledge is, if not impossible, highly restricted. One popular response to this argument has been the appeal to the so-called 'theoretical virtues'. These are properties of our scientific theories that scientific realists take to be epistemic in nature: if they are had by our theories, they make it more likely for those theories to be true. As a result, they are said to be capable of conferring extra epistemic power on our theories, and, so, able to break ties in underdetermination scenarios.

D. Tulodziecki (✉)
Department of Philosophy, Purdue University, 7132 Beering Hall,
100 N. University Street, IN 47907, West Lafayette
e-mail: tulodziecki@purdue.edu

A. Fairweather (ed.), *Virtue Epistemology Naturalized: Bridges Between Virtue Epistemology and Philosophy of Science*, Synthese Library 366, DOI 10.1007/978-3-319-04672-3_15,

The main challenge faced by proponents of this view is to establish a link between the theoretical virtues and truth (see, for example, Psillos 1999: 171ff). And, curiously, despite the fact that the virtues are frequently invoked in response to the underdetermination argument, how to establish this link, even in principle, has not really been addressed.[1] My goal in this paper is to spell out what I think is the most promising approach in making this link: an account according to which the question of whether the virtues (and what virtues, if any) have epistemic import is answered empirically, through an examination of cases of epistemically successful (and not so successful) theories.[2] As a concrete example of how this approach works, I will discuss in some detail a case-study from the history of medicine, concerning the mid-nineteenth century debate about the transmissibility of puerperal fever. After showing how some specific virtues are put to work in this particular case, I will explain, using the case-study as a basis, what is required in order to make a more general argument for the epistemic potential of the theoretical virtues along these lines. I will then go on to argue that putting the virtue-question on empirical grounds is enough to undermine the anti-realist position on underdetermination – regardless of whether, in fact, the virtues turn out to be epistemically potent or not. More importantly, however, thinking about the virtues in the way proposed in this paper is a beginning for a more systematic and extensive account of how to discover and put to use factors other than the empirical evidence in our thinking about the epistemic ingredients and justification of our scientific theories.

I will proceed as follows: in Sect. 2, I will outline the scientific-realism debate and explain the role that the underdetermination argument and the theoretical virtues play in it. Section 3 focuses on some competing claims from early nineteenth-century Britain concerning the transmissibility of puerperal fever, showing how it is, at least in principle, possible to link the theoretical virtues to empirical data. In Sect. 4, I discuss the upshot of the case-study and argue that it undermines the anti-realist argument for underdetermination. In Sect. 5, I move on to discuss some objections to the approach outlined in this paper. I end by highlighting some of the challenges that this approach faces, but conclude, that, nevertheless, it is our best option for getting a more robust grip on epistemic virtues and their role in science.

2 Epistemic Virtues, Underdetermination, and the Debate About Scientific Realism

One of the main venues for discussion of potentially epistemic virtues in (philosophy of) science is the debate about scientific realism, and, in particular, the debate about underdetermination, so it's useful to start off with a brief overview of the status of

[1] Two exceptions are McMullin, according to whom some properties, such as explanatory power, are constitutive of the aim of science (see McMullin 1987) and Psillos (1999, chapter 8, 2004: 404ff). There are also some notable discussions of individual virtues, especially simplicity (cf. note 7).

[2] For a defence of the view that there are other factors that make epistemic contributions to our scientific theories – specifically, methodological rules and practices – see Tulodziecki (2013a).

this debate. Following Psillos, we might think of scientific realists as being committed to the following three claims:

> (1) The metaphysical stance asserts that the world has a definite and mind-independent natural-kind structure.
> (2) The semantic stance takes scientific theories at face-value, seeing them as truth-conditioned descriptions of their intended domain, both observable and unobservable. Hence, they are capable of being true or false. Theoretical assertions are not reducible to claims about the behaviour of observables, nor are they merely instrumental devices for establishing connections between observables. The theoretical terms featuring in theories have putative factual reference. So, if scientific theories are true, the unobservable entities they posit populate the world.
> (3) The epistemic stance regards mature and predictively successful scientific theories as well-confirmed and approximately true of the world. So, the entities posited by them, or, at any rate, entities very similar to those posited, do inhabit the world. (1999: xvii)

The main lines along which different anti-realisms divide is according to which of the three realist theses they attack or deny. The thesis that takes centre-stage in the scientific realism debate is the epistemic thesis.[3] Since the argument from underdetermination directly targets the realist's epistemic commitments, this is also where discussions of underdetermination fall; in line with common usage, I will, from now on, have in mind deniers of thesis (3) when talking about anti-realists. Anti-realists targeting thesis (3) do not deny that science might attain truth, or even that claims about unobservables have truth-values. Rather, what they deny is that we can ever be justified in believing either claims about such unobservables, or that science has achieved truths involving them.[4] The underdetermination argument is designed precisely to call into question the claim that successful theories in mature science are approximately true. Here is a pretty standard formulation of the argument:

1. Empirical Equivalence Thesis (following Kukla 1998; Psillos 1999 in terming it thus): any theory has logically incompatible and empirically equivalent rivals.
2. Entailment Thesis (cf. Psillos 1999): entailment of the evidence is the only epistemic constraint on theory-choice.
3. Conclusion: Believing one theory over its rivals "must be arbitrary and unfounded" (Kukla 1998: 58).[5]

According to the anti-realist, underdetermination demonstrates that there are no reasons (and that, in fact, there cannot be any such reasons) to accept those parts of a theory that transcend the data, and that, therefore, we cannot have any reasons to believe theories that transcend the empirical evidence. For, so the argument goes, underdetermination teaches us that, for any currently accepted theory T, all the

[3] Since scientific realists and anti-realists agree on theses (1) and (2), I will not discuss these further.

[4] The most influential anti-realist account has been the constructive empiricism developed by van Fraassen in *The Scientific Image* (1980).

[5] For the purposes of this paper, I adopt a broad notion of both 'theory' and 'empirical equivalence' in order to not beg the question against any particular version of the underdetermination argument. For some different notions of empirical indistinguishability, see Earman (1993).

evidence that is at present taken to support T (or even alleged to favour T over its rivals), is also compatible with another theory T* that posits a theoretical structure that is both different from and logically incompatible with that of T. The anti-realist's claim then is that, from an empirical standpoint, it is completely arbitrary that we happen to believe in the approximate truth of T rather than in the approximate truth of T*. What explains our preference for T instead of T* are merely pragmatic criteria, whereas, from a doxastic point of view, we ought to regard both T and T* as equivalent or equally believable. Clearly, if this were true in all cases, realists would be in trouble, for their position is precisely that there are epistemic criteria for preferring certain theories (usually current ones) to others. A central part of the realism debate, then, focuses on whether the criteria that are responsible for theory choice are purely or primarily epistemic, as the realist would have it, or whether they are merely pragmatic, as the anti-realist would have it. While both agree that empirical evidence alone cannot single out a preferred theory, they disagree on whether there are other epistemically relevant criteria for theory-choice.[6]

The most prominent class of such criteria is that of theoretical virtues: according to the realist, while it might be true that there are empirically equivalent and logically incompatible rival theories, such theories are not *epistemically* equivalent, since they are unequal with respect to some other theoretical virtues relevant for theory-choice. Among the candidates that have been proposed to fill this role are coherence with other (established) theories, unifying power, consilience, generation of novel predictions, explanatory power, simplicity, elegance, parsimony, lack of ad hoc features, and fruitfulness. While realists think that (some of) these virtues have epistemic power, anti-realists think they are merely pragmatic, and, believe that, even if these factors play a role in theory-choice, they fail to be epistemically significant. Van Fraassen, for example, explicitly claims that the theoretical virtues "cannot rationally guide our epistemic attitudes and decisions" (1980: 87). Thus, part of the question that is at stake in the debate about underdetermination is the more general question of what constitutes an epistemically relevant factor for theory-choice, with anti-realists claiming that only the empirical evidence (or entailment thereof) counts, while realists hold either that the theoretical virtues themselves are somehow evidential, or, else, that, at the very least, they are capable of breaking the alleged epistemic tie between the rival theories in question. Realists claim that theories having these virtues are more likely to be true and that, therefore, these virtues provide extra reasons for believing in the approximate truth of those theories, while this is exactly what is denied by anti-realists.

However, despite the fact that the virtues frequently feature in realist responses to the underdetermination argument, there have not been any real attempts to develop substantial accounts of the virtues that could make them function robustly

[6] There are also some other points of contention, such as questions about what counts as a 'proper' rival, questions about the different relationships evidence and theory may bear to each other, etc. Since these discussions don't bear directly on my argument in this paper, I leave them aside here. For further detail, see Laudan and Leplin (1991), Hoefer and Rosenberg (1994), and Kukla (1998, Chapters 5 and 6).

either in debates about the argument itself, or in debates about the dimensions of justification of our scientific theories more generally. There are some accounts of individual virtues[7]; however, these accounts focus on specific virtues and, as such, do not aim to be more general accounts of how epistemic virtues function in science and do not address questions about how we might think about the epistemic function of various theoretical properties as they feature more generally in contexts of scientific justification.[8] Psillos (1999: chapter 8) perhaps comes closest to such an account, by proposing "a combination of the insights of Boyd and Salmon" (165). However, even Psillos admits that "[i]t hardly needs stressing that we are far from having such a theory [of theory-appraisal]" (2004: 405).

My paper can be read in the spirit of providing the beginnings of just such an account. Specifically, my aim is to show that it is – at least in principle – possible to uncover both theoretical virtues and their role in scientific justification through engaging in detailed scientific case-studies. Whether these virtues are ultimately to be regarded as epistemic or pragmatic is, at this point, an open question; however, the main point of this paper is not to show that specific virtues, or even the virtues in general *are* epistemic in nature (although I think its conclusions are suggestive of such a view), but, rather, to make plausible the view that the virtue-question is one that can be settled empirically, and to show that answers about the epistemic standing of the virtues will involve a robust engagement with contexts in which we see these virtues at play. Moreover, as I will argue in Sect. 4, putting the virtue-question on empirical grounds is actually *enough* to undermine the anti-realist argument, regardless of whether or not it will *in fact* turn out that the virtues are epistemic in nature: putting the virtue-question on empirical grounds is sufficient to undermine the argument from underdetermination, and accomplishing this shift to empirical grounds is the purpose of the extended case-study at hand. While one case-study is, of course, not enough to establish conclusions about the epistemic nature of the virtues, it is, I will argue, enough to make plausible the view that the virtue-question ought to be approached empirically.

3 Puerperal Fever in the Mid-1800s

If the virtue-question is to be thought of as an empirical one, the question arises as to what sort of empirical data is required to make the link between virtues and epistemic success. My goal in this section is to provide an example of this sort of data,

[7] Simplicity (and, relatedly, parsimony) are particularly notable in this respect. See, for example, Sober (1988, 1996, 2002a, b), Forster (1995a, b), Forster and Sober (1994), and Kelly (2007a, b).

[8] A noteworthy exception is McAllister (1989, 1996), who draws a distinction between virtues as indicators of truth and virtues as indicators of beauty, the former being formulated a priori, while "aesthetic criteria are inductive constructs which lag behind the progression of theories in truth-likeness" (1989: 25).

both to show in more detail what form it might take, and also in order to show that it is, in fact, possible to obtain it. Specifically, I want to show that it is possible to obtain it through the type of case-study that I am offering here (and, by extension, through others like it). One case-study by itself cannot establish the epistemic importance of anything; however, it does suggest that the virtues can function in an epistemic role, by showing just how they were made use of by the relevant practitioners at the time, who appealed to them as considerations in favour of their hypotheses over others. Moreover, regardless of the actual epistemic standing of the virtues, doing so is enough to undercut the main line of anti-realist defence in favour of the underdetermination argument. This argument will come in Sect. 4.

The case-study I want to look at is that of puerperal fever in the mid-1800s. This case was made famous by Hempel's discussion of Semmelweis in the Philosophy of Natural Science (1966), and has been taken up by a number of philosophers since (Lipton 2004; Gillies 2005; Bird 2007, 2010; Scholl 2013). Here, however, I am not interested in the various controversies surrounding Semmelweis (for a detailed discussion of these, see Tulodziecki (2013b)), but, rather, in an examination of the different hypotheses that were put forward with respect to puerperal fever, and with the sorts of considerations that were adduced in their favour at the time.

Puerperal fever was the most common cause of maternal mortality in the nineteenth century. Appearing after having given birth, its symptoms included shivering, a high pulse with accompanying fever, and extraordinary abdominal pain (to the extent that even being covered with sheets or blankets could not be tolerated). The sporadic version had a mortality rate of about 35 %; in epidemics, the mortality rate was as high as 80 %.[9] The medical situation with respect to puerperal fever (and most other diseases, for that matter) was rather complicated: there was no consensus on virtually any aspect of the disease – its symptoms, its pathology, whether slightly different sets of symptoms ought to all be classified as puerperal fever or different diseases, whether there was one type or many, whether it was transmissible from person to person (and, if so, in what ways), whether it had one cause or many, what the right treatments were, whether different treatments were called for under different circumstances, what factors would exacerbate prevalence of the disease, and who was particularly vulnerable to it. The only thing there was agreement on was that it caused a lot of deaths.

Since it is impossible to do justice to all of these issues, in what follows, I will focus on one specific issue with respect to which things were relatively more straightforward than they were with respect to other issues: that of the transmissibility of puerperal fever. For our purposes, we can distinguish the following two views: first, the view that puerperal fever could not be transmitted at all; second, the view that it could and, specifically, that it could be so transmitted through the hands of medical practitioners, such as doctors and midwives, from whom it would be

[9] There are widely diverging accounts as to what percentage of women actually contracted the disease, however. For more details on the disease and its history see Loudon (1986, 1992, 2000).

introduced into the birth canal and make women sick. An additional question concerned the issue of whether there were other routes of transmission besides the practitioners' hands, such as 'infected air' that might emanate from sick people. Since, however, adherents of the transmissibility hypothesis thought that the main route of transmission was via hands, regardless of whether there were other routes besides, and since this was the main focus of the debate, I will leave this additional complication aside and focus solely on the hand-hypothesis. According to this hypothesis, puerperal fever was propagated "through the medium of a third person; and that person generally the medical attendant or nurse" (Simpson 1851: 507).

In opposition to this was the telluric hypothesis, according to which the main cause of puerperal fever were telluric influences rising from the ground (a type of miasma theory).[10] These were thought to be the main cause of puerperal fever, which is why it was, for example, that such emphasis was put on the locations of hospitals (such as whether they were close to a swamp, on top of a hill, etc.). There were also certain other factors that were thought to exacerbate prevalence and intensity of the disease, such as overcrowding or bad ventilation and, in addition, it was thought that people of certain predispositions – for example women who had a 'bad constitution' or gone through difficult labour – were particularly prone to falling ill. While these other factors were thought to play a role, however, by themselves they were not taken to be sufficient for causing the disease, and some telluric influence or other was always thought to be required for doing so.[11] It was this last part in particular that was at odds with the hands-hypothesis, since proponents of the latter view thought that infection through matter on practitioners' hands, whence it was introduced into the birth canal, was often sufficient to cause the disease.[12]

What sorts of considerations, then, did proponents of the hand-hypothesis appeal to? Why did they think it was possible for puerperal fever to be transmitted via the hands of doctors and midwives? In what follows, I will discuss three theoretical virtues that are prominent on realists' lists – explanatory power, consilience, and the generation of novel predictions – and show that these were indeed considerations that played an important and, moreover, epistemic role in the arguments for the hand-hypothesis.[13]

[10] For details on different types and versions of miasmatic theories, see Baldwin (1999) and Worboys (2000).

[11] For a discussion of the distinction between predisposing and exciting causes, see Hamlin (1992).

[12] Note, however, that even adherents of the hand-hypothesis did not claim that puerperal fever could *only* be caused via hands; indeed, the hypothesis was perfectly compatible with puerperal fever also – just not primarily – being caused by telluric influences.

[13] This choice is motivated by the fact that these three virtues are ones on which there seems to be agreement among most realists (as opposed to simplicity or elegance, say); however, plenty of others can be found.

3.1 Explanatory Power

With this in mind, let's start looking at the virtues. One consideration that we find was consistently appealed to by virtually every proponent of the transmissibility-hypothesis was that this hypothesis could explain a number of phenomena the telluric hypothesis failed to explain. For example, it was pointed out that puerperal fever often followed a single practitioner "with the keenness of a beagle" (Holmes 1892: 157). Here is a typical passage from James Young Simpson, a mid-nineteenth century Scottish physician (also famous for introducing chloroform into anaesthesia during birth) that describes this phenomenon, which was a frequent occurrence:

> [T]hat it [puerperal fever] was so propagated by the medical attendant or nurse, we further believe upon the following species of evidence –viz. that it was..... distinctly and precisely limited to the practice of one or two practitioners only, out of a large number of medical practitioners, practicing in a large community. Many examples were recorded, and many more unrecorded were known to the profession, of the disease being this limited to the practice of a single practitioner in a town or city; all, or almost all, the patients of that practitioner being affected with it, where none of the patients of other practitioners were seized with any attack of the disease. In these cases we could not believe it to be owing to any morbific influence present in the air, or emanating from the locality in these cities or towns. For is so, it would affect indiscriminately the patients of all practitioners. But it had been often seen, as it was just now remarked, to haunt the steps of a single practitioner, and a single practitioner only, in a community. (1851: 507–508)

The first thing to note is that it is clear that the fact that the transmissibility hypothesis can explain what the telluric hypothesis cannot is considered to be "a species of evidence" for the transmissibility view. What is being pointed out by Simpson is that the telluric hypothesis simply cannot account for the existing patterns of infection and the transmissibility hypothesis can. If the telluric hypothesis had been correct, for example, one should have found a more uniform pattern of infection in areas similar to each other; in this case, if the telluric hypothesis were true, most women in a certain area ought to have been affected by the disease, not just specific individuals. The telluric hypothesis might have tried to account for this pattern by appealing to a combination of other factors (poor constitution, difficult labour, etc.). However, even in that case, there was no reason for the link between the disease and specific doctors or midwives, such as that in the following passage, an example of a common type of story:

> Dr. Roberton, of Manchester, tells us, that in 1840 upwards of 400 women were delivered by different midwives in connection with the Lying-in Hospital in Manchester. These 400 women were delivered in different parts of the town at their own houses: 16 of them died of puerperal fever; all the others made good recoveries. The production of this could not have arisen from any general epidemic, or atmospheric or telluric influence; for the fatal cases occurred in no one particular district, but were scattered through different parts of the town. Now, these 400 women and more were attended in their confinements by twelve different midwives. Eleven of these twelve midwives had no puerperal fever amongst their patients. The sixteen fatal cases had occurred in the practice of one only of the twelve. The disease, in fact, was limited entirely to her patients. There must have been something, then, connected with that one midwife, in which she differed from the other midwives, inasmuch as all her

> patients took the disease, whilst the patients of all the other midwives escaped from it. And in medical philosophy we cannot fancy that this something consisted of aught else than some form of that morbific principle or virus to which pathologists give the name of contagion. (1851: 508)

The telluric hypothesis, even one taking into account a number of modifying factors, such as individual predispositions and circumstances, simply had nothing to say about why the disease followed specific physicians and midwives. If this phenomenon had been an isolated one, one might have put it down to coincidence – a practitioner might have been unlucky in happening to attend women with particularly weak constitutions, for example – but the point being made is precisely that this phenomenon is widespread, and the norm rather than the exception, and that *that* is something for which the telluric hypothesis cannot provide an explanation.

There were also additional patterns that the telluric hypothesis lacked any sort of account of and that had a perfectly good explanation according to the transmissibility hypothesis, according to which such patterns were to be expected. These additional patterns were constituted, not by cases in which the disease followed a particular individual, such as above, but by cases in which it was possible to trace 'paths of infection': cases in which it was possible to track the spread of puerperal fever across a number of different practitioners who gave the disease to certain women only to be picked up by others who would in turn spread it among their patients and others who treated those patients. A striking example of this sort of data comes from Alexander Gordon, as early as 1795 (cf. Figure 1):

> The midwife who delivered No. 1 in the table [Fig. 1], carried the infection to No 2, the next woman whom she delivered. The physician who attended Nos. 1 and 2, carried the infection to Nos. 5 and 6, who were delivered by him, and to many others. The midwife who delivered No. 3 carried the infection to No. 4, from No. 24 to Nos. 25, 26, and successively to every woman whom she delivered. The same thing is true of many others, too tedious to be enumerated. (471)

Thus, whereas before, it remained mysterious why particular women would contract the disease, the transmissibility hypothesis could explain *who* got infected and how. The only option for the telluric hypothesis was, once again, to appeal to additional criteria besides the cosmic-telluric influence. However, even if such a story could be told, the different cases above would still have required separate and different explanations, whereas the transmissibility-hypothesis could supply a unified explanation for all cases in one stroke. Moreover, in the same way, it could also account for new (and previously puzzling) epidemics: "The midwife who delivered Mrs. K – carried the infection to No. 55 in Nigg, a country parish not far from Aberdeen, from whom it spread through the whole parish" (471).

Lastly, the transmissibility-hypothesis could explain the existence of anomalous regions that, mysteriously, were free from the disease. On the telluric hypothesis, there was no explanation for the existence of these regions; after all, if telluric conditions were responsible, they should have affected areas in their entirety, without pockets free from the disease. Thus, for example, it seemed "remarkable, that the puerperal fever should prevail in the new town, and not in the old town of Aberdeen, which is only a mile distant from the former" (472). Once again, however, Gordon has an explanation: “the mystery is explained when I inform the reader, that the midwife, Mrs. Jeffries, who had all the practice of that town, was so very fortunate as not to fall in with the infection, otherwise the women, who she delivered, would have shared the fate of others” (472). Gordon concludes that all this shows that “the cause of the puerperal fever, of which I treat, was a special contagion or infection, altogether unconnected with a noxious constitution of the atmosphere” (472).

To sum up: the fact that the transmissibility hypothesis could explain these various patterns and that the telluric hypothesis could not, was taken to count in favour of the transmissibility hypothesis. Among the phenomena that lacked an explanation on the telluric view, but that were explained and, indeed, expected, on the transmissibility view were: (i) the fact that the disease tended to follow specific practitioners, (ii) that it was possible to trace specific and detailed paths of infection both locally and across geographical areas, (iii) that it could explain why exactly those people who fell sick fell sick, (iv) how new epidemics came about, (v) why there existed ‘anomalous’ regions free from the disease, and (vi) how, precisely, the disease was spread through different parts of town, different towns, and, indeed, different geographical regions.

A TABLE,—Containing an account of those Patients affected with the Puerperal Fever, who were attended by Dr. Gordon, from December 1789 to October 1792.

When taken ill.	No.	Name.	Age.	Residence.	Cured.	Dead.	By whom delivered.
1789.							
December	1	James Garrow's wife	27	Woolman-hill		5th day	Mrs. Blake.
Ditto	2	James Smith's wife	30	Ditto		23d „	Ditto.
Ditto	3	John Smith's wife	34	Green		11th „	Mrs. Elgin.
Ditto	4	Al. Mennie's wife	25	Hardgate		11th „	Ditto.
1790.							
January	5	John Anthony's wife	25	North-street		3d „	Dr. Gordon.
February	6	Christian Durward	36	Rottenholes		3d „	Ditto.
April	7	Al. Stuart's wife	30	Denburn	1		Mrs. Philp.
May	8	William Elrick's wife	34	Exchequer-wynd	2		Mrs. Blake.
Ditto	9	Elizabeth Murray	28	North-street		7th „	Ditto.
Ditto	10	Helen Mitchell	30	Ditto	3		Ditto.
Ditto	11	Janet Wier	34	Denburn	4		Mrs. Elgin.
August	12	Mrs. Johnston	36	Littlejohn's-st.	5		Mrs. Smith.
Ditto	13	Geo. Webster's wife	38	Fowler's-wynd	6		Mrs. Blake.
Ditto	14	Peter Paul's wife	32	Windmill-brae	7		Ditto.
Ditto	15	John Low's wife	25	Justice-mills		5th „	Mrs. Smith.
Ditto	16	Mrs. Milne	27	North-street	8		Mrs. Blake.
Septemb.	17	Isabel Allan	36	Birnie's-close		5th „	Mrs. Coutts.
Ditto	18	Robert Burr's wife	30	Gallowgate		2d „	Mrs. Irvine.
October	19	Al. Eddy's wife	36	Ditto		3d „	Mrs. Clark.
Ditto	20	Agnes Milne	24	Putachie-side	9		Ditto.
Ditto	21	Al. Stuart's wife	26	Green	10		Mrs. Blake.
Ditto	22	Elizabeth Jamieson	25	Windmill-brae		5th „	Dr. Gordon.
Ditto	23	Dundas Nicol's wife	25	Green	11		Mrs. Philp.
Ditto	24	Al. Brown's wife	27	Loan-head		5th „	Mrs. Elgin.

Fig. 1 Taken from Gordon 1795, pp. 452–453

CASES AND DISSECTIONS. 453

TABLE (*continued*).

When taken ill.	No.	Name.	Age.	Residence.	Cured.	Dead.	By whom delivered.
1790.							
October	25	Anne Smith	24	Denburn		5th day	Mrs. Elgin.
Ditto	26	Mrs. Malcolm	25	Green		1st „	Ditto.
Ditto	27	Wm. Robertson's wife	30	Gilcomston		5th „	Mrs. Emslie.
Ditto	28	Jean Webster	17	Justice-port	12		Mrs. Anderson.
Novemb.	29	Anne Cumming	29	North-street	13		Ditto.
Ditto	30	Margaret Still	25	Ditto	14		Ditto.
Ditto	31	Janet M'Kay	38	Gallowgate	15		Mrs. Clark.
Ditto	32	Jean Laing	32	Ditto		7th „	Dr. Gordon.
Ditto	33	Mrs. Leitch	40	Carnegie's-brae	16		Ditto.
Ditto	34	Anne Barclay	20	Tannery-street	17		Mrs. Clark.
Decemb.	35	Mrs. Muffart	36	Hardgate	18		Mrs. Davidson.
Ditto	36	Jean Galloway	27	North-street	19		Mrs. Anderson.
Ditto	37	Janet Anderson	25	Putachie-side		5th „	Mr. Harvey.
Ditto	38	Mrs. ——	25	....		5th „	Dr. Gordon.
1791.							
January	39	Al. Main's wife	40	Poinernook		1st „	Mrs. Henderson.
February	40	Violet Thom	25	Green	20		Dr. Gordon.
Ditto	41	Mrs. Home	22	Carnegie's-brae	21		Mrs. Ogilvie.
Ditto	42	Mrs. Walton	25	North-street		11th „	Ditto.
Ditto	43	Elspet Riach	25	Ditto		5th „	Mrs. Balfour.
March	44	Janet Cormack	25	Back-wynd	22		Ditto.
Ditto	45	Andrew Duncan's wife	26	Ditto		5th „	Mrs. Blake.
Ditto	46	Anne Davidson	34	Justice-port	23		Mrs. Anderson.
Ditto	47	Elspet Fife	30	Windmill-brae	24		Mrs. Keith.
Ditto	48	Margaret Forbes	40	Footdee	25		Mrs. Anderson.
April	49	Janet Robertson	36	Correction-wynd	26		Mrs. Coutts.
Ditto	50	Wm. Gibbon's wife	27	Ditto	27		Dr. Gordon.
Ditto	51	John Duncan's wife	26	Woman-hill		7th „	Mrs. Keith.
Ditto	52	James Davidson's wife	25	Castle-street	28		Dr. Gordon.
Ditto	53	Rachel Gordon	36	Ditto	29		Mrs. Mitchell.
May	54	Mrs. Clark	25	Gallowgate	30		Dr. Gordon.
Ditto	55	George Duthie's wife	30	Torry		5th „	Mrs. Philp.
June	56	Anne Molison	27	Windmill-brae	31		Mrs. Emslie.
Ditto	57	Mrs. Henrie	30	Lodge-walk	32		Mrs. Elgin.
Septemb.	58	Elspet Robertson	25	Shoe-lane	33		Mrs. Blake.
Ditto	59	Rachel Leith	25	Back-wynd	34		Mrs. Taylor.
Ditto	60	Mrs. Thomson	25	Lodge-walk	35		Dr. Gordon.
October	61	Mrs. Ligertwood	30	Queen-street	36		Ditto.
Ditto	62	Widow Forbes		Printfield	37		Mrs. Taylor.
Novemb.	63	Mrs. Brown	42	Fintray		5th „	Mrs. Mitchell.
Ditto	64	Mary Meldrum	32	Windmill-brae		5th „	Mrs. Chalmers.
Decemb.	65	Jean Brown	36	Vennel	38		Mrs. Anderson.
Ditto	66	Margaret Yull	23	Castle-street	39		Dr. Gordon.
Ditto	67	Anne Hervie	23	Woman-hill	40		Mrs. Keith.
Ditto	68	Isaac Allan's wife	22	Windmill-brae	41		Mrs. Emslie.
1792.							
January	69	Mrs. White	30	Printfield		5th „	Mrs. Keith.
Ditto	70	Mrs. Byrn	27	Broadgate	42		Mrs. Philp.
Ditto	71	Christian Sangster	30	Green	43		Mrs. Ogilvie.
February	72	Al. Sim's wife	27	Printfield	44		Mrs. Chalmers.
Ditto	73	James Gordon's wife	28	Ditto	45		Dr. Gordon.
Ditto	74	Mrs. Mather	26	Drum	46		....
March	75	Tho. Wallader's wife	36	Printfield	47		Mrs. Keith.
Ditto	76	Mrs. Imlach	24	Pesly	48		Dr. Gordon.
October	77	Anne Skinner	36	Gallowgate	42		Ditto.

Fig. 1 (continued)

3.2 *Consilience*

Another much cited virtue is consilience. The idea behind this virtue is that it speaks in favour of a hypothesis if that hypothesis can account for types of phenomena that did not play a role in the formation of the original hypothesis. The term 'consilience' goes back to Whewell, who thought that this was a property exhibited by "only the best established theories which the history of science contains" (1858: 88):

> No accident could give rise to such an extraordinary coincidence. No false supposition could, after being adjusted to one class of phenomena, exactly represent a different class, where the agreement was unforeseen and uncontemplated. That rules springing from remote and unconnected quarters should thus leap to the same point, can only arise from *that* being the point where truth resides. (ibid.)

It turns out that this is the case for the transmissibility hypothesis: the hypothesis that puerperal fever is transmissible through the hands of medical practitioners can explain a number of phenomena "different from those which were contemplated in the formation of our hypothesis" (ibid.). For example, the transmissibility hypothesis can explain a variety of phenomena concerning erysipelas (a streptococcal rash) even though considerations about erysipelas were not part of the original evidence for the transmissibility hypothesis. Among the claims being made by proponents of the transmissibility hypotheses was

> that when the fingers of medical men were impregnated with the morbid secretions thrown out in erysipelatous inflammation, the inoculation of these matters into the genital canals of parturient females produced puerperal fever in them in the same way as the inoculation of the secretions from patients who had died of puerperal fever itself. The effused morbid matters in the one disease, as in the other, were capable of producing the same effect when introduced into the vagina of a puerperal patient. (Simpson 1851: 516)

As before, there are plenty of stories to support this conclusion. Here is a representative example:

> In an instance recorded by Mr. Hutchinson, two surgeons, living at ten miles' distance from each other, met half-way to make incisions into a limb affected with erysipelas and sloughing. Both practitioners touched and handled the inflamed and sloughing parts; and the first parturient patients that both practitioners attended within thirty or forty hours afterwards, in their own distant but respective localities, were attacked with, and died of, puerperal fever. The late Mr. Ingleby mentions an instance of a practitioner making incisions into structures affected with erysipelas, and going directly from this patient to a patient in labour. This patient took puerperal fever and died. And within the course of the next two days, seven cases of puerperal fever occurred in the practice of the same practitioner, almost all of them proving fatal. And various other cases, similar to the preceding, were well known to the profession. (Simpson 1851: 516)

In addition, it was also observed that the connection between puerperal fever and erysipelas did not just go one way, but both, that is, "[n]ot only was the morbid matter in erysipelas apparently sometimes capable of producing puerperal fever, but the secretions and exhalations form puerperal fever patients seemed, on the other hand, sometimes capable of producing erysipelas" (Simpson 1851: 516–517). In fact, it was pointed out not just that erysipelas produced puerperal fever in patients of doctors who had treated erysipelas, but, in addition, that it was found that the patients' secretions "produced also erysipelas in several of the nurses, relations, and attendants upon the patients" (Simpson 1851: 516/517). For example, the doctor Mr. Sidey had a patient die of puerperal fever, and it was found, in the week following the patient's death, that

> [t]he patient's mother-in-law, who was in constant attendance upon her, was attacked with fever and erysipelas of the face and head. One of the patient's sons, a boy five years of age, was attacked with erysipelas of the face; a daughter was seized with fever and sore throat, with dusky redness, which continued for some time; and the patient's sister-in-law was attacked with acute gastric symptoms, and great abdominal irritation, under which she sank in a few days. Here we have apparently the same focus on contagion producing puerperal fever in puerperal patients, and erysipelas, inflammatory sore throat, etc., in patients who were not in a puerperal state. (Simpson 1851: 516–517)

Lastly, it was pointed out that "[t]he two diseases had in Britain been repeatedly observed to prevail at the same time, in the same town, in the same hospital, or even in the same wards" (Simpson 1851: 515–516). All these phenomena would have been puzzling on the assumption that the telluric hypothesis was true, yet, once again, they are exactly what is to be expected on the transmissibility hypothesis. Moreover, the original formulation of the hand-hypothesis was not based on evidence about erysipelas, but, rather, on phenomena concerning the connection between the onset of puerperal fever and treatment by specific individuals who had previously been associated with the disease in one way or another. Yet, the transmissibility hypothesis could "explain and determine" these different types of cases, specifically, (i) the observation that puerperal fever was often contracted by patients whose doctors had previously treated cases of erysipelas, (ii) the reverse, i.e. that cases of erysipelas often followed incidents of childbed fever, and (iii) the coinciding of epidemics of erysipelas and puerperal fever.

3.3 The Generation of Novel Predictions

To end, I want to briefly draw attention to one last virtue that is to be found on virtually every list: the generation of novel predictions.[14] Again, this virtue is one we can clearly see exemplified by the transmissibility hypothesis and not by the telluric hypothesis: while the telluric hypothesis was able to tell, in certain cases, a story of how the disease might have come about in certain areas (cosmic-telluric conditions) and in certain people (a combination of conditions and predispositions), it failed to provide any sort of systematic account of why particular individuals fell sick. With *hindsight*, it was possible for the theory to look at a sick person, and invoke various factors that might have contributed to that person's falling sick; however, the reasons were different every time, and while it was always possible to appeal to certain contributing factors, say, a weak constitution, even this could

[14] An exception is Solomon (2001: Chapter 2).

not be done systematically, since it just wasn't the case that there was an actual correspondence between these properties and incidences of the disease. For example, it simply was not the case that, generally speaking, people with sickly dispositions, those who had gone through difficult labour, or those who had conceived out of wedlock, were more prone to the disease, even though these factors would with hindsight be invoked as explanations for why specific women fell ill. Indeed, the list of potential factors was so long that *some* explanation along the above lines could always be found, since the number and combination of different factors was simply huge. Despite this, however, the telluric hypothesis was in no position to make any predictions whatsoever: it could not predict who would fall ill or what groups of people might fall ill, since no cause by itself was deemed sufficient, and any combination of factors might or might not actually bring about the disease. The case was quite different for the transmissibility hypothesis. It could predict, quite neatly, that if a certain practitioner, for example, had been in touch with victims of childbed fever, there would likely be death among his or her patients in the immediate future; it could predict that 'thorough cleansing' on the practitioner's part would diminish this possibility; it could predict that if there were cases of erysipelas, childbed fever would soon follow (and the other way round), that outbreaks of surgical fever in hospitals would usually be followed by outbreaks of puerperal fever in the nearby maternity wards (due to cross-contamination and lack of hygienic measures on the part of surgeons who would move frequently between the maternity wards and operating theatres), and so on. In short, it made predictions about who the likely next victims would be. As Gordon puts it:

> I could venture to foretell what women would be affected with the disease, upon hearing by what midwife they were to be delivered, or by what nurse they were to be attended during their lying-in; and almost in every instance my prediction was verified. (1795: 447)

So, as we can see, in the case of the debate about the origin of puerperal fever, some of the most prominent theoretical virtues – explanatory power, consilience, and generation of novel predictions – were invoked in favour of the transmissibility hypothesis, while they do not feature similarly in the telluric hypothesis.

4 Epistemic Virtues and the Argument from Underdetermination

The case-study shows that there are at least some instances in which it is plausible to think that the virtues make an epistemic contribution to our theories and hypotheses. Moreover, instead of abstractly suggesting the possibility of epistemic tie-breakers, it offers a concrete scenario in which we can see specific virtues doing specific epistemic work. By showing *how* empirical ties may actually be broken by specific considerations, it severely calls into question premise 2 of the original underdetermination argument, the premise that claims that the only epistemic constraint on theory-choice is a theory's empirical evidence. It does so precisely because it shows that and how other criteria were in fact invoked in a case of real-life theory-choice, and by showing that they were used in order to argue for the *truth* of the transmissibility hypothesis – a distinctly epistemic context.

Of course, the mere fact that they were invoked in such a context does not make them generally epistemic; however, the case-study makes plausible the view that they might be, by suggesting that, contra premise 2, there are at least some cases of empirically equivalent theories that are not equally believable – certainly, it shows that this was so for the practitioners in the case above. Standing on its own, the case-study does not – and cannot – establish an epistemic connection between the virtues and truth or epistemic success more generally.[15] However, what it does do is show us both what sort of data is required to do so, and also that it is possible to obtain this kind of data by conducting case-studies of a particular kind. Thus, by pointing to the (potential) connections between specific theoretical virtues and epistemic success, the case of puerperal fever shows us how it is in principle possible to make the connection between virtues and epistemic success that scientific realists require.[16] Moreover, the case shows that whether or not the virtues are ultimately to be regarded as epistemic is an *empirical* question that can only be settled by more extensive, systematic, and detailed examination of relevant data.

Thus, once virtues such as the above are taken into account, anti-realists need to do more than establish mere empirical equivalence: they also need to establish equivalence with respect to the virtues. And whereas, in the context of the old underdetermination argument, it was possible for the anti-realist to appeal to algorithms or skeptical hypotheses in order to do so, this is no longer the case once the lessons of the case-study are taken into account. The anti-realists' argument for underdetermination, in order to successfully target the epistemic thesis, needs to be a general and in-principle argument working against any theory whatsoever, but putting the virtue-question on empirical grounds takes away the *guarantee* that anti-realists need in order to show that there is always underdetermination of the required kind. Crucially, the case actually undercuts this in-principle strategy *regardless* of whether, in fact, the virtues make epistemic contributions to our theories or not: even if they don't, the case-study establishes that it's plausibly *possible* for them to do so, and that's all that realists require. By moving the discussion away from in-principle arguments to discussions of real and individual cases, the case-study changes the game about underdetermination: what it makes clear is that it's virtually impossible to figure out, by means of a general argument, in advance, and without examining in detail the empirical data, whether a certain tie can or ought to be broken (much less whether this is the case for all and any ties). It is in this sense that the case-study is bad news for the anti-realist. Unless the anti-realist can show that underdetermination is universal, the realist can maintain that we are, at least sometimes, justified in picking certain theories over others on the basis of epistemic

[15] What connection exactly is required depends on one's understanding of epistemic success, and on whether one is a monist or pluralist about this notion.

[16] I discuss this point in more detail in Tulodziecki (2013a).

criteria, and proceed to argue that our current theories fall into this class (as many prominent realists have done).

A favourite recourse in the context of the old underdetermination argument was the idea that empirical equivalence could always be maintained or produced, if need be, by appealing to skeptical hypotheses (see, for example, Kukla (1993, 1996), or algorithms (van Fraassen (1980)). As we have already seen, in the context of a new underdetermination argument that includes the virtues, anti-realists need to produce virtue-equivalence on top of empirical equivalence. Initially, one might think that all we are doing is moving underdetermination one level up: all we need to do is to invoke a higher-level underdetermination argument that includes the virtues, and, as a result, our basic predicament is the same and we simply exchanged the old version for a new. However, the case-study undercuts this response by example, by highlighting that anti-realists can no longer appeal to a mechanism for establishing virtue-equivalence in the new argument whose role is equivalent to that of sceptical hypotheses and algorithms in the old.

The anti-realist might claim that, surely, any theory will possess *some* virtues, and that that's all that is needed to establish empirical-cum-virtue-equivalence. However, once we consider the anti-realists' options in more detail, we see that that is not so. The anti-realists' first option is to show that for every theory T there is a theory T* that has the same virtues as T. The only way in which this is possible would be to show – for every single virtue – that it is possible to engineer it into any given theory, in just the same way in which empirical equivalence can be engineered into T* through sceptical hypotheses. If this was possible, one could then show that any theory would have a rival theory T* that ties with respect to the theoretical virtues, because we could selectively engineer into T* exactly the virtues that are required for the equivalence at hand. So, for example, if our original theory was found to possess the virtues of simplicity and coherence with other theories, we could simply engineer these virtues into T*. However, even if this was possible for *some* of the virtues, it's hard to see how this would be possible for *all* of the virtues (how, for example, could fruitfulness or consilience be engineered?). This is especially so in view of the fact that realists don't make any claim to the exhaustiveness of their current list. Indeed, on this approach, it's quite likely that further detailed studies will turn up various properties not currently on the list, and it's unclear how anti-realists could already now give arguments for virtues we don't even know about at this point.

A slightly less demanding option would require only that anti-realists show that it's possible for *some* virtues to be had by any theory, and to argue that this is enough. For example, they might argue that, unless realists can produce a ranking of how the different virtues compare to each other, as long as a theory has some virtue or other, there is simply no way to argue for the epistemic superiority of any one theory over another. So, while we might be able to pick theories having virtues over those that lack them, once virtues are at play in both theories, we don't really have the resources to make comparisons.[17] In this vein, anti-realists might for example

[17] For some details on how such an argument might go, see Tulodziecki (2012).

argue that we can always add unifying power to a theory by proposing a single cause for a variety of phenomena, or simplicity by invoking demons. The case-study, however, makes clear that even if this was possible, this would no longer be enough, because what it generates is an account of *how* certain virtues function in epistemic contexts, and not just *that* certain hypotheses might be linked with specific virtues. For anti-realists to achieve something comparable, thus, would involve being able to show not just that we can add some 'virtue' to our theories, but also showing how it is that that virtue plays a role in T* that is analogous to the role it or some other virtue plays in T. And, once again, it's unclear how *that* could be established by means of an all-encompassing and general argument that applies to any possible theory in advance.[18]

5 Objections

Let's now move on to some worries one might have about the general strategy in this paper. Even if one agrees that anti-realists would have trouble establishing a new form of underdetermination argument that both takes into account the virtues and also remains as strong as the old, there might be other concerns one might have, the most obvious one, perhaps, being connected to a potential circularity.

Specifically, one might worry about how we can ever be in a position to judge whether a virtue is linked to epistemic success or not. After all, we cannot appeal to the virtues themselves in order to judge success, since doing so would involve us in a circularity: if we judge to be successful those theories that exemplify (some of) the virtues, we cannot then invoke that very same success in order to determine what constitutes a virtue in the first place. Unbridled access to the truth, of course, would solve this problem, since this would allow a proper and independent check on whether the virtues were, in fact, successful; however, defending the view that science gives us this sort of unbridled truth-access is a view few would be willing to defend or even believe (certainly I'm not willing to do either). So, then, how do we judge whether the virtues do any epistemic work without either appealing to the virtues themselves or invoking otherwise implausible views of science?

I think there are two main lines of response to this. The first is that in order to judge success, we don't need anything like unbridled access to the truth. What we are interested in are theories and hypotheses that we judge to be epistemically successful, but making such judgements needn't involve truth (even though, of course, it can). Instead, it is possible to hold that there are many different kinds of epistemic success, and while some of them might involve truth, they need not do so. We might recognise other legitimate epistemic goals, and do so in a way that goes beyond the rather limited anti-realist conception of epistemic success as only empirical adequacy; for example, we might be interested in gaining understanding, providing solutions to certain puzzles, adhering to certain standards (rigorous testing,

[18] For further implications, see Tulodziecki (2013a, especially Section 6).

well-calibrated instruments, etc.), and so on. And, regardless of whether one holds these goals to be only derivative on our way to truth or to be epistemically valuable in their own right, they can certainly function as examples of epistemic success that do not involve independent access to the truth.[19] So, one might claim, for example, that the Bohr-Rutherford model of the atom provides understanding of the (structure of) the Rydberg formula about the wavelengths of spectral lines, even though the model is no longer taken to be correct (and, indeed, was inconsistent, so could not have been right at any rate) and has been replaced by atomic orbitals.[20] We could – and still can – view this as an instance of epistemic success, despite the fact that we have no access to the 'real truth' about electrons (or whether there even are any). Certainly, it would be helpful at this point to have a more precise notion of different kinds of epistemic success (a project that is worth pursuing and about which the literature is curiously silent); however, we can make quite a lot of headway even without this. Regardless of what sorts of details about epistemic success we might end up with, realists and anti-realists agree when it comes to picking out 'better' over 'worse' theories, despite the fact that we lack a clear definition of what exactly constitutes a theory's good- or badness. As long as there is substantial agreement on which theories are generally regarded as better or worse – and there is – the fact that our notion of epistemic success is vague, and perhaps even subjective, is not an obstacle. After all, we are interested precisely in what distinguishes those theories that we consider 'good' from those we consider inferior in certain ways.[21]

A second point worth stressing is that, even if we eschew the notion of unbridled access to the truth in general, we often *do* have access to our paradigm standard of success: there often *are* instances in which we know that we got things right, and instances in which we know that we didn't (at least to the extent that scientific knowledge is possible at all). For example, we now know that the earth goes around the sun, and not the other way round, we now know that certain diseases are caused by specific microorganisms (indeed, this is what it means to have such diseases), and we now know that the planets move neither in circles, nor perfect ellipses, that there is no phlogiston or luminiferous aether, and that descent with modification is a fact. If we believe in scientific knowledge at all, this is as good as it gets. And, for our purposes, that is enough: we can examine how people arrived at conclusions that we now take to be established, and we can examine what sorts of arguments they were putting forward in favour of their views, what sorts of properties they appealed to in the course of doing so, and what sorts of considerations went into their accepting a number of conclusions that others rejected.

Another worry grants that we have a way (even if it's rough) of distinguishing more from less successful theories, and focuses on how we can actually distinguish properties that are genuinely epistemic from those that aren't. After all, our theories

[19] If one thinks truth is the only intrinsically epistemically valuable goal in science (as I don't), one also faces the task of explaining how the derivative goals are related to truth.

[20] For more details, see Norton (2000).

[21] Note also that we don't need perfect agreement; a number of agreed-upon cases is enough to get this project off the ground.

have all kinds of prima facie non-epistemic properties, such as having been conceived by someone with a certain shoe size, having been thought of on a certain weekday, and so on. So, given any property at all, how do we go about ascertaining whether that property is epistemic or not?

Recall that what we are interested in is whether having certain properties may in any way legitimately be viewed as making an epistemic difference to our theories. Clearly, examining a number of successful theories and finding that they all share a property is insufficient. Rather, we need to examine *both* cases of success and cases of failure. After all, we are interested in finding out whether successful theories have properties that unsuccessful ones lack, and simply examining one of the two classes is not enough to help us answer this question. Take the case, for example, in which we are interested in ascertaining whether a certain property is linked to a certain eye-colour, say, having blue eyes. Examining only blue-eyed people will not tell us what we are interested in (unless they lack the property, in which case we need not enquire further), since, even if it turns out that all blue-eyed people have the property, this will not tell us that the property is, in fact, associated with being blue-eyed unless we also know that non-blue-eyed people lack it. Failing to examine both blue-eyed and non-blue-eyed people and focusing on blue-eyed people only will presumably turn up all sorts of spurious associations. To see this, assume we find, after careful and thorough examination, that a very high percentage of blue-eyed people own umbrellas. It would be a mistake to conclude from this that umbrella-owning is associated with blue eyes, since (presumably) examining people with other eye-colours would make clear that there is no difference along these lines. Similarly, in the case of scientific theories, it is important to examine both more and less successful theories, in order to tell whether they exhibit systematically different properties. Our concern is whether there is a higher incidence of a given property in one class as opposed to the other, and this is something we can only determine by looking at both.

However, it is worth nothing that even if we do find that certain properties are systematically associated with epistemic success, merely having an association is, of course, not enough to tell us whether the property is in any way *responsible* for this success.[22] It might be, for example, that certain properties are only 'derivatively' associated with success, in the sense that there is nothing about the property itself that somehow brings about the success; yet we might be able to give an explanation of why it is so associated. The place of the above case-study in this endeavour is to show how, in principle, one might conduct such studies.

To end, one last potential worry I want to address is one that arises in any analysis that appeals to case-studies for philosophical purposes – that of selection bias. The worry is that we might get skewed results, because we are more (or less) likely to select certain theories rather than others. For example, one might worry that we tend to select cases of spectacular scientific successes, simply because those cases are more famous and more likely to come to mind, and, as a result, we might tend to identify plenty of virtues in those theories, despite the fact that those virtues are

[22] The usual worries about causation vs. correlation and related issues arise at this point.

not typical. Perhaps those cases are so exceptional that they tend to have virtues when, in fact, having those virtues is not a feature of successful theories in general. Likewise, when examining failures, the first cases that spring to mind might be spectacular failures, lacking any kind of virtue, despite the fact that 'average failures' might exemplify some of them. In short, the worry is that we are generating artificially produced contrasts by tending to select only cases at the extreme ends of the spectrum.

While this is an important issue, I do not think it is a reason for deterrence (although it is something to be aware of and careful about). This is so especially in view of the fact that there are plenty of things that can be done to alleviate this concern. For one, as I have already emphasised, we need to look at both cases of success and cases of failure. In order to avoid the worry of picking theories from either extreme, a good strategy is to look at episodes as a whole, not just one side of a given debate. This means looking at both those who got it right and those who didn't, and engaging in a detailed analysis of the sorts of criteria different camps appealed to in establishing their views, regardless of outcome. In this context, moreover, it is also useful to examine specific hypotheses rather than whole-sale theories: by focusing on more localised instances of (non-)success and by determining the epistemic factors that were involved in the generation of specific claims, as opposed to large-scale theories, we can get a more balanced picture of the hypotheses in question. The focus, in this case, will be not just on the best of the best or the worst of the worst, but, instead, on the most successful parts of less successful theories and on the least successful parts of more successful theories, thus ensuring that we are examining a variety of 'good' and 'bad' claims across a number of different contexts. Lastly, we ought to examine a succession of theories or claims in a certain domain, regardless of their degree of success; there are high and low points in the history of any domain and examining claims chronologically, without gaps, ensures a more balanced analysis.[23]

6 Conclusion

The case of the debate about the transmissibility of puerperal fever in the mid-nineteenth century shows that no good answer to the virtue-question can be had without taking into account detailed empirical studies of the virtues in action. In order to ascertain whether there are virtues that contribute to the epistemic standing of our scientific theories, we need to examine cases in which we can observe the virtues at work. Putting the virtue-question on empirical grounds in this way – regardless of whether it turns out that the virtues are actually epistemically potent – is, as we have seen, sufficient to undermine the general anti-realist position on underdetermination. If it turns out that there really are properties that are

[23] For discussions of a variety of issues arising at the intersection of epistemology and history of science, see Feest and Sturm (2011).

systematically associated with epistemic success, those virtues bear close scrutiny. As I have noted, merely being associated with success does not tell us whether the virtues are, in any way, responsible for said success; however, even finding out whether there are any such associations (and, if so, what they are) would be an important step towards an account that seeks to put the virtues to real use in contexts of actual, real-life scientific justification. As such, thinking about epistemic virtues – and other epistemic properties that scientific theories might possess – goes much beyond resolving issues to do with underdetermination, or even the realism debate. Rather, it is the beginning of an account that seeks to shed light on more general questions about the epistemic status of scientific theories, such as questions about what sorts of factors make epistemic contributions to our scientific hypotheses and in what ways they do so.

Acknowledgements I would like to extend my thanks to John Norton for helpful conversations, and, especially, to Uljana Feest, both for very helpful discussions and also for many thoughtful comments on a previous version of this paper.

References

Baldwin, P. 1999. *Contagion and the state in Europe, 1830–1930*. Cambridge: Cambridge University Press.

Bird, A. 2007. Inference to the only explanation. *Philosophy and Phenomenological Research* 74(2): 424–432.

Bird, A. 2010. Eliminative abduction: Examples from medicine. *Studies in History and Philosophy of Science* 41: 345–352.

Churchill, F. (ed.). 1849. *Essays on the puerperal fever and other diseases peculiar to women*. London: The Sydenham Society.

Earman, John. 1993. Underdetermination, realism, and reason. *Midwest Studies in Philosophy* 18(1): 19–38.

Feest, U, and T. Sturm (eds.). 2011. What (good) is historical epistemology? *Erkenntnis* 75(3): 285–302.

Forster, M.R. 1995a. The golfer's dilemma: A reply to Kukla on curve-fitting. *British Journal for the Philosophy of Science* 46(3): 348–360.

Forster, M.R. 1995b. Bayes and bust: Simplicity as a problem for a probabilist's approach to confirmation. *British Journal for the Philosophy of Science* 46(3): 399–424.

Forster, M.R., and E. Sober. 1994. How to tell when simpler, more unified, or less ad hoc theories will provide more accurate predictions. *British Journal for the Philosophy of Science* 45(1): 1–35.

Gillies, D. 2005. Hempelian and Kuhnian approaches in the philosophy of medicine: The Semmelweis case. *Studies in History and Philosophy of Biological and Biomedical Sciences* 36(1): 159–181.

Gordon, A. 1795. A treatise on the epidemic puerperal fever, etc. In *Essays on the puerperal fever and other diseases peculiar to women*, ed. F. Churchill (1849). London: The Sydenham Society.

Hamlin, C. 1992. Predisposing causes and public health in early nineteenth-century medical thought. *Social History of Medicine* 5(1): 43–70.

Hempel, C.G. 1966. *Philosophy of natural science*. Englewood Cliffs: Prentice-Hall.

Hoefer, C., and A. Rosenberg. 1994. Empirical equivalence, underdetermination, and systems of the world. *Philosophy of Science* 61: 592–607.

Holmes, O.W. 1892. *The works of Oliver Wendell Holmes: Medical essays 1842-1882*, vol. 9. Boston/New York: Houghton, Mifflin, and Company.
Kelly, K. 2007a. A new solution to the puzzle of simplicity. *Philosophy of Science* 74: 561–573.
Kelly, K. 2007b. How simplicity helps you find the truth without pointing at it. In *Induction, algorithmic learning theory, and philosophy series: Logic, epistemology, and the unity of science*, vol. 9, ed. M. Friend, N. Goethe, and V. Harizanov, 111–143. Dordrecht: Springer.
Kukla, A. 1993. Laudan, Leplin, empirical equivalence, and under- determination. *Analysis* 53: 1–17.
Kukla, A. 1996. Does every theory have empirically equivalent rivals? *Erkenntnis* 44(2): 137–166.
Kukla, A. 1998. *Studies in scientific realism*. New York: Oxford University Press.
Laudan, L., and J. Leplin. 1991. Empirical equivalence and underdetermination. *Journal of Philosophy* 88: 449–472.
Lipton, P. 1991/2004. *Inference to the best explanation*. London/New York: Routledge.
Loudon, I. 1986. *Medical care and the general practitioner: 1750–1850*. Oxford: Oxford University Press.
Loudon, I. 1992. *Death in childbirth*. Oxford: Clarendon.
Loudon, I. 2000. *The tragedy of childbed fever*. Oxford: Oxford University Press.
McAllister, J.W. 1989. Truth and beauty in scientific reason. *Synthese* 78(1): 25–51.
McAllister, J.W. 1996. *Beauty and revolution in science*. Ithaca: Cornell University Press.
McMullin, E. 1987. Explanatory success and the truth of theory. In *Scientific inquiry in philosophical perspective*, ed. N. Rescher. Lanham: University Press of America.
Norton, J. 2000. How we know about electrons? In *After Popper, Kuhn and Feyerabend; Recent issues in theories of scientific method*, ed. R. Nola and H. Sankey, 67–97. Dordrecht: Kluwer.
Psillos, S. 1999. *Scientific realism: How science tracks truth*. London/New York: Routledge.
Psillos, S. 2004. Tracking the real: Through thick and thin. *British Journal for the Philosophy of Science* 55(3): 393–409.
Scholl, R. 2013. Causal inference, mechanisms, and the Semmelweis case. *Studies in History and Philosophy of Science* 44: 66–76.
Simpson, J.Y. 1850. Chapter 10: The analogy between puerperal and surgical fever. In *Childbed fever: A documentary history*, ed. I. Loudon. New York: Garland Publishing: 1995.
Simpson, J.Y. 1851/1871. *Selected obstetrical & gynaecological works of Sir James Y. Simpson: Containing the substance of his lectures on midwifery*, ed. J. Watt Black. Edinburgh: Adam and Charles Black.
Sober, E. 1988. *Reconstructing the past – Parsimony, evolution, and inference*. Cambridge: MIT Press.
Sober, E. 1996. Parsimony and predictive equivalence. *Erkenntnis* 44(2): 167–197.
Sober, E. 2002a. Instrumentalism, parsimony, and the Akaike framework. *Proceedings of the Philosophy of Science Association* 69(S3): S112–S123.
Sober, E. 2002b. What is the problem of simplicity. In *Simplicity, inference, and modelling*, ed. A. Zellner, H. Keuzenkamp, and M. McAleer, 13–32. Cambridge: Cambridge University Press.
Solomon, M. 2001. *Social empiricism*. Cambridge: MIT Press.
Tulodziecki, D. 2012. Epistemic equivalence and epistemic incapacitation. *British Journal for the Philosophy of Science* 63(2): 313–328.
Tulodziecki, D. 2013a. Underdetermination, methodological practices, and realism. *Synthese* 190(17): 3731–3750.
Tulodziecki, D. 2013b. Shattering the myth of Semmelweis. *Philosophy of Science* 80(5): 1065–1075.
van Fraassen, B. 1980. *The scientific image*. Oxford: Clarendon.
Whewell, W. 1858. *Novum Organon Renovatum*. London: John W. Parker.
Worboys, M. 2000. *Spreading germs: Diseases, theories, and medical practice in Britain, 1865–1900*. Cambridge: Cambridge University Press.

Experimental Virtue: Perceptual Responsiveness and the Praxis of Scientific Observation

Shannon Vallor

1 Introduction

A proper account of the scientific virtues must perform a delicate task: it must articulate and explain the genuinely *normative* implications of each of those virtues, while simultaneously ensuring that the virtues described do, in fact, characterize ideal scientific practice and are not simply carted over from the domain of conventional moral action without concern for their appropriateness to the scientific enterprise. It follows from this that any scientific virtue properly identified by such an account ought to be discoverable from historical examination of the habits, dispositions and attitudes of scientists widely credited with a lifetime of excellent scientific practice. While such a virtue need not be manifest in all scientific practice, and in fact as an 'excellence' we should expect that it will *not* be so manifest in the ordinary course of scientific work, a purported scientific virtue ought to been seen at work in cases of *exceptional* or distinctly *praiseworthy* lifetimes of scientific practice. Furthermore, we should expect any such virtue to be recognized and acknowledged as normatively binding by exceptional and praiseworthy scientists themselves.

In this chapter I will develop and defend an account of one particular scientific virtue, one not easily identifiable among traditional lists of the epistemic *or* the moral virtues, though components or preconditions of this virtue are found in most such accounts. Although my special focus here will be the manifestation of this virtue of scientific character in experimental/observational praxis, I will show how this virtue functions in both experimental and theoretical contexts, and is in fact critical to the excellent function of each as a guide and constraint for the other. While there is no English term that captures precisely the meaning of the virtue I shall emphasize, the nearest approximation would be *perceptual responsiveness*. The virtue of being

S. Vallor (✉)
Santa Clara University, Santa Clara, CA, USA
e-mail: SVallor@scu.edu

A. Fairweather (ed.), *Virtue Epistemology Naturalized: Bridges Between Virtue Epistemology and Philosophy of Science*, Synthese Library 366, DOI 10.1007/978-3-319-04672-3_16,

perceptually responsive is conceptually complex, and will require precise definition and clarification in Sect. 2 to remove any confounding ambiguities.

It will also be necessary to establish the appropriateness of this virtue as a norm of scientific practice. In order to accomplish this, I draw in Sect. 3 upon the phenomenological philosophy of Edmund Husserl, which helps us to connect the virtue of perceptual responsiveness in experimental/observational praxis back to normative considerations of scientific rationality and truth. We first see such a link near the end of Husserl's *Ideas I,* where he defines rationality in the perceptual situation as a proper responsiveness to the 'motivating' content of a phenomenon (1982, p. 328). The structure of motivation and response forms the core of any phenomenological understanding of perception, and extends to scientific observation and experiment. This phenomenological account, with its emphasis on scientific inquiry as a reciprocal and communicative interaction with nature, will also help us to draw out some instructive parallels between scientific/epistemic virtue and moral virtue. Maurice Merleau-Ponty described perception as the "art of interrogating [the phenomenon] according to its own wishes" (1968, p. 133). While most philosophers of science would blanch at any suggestion that natural phenomena have their own "wishes" about how the scientist views or interacts with them, I will employ Joseph Kockelmans' phenomenology of the physical sciences to reconstruct the point in a fashion that even the most hardheaded philosopher of science can appreciate.

In order to complete this reconstruction, I show in Sect. 4 how the virtue of responsiveness can be read from the experimental practices and explicit normative commitments of two historical models of scientific excellence, each of whom earned an enduring reputation for a lifetime of extraordinary scientific achievement: seventeenth century experimenter Robert Hooke and twentieth century cytogeneticist Barbara McClintock. While Hooke and McClintock otherwise possessed markedly different personalities and strengths, evidence from a brief study of their characteristic habits, dispositions and attitudes suggests that they both shared and regarded as normatively binding the form of excellence I have termed *perceptual responsiveness.*

Finally, Sect. 5 will explore how the scientific virtue of perceptual responsiveness to motivating appearances of natural phenomena invites comparison with moral contexts of virtue, where the *practical wisdom* of a moral agent entails a disposition to perceive and enact the appropriate response motivated by the moral phenomena embedded in the particulars of the practical situation. While noting significant obstacles to a true identification of perceptual responsiveness with practical wisdom, I maintain that the parallel structures of these virtuous habits of mind are not accidental. Here I take a cue from Zagzebski's (1996) neo-Aristotelian account of epistemic virtue, suggesting that experimental/observational praxis and moral praxis have more in common than we might realize.

2 Perceptual Responsiveness as a Scientific Virtue

Accounts of the scientific virtues vary considerably, but there is an impressive list of plausible candidates for such virtues, among them *honesty, fairness, humility, open-mindedness, perseverance, diligence, intellectual courage, adaptability, sensitivity, insight* and *creativity* (Zagzebski 1996). Some of these, such as honesty and fairness, seem quite evidently to encompass both moral and intellectual praxis, while others such as perseverance and creativity may seem to be more at home in the intellectual domain. I will address the question of how perceptual responsiveness ought to be understood in relation to the moral virtues in Sect. 5.

Provisionally, let us define the scientific virtue of *perceptual responsiveness* as *a tendency to direct one's scientific praxis in a manner that is motivated by the emergent contours of particular natural phenomena and the specific form(s) of practical and theoretical engagement they invite.* The conceptual content of this definition will be clarified in the third section below, where its phenomenological roots are laid open. The practical scope of the definition will be more fully articulated by the historical examples discussed in the fourth section. My first task is to explain how perceptual responsiveness is related to the list of scientific virtues sketched at the beginning of this section, and particularly to those virtues which seem to be closely related to, or preconditions for such responsiveness. For the provisional definition I have given already allows us to see that the practical cultivation and habitual expression of perceptual responsiveness would entail, among other things, a perceptual *sensitivity* to the emergent contours of phenomena, *perseverance* and *diligence* in the continued exploration of those contours, *insight* into the practical and theoretical possibilities they open up, *open-mindedness* and *adaptability* with respect to the way these contours and possibilities may challenge or violate prior cognitive expectations, and *creativity* in finding ways to take up the practical and theoretical possibilities opened by the phenomena.

While I will not assert that *all* scientific virtues are conditions for or implications of perceptual responsiveness, it does seem that such responsiveness can serve as a sort of 'umbrella' scientific virtue that expresses the practical union of several others. In fact, perceptual responsiveness functions in a manner markedly similar to *phronesis* or *practical wisdom* in the moral domain. Just as *phronesis* is a virtue that implies the presence and unified practical expression of a number of other intellectual *and* moral virtues, *perceptual responsiveness* implies the presence and unified practical expression of a number of other scientific virtues. I will stop short of identifying them as one and the same virtue, for reasons explained in Sect. 5, but I shall suggest that the parallel is not accidental, and that in fact it supports the intuition of Zagzebski (1996), who argues that a sharp separation between the intellectual and moral virtues is lacking in justification.

3 The Phenomenological Roots of Responsiveness as a Scientific Virtue

I have defined the scientific virtue of perceptual responsiveness as *a tendency to direct one's scientific praxis in a manner that is motivated by the emergent contours of particular natural phenomena and the specific form(s) of practical and theoretical engagement they invite.* This conception of virtuous scientific practice involves both a responsibility and a reliability component – experimental or observational responsiveness entails a perceptual 'openness' on my part to the target phenomenon's *actual* contours, a form of motivated responsibility to 'the things themselves', but equally entails a reliable praxis for successfully discriminating those contours from confounding influences, artifacts of my sensory system or instrumentation, background assumptions, projected bias, expected or desired outcome, and so on.

Several elements of this definition, however, require clarification by means of an exposition of their conceptual roots in Husserlian phenomenology (one aim of which was the clarification of the legitimizing basis of scientific knowledge). The elements requiring phenomenological elucidation are: (1) the concept of *motivation*; (2) the notion of phenomenal features as *emergent contours* and (3) the concept of an *invitation* issued by a phenomenon to an observer or inquiring scientific agent.

Let us begin with the concept of *motivation*. It is important to distinguish the phenomenological meaning of this term from its broader philosophical use, especially within the internalism/externalism debates in virtue epistemology.[1] The phenomenological concept will, I think, serve to establish that epistemic and scientific virtue is not the product of a purely psychological motivation for truth-seeking (as internalists suggest), nor merely a reliably truth-conducive mechanism for forming beliefs (as externalists have claimed), or even a combination of these as two conceptually distinct conditions. Rather, epistemic virtue is the product of a form of motivation in which these two conditions *are conceptually unified*, constituting a rational-perceptual praxis of responsiveness to what Husserl terms 'originary evidence'.

Husserl defines the phenomenological concept of motivation in the first volume of his *Ideas*, published in 1913, and elaborated on its scope in the second volume. Motivation is first described by Husserl in connection with the fundamental 'experienceableness' of any perceptual phenomenon, which means for him that such a phenomenon or object always bears within its present appearance a future horizon of "open possibilities of fulfillment" that may be actualized in a manner contingent upon the subject's response to the phenomenon (1982 [1913], p. 107).[2] Husserl notes that these possibilities, while open, "are by no means completely undetermined but are, on the contrary, *motivated* possibilities predelineated with respect to their essential type." (*ibid.*, emphasis modified)

To say that these perceptual possibilities are *motivated* is to say that they are in some way prefigured by and communicated in the present experience. A very simple example might be the way in which a tomato perceived hanging on the vine presents the possibilities of being turned or lifted on the vine (perhaps to inspect the other sides), being plucked from the stem (and perhaps eaten, or dropped and discarded, or thrown), being smelled for ripeness by leaning in close, etc. Certain other possibilities seem far more remote, or even excluded (e.g., using the tomato as a cutting instrument). Importantly, as the tomato example conveys, motivation is a

[1] See Fairweather and Zagzebski (2001) for an excellent account of these debates.

[2] A limit-case would be rare and ephemeral experimental phenomena such as those produced by high-energy particle collisions. Many of these are not stable enough to explore, manipulate or otherwise engage in a temporally extended interaction, yet even these project a future horizon of motivated possibilities (e.g., their possible re-creation under similar experimental conditions).

structure of perceptual experience that bears within itself an exceedingly rich meaning-content, involving both value-meanings (the practical goals and desires I have with respect to the object) *and* kinesthetic meanings (the bodily potentials I need to enact in order to achieve or maintain a desired relation to the object). Even theoretically meaningful possibilities may be perceptually motivated (e.g., seeing the genetic implications of a tomato's novel phenotypic profile). Husserl's phenomenology of perception, later elaborated by Merleau-Ponty, reminds us that observation is never a passive or value-free inspection of a thing but an active and practically motivated response, an exploration situated within a specific context of human meaning. In fact, Husserl claims that the ordinary conception of practical motivation as the willing of a particular means to an end is simply a species of this more universal phenomenological concept of perceptual motivation (1982 [1913], p. 107ff).

In order to see the relevance of phenomenological motivation to scientific virtue, it is necessary to show how Husserl uses this concept as a foundation for *rationality* as a normative demand. In the section of *Ideas I* entitled 'The Phenomenology of Reason', Husserl describes how a perceptual position-taking is only rational to the extent that it is grounded in the fulfillment of a motivated possibility given in an 'original' perceptual experience (1982 [1913], p. 327). Put another way, a perceptual judgment depends for its rational legitimacy on the ongoing fulfillment in 'originary evidence' of one or more of the possibilities motivated by a perceptual encounter. For example, if I judge that the table is solid, the judgment is rationally legitimized only to the extent that I (or others with whom I share the perceptual horizon) am able to fulfill certain possibilities motivated by the table's perceptual appearance (resting objects on its surface, feeling the table's resistance with one's hand, etc.).

"Evidence" (including but not limited to scientific evidence) therefore represents for Husserl the "unity of a rational position with that which essentially motivates the position" (1982 [1913], p. 328). This evidence occurs in two species, each relevant to scientific praxis. *Assertoric* evidence refers to the unity of a judgment about a particular phenomenon or state of affairs with the fulfillment of that judgment by a *physically* perceived appearance that motivates it (example: correctly measuring the angle subtended by a particular ray of light actually observed in an experiment). *Apodictic* evidence, on the other hand, is described by Husserl as the result of the unity of a judgment concerning some general structure, essence or 'predicatively-formed essence complex' with the motivated *intellectual* 'seeing' of that structure, essence or complex (example: correctly judging, on the basis of a given mathematical intuition, that the members of a numerical set are all prime numbers). Rational motivation is then described by Husserl as the common legitimizing standard for all sorts of positings: theoretical, axiological and practical (1982 [1913], p. 333).[3] Whether the truth-value of the positing is apodictically demonstrated or merely assertorically plausible, the rationality of the positing is contingent on its having its

[3] This helps us to conceive of how scientific and moral virtues could be genuinely related and yet distinguishable; that is, if they each entail forms of rationally motivated action and judgment, albeit in different spheres of praxis.

'cash backing', so to speak, in motivating content actually given in perceptual experience, or put another way, the fact that the experience itself "speaks on behalf of" the objective content of that judgment (1982 [1913], p. 334).

For Husserl, rationality as a property of judgments *and* persons is anchored in the motivation-structure of perceptual intuitions, whether of a physical or intellectual kind. Judgments not based on perception (for example, judgments based on memory or testimony) are rational only in a derivative and weaker sense. It is, then, a necessary characteristic of virtuous scientific activity that our judgments or conclusions are *habitually responsive* to the rationally motivating evidence given in contexts of scientific perception. A virtuous scientist, then, will be one who *in general* does not, for example, ignore pertinent and motivating perceptual content due to considerations not in evidence, or worse, for reasons robustly incompatible with the given perceptual evidence. Nor will a virtuous scientist be in the habit of forming unmotivated judgments, as even her hypotheses and guesses are inclined to be motivationally guided by the available perceptual evidence (even if highly underdetermined by it).

It is essential to stress, however, that due to the assertoric rather than apodictic character of physical evidence in the domain of experimental scientific praxis, an experimentalist's rationally motivated judgment always remains contingent for its validity on the ongoing harmonious fulfillment of those motivated possibilities; the experiment must be, if possible, repeated with consistent results, and ideally, the new experimental implications an observed result sets up will themselves be tested and fulfilled. Establishing the rational character of an experimental conclusion therefore is not a singular event; it implies "a thoroughgoing harmonious fulfilling with a steadily increasing rational power" (Husserl 1982 [1913], p. 332). Another way of saying this is that the scientific virtue of responsiveness is typically displayed not in discrete experimental instances but in a scientist's ongoing and holistic assessment of the total experimental evidence at hand.[4]

It is this epistemic duty to cultivate a temporally extended and holistic assessment of the evidential dynamic that is indicated by our definition's emphasis on a scientist's responsiveness to the *emergent contours* of experimental phenomena. A virtuous scientist expects the character of a physical phenomenon to emerge clearly only in time, through extended and repeated experimental interactions, and does not overestimate the epistemic import of isolated or singular appearances. Nor does the virtuous scientist expect to render the phenomenon fully transparent, that is, to wholly exhaust the truth-content of its appearances in her experimental activity or her theoretical formulations. Instead, the virtuous scientist understands the experimental interactions to effectively and systematically palpate the *contours* of a phenomenon, where 'contours' refers to those positive points of contact between

[4]The distributed means of evidence acquisition in contemporary scientific praxis, where experimental tasks are often divided among teams of sub-specialists and where most scientists rely heavily on peers at other institutions to confirm their results, suggests that this holism also has a strong social dimension. This is addressed further in Sect. 4 and shown to pose no difficulty for my view.

phenomenon and experimenter, or between data and theory, that can never exhaust the full phenomenal possibilities of a thing.[5]

As an instructive parallel, consider what responsiveness entails taken as a moral or social virtue. A virtuous agent in the moral domain understands and has due respect for the *depth* of the human personality, and does not reduce the personhood of another human being to the total significance of their interactions with *her*. Yet she takes those same interactions as an opportunity to gradually draw out the stable contours of the other's person, so that over time she may better know, anticipate and effectively respond to the other's evolving moral needs. Contrast this virtuous moral agent with an individual who habitually and falsely believes herself to have wholly fathomed the personality of others and discovered all of the other's morally relevant needs/desires, often from just a single encounter or a few communicative transactions. Such an agent lacks a virtuous appreciation of the *depth* and dynamic *complexity* of the human personality, and her moral interactions are characteristically less successful as a result of her reductive and presumptuous attitude to persons. Likewise, a scientist who tends to overestimate the rationally motivating weight of singular experimental results lacks the virtuous scientist's appreciation for the depth and dynamic complexity of nature.

The above example sets up the phenomenological model of perception, and of natural-scientific observation more narrowly, as a two-way communicative transaction between the subject and a part of nature. This brings us to the third element of our provisional definition of perceptual responsiveness. For Husserl, the natural-scientific inquirer responds to *invitations* of a plurality of theoretical and practical themes presented to her *by* objects in the surrounding world: (1989 [1952], p. 230).

> The Object 'intrudes on the subject' and exercises stimulation on it (theoretical, aesthetic, practical stimulation). The Object, as it were, wants to be an Object of advertence, it knocks at the door of consciousness...it attracts, and the subject is summoned until finally the Object is noticed. Or else it attracts on the practical level; it, as it were, wants to be taken up, it is an invitation to pleasure, etc. (1989, p. 231)

This characterization of phenomena may initially give us pause. How literally are we to take Husserl's references to the Object 'wanting' to be taken up or attended to? It should be obvious that Husserl does not mean to attribute conscious intentionality to natural phenomena. Rather, the point is that when natural phenomena appear to us they always do so within a practical and theoretical situation of meaning and value that pre-conditions the experience; it is within and because of this lived situation that observed phenomena affect us in particular ways, and *invite* as appropriate a range of cognitive, kinetic, aesthetic and/or volitional responses on our part. Being a responsive observer of natural phenomena, then, entails a tendency to successfully 'read' the epistemically salient features of the observational or experimental situation, and to respond in a way that is appropriate

[5] For a fuller account of this phenomenological model of scientific experimentation rooted in Maurice Merleau-Ponty's 'reversibility thesis,' see Vallor (2010).

to those features and that wisely selects among those possibilities of further interaction that the phenomenon invites.[6]

How, then, does the virtue of perceptual responsiveness manifest itself in modern scientific activity? Joseph Kockelmans has described it this way, with respect to the domain of physics:

> The physicist, therefore, knows that in his science he is in real contact with the real world. The world invites him to assume a determined attitude toward itself. The physicist accepts this invitation and begins by asking questions. He selects his questions in such a way that nature is forced to reply in a determined way. These replies invite him to ask further questions, destined to lead to well-defined replies because they force nature to reply in such a fashion. In this way a certain field of meaning develops, and within this field the world of physical science receives its form and structure. (1966, p. 164)

The physicist's replies can come in different forms, as can the invitations. The invitation to which the theoretical physicist replies is typically a definite set of possibilities opened up by a specific group of mathematical formalisms taken to model some domain of physical phenomena in the real world. The physicist may explore one of those possibilities with the aim of putting a certain mathematical question back to the model; for example, asking whether the model under a given mathematical transformation will still preserve those features which currently allow it to serve as a successful model, or whether the model is 'broken' by such a transformation. The virtuous theoretical physicist is one who 'sees' as fully as possible the range of mathematical possibilities the model offers, and who wisely selects those possibilities to pursue that are most likely to produce illuminating results, for example results that definitively establish whether the theoretical model is consistent or inconsistent with certain physical possibilities in nature. Here, it is evident that the physicist's interrogation of nature is not direct but passes through an intermediary; a given mathematical model of nature (which may or may not be adequate).

The experimental physicist is, on the other hand, more commonly concerned with the invitations to measurement offered by physical phenomena themselves. Yet even invitations to experimental measurement are typically not 'read' directly from pure phenomena, but also (and sometimes solely) from theoretical models. Yet unlike the theorist, the experimenter has the opportunity to address her responding question, not back to the model alone, but also to the physical phenomenon created in the experimental situation. The virtuous experimenter is one who, among other things, properly reads or 'decodes' all of the salient invitations to measurement implied by the phenomenon and/or model, and creatively finds a way to take up just those invitations whose answer may shed the most light: either on the model, the experimental phenomenon itself, or some other, related physical phenomenon.

[6] Scholars of virtue ethics will see the parallels with the virtue of Aristotelian *phronesis* or practical wisdom quite clearly here, but let us make them explicit: being a practically wise moral agent entails a tendency to successfully 'read' the morally salient features of each practical situation, and to respond in a way that is appropriate to those features and that wisely selects among those possibilities of further human interaction that the moral situation invites (Aristotle 1999, Book IV).

Again, what reason do we have to take such appropriate responsiveness as a *virtue*? And from what does it derive its normative force? Appropriate responsiveness to natural phenomena is a scientific *virtue* because it is not a rule-driven technique or procedure that can be operationalized or directly taught. Instead:

1. It is a cultivated praxis or habit of *seeing* and *acting* in particular theoretical and experimental situations that tends to bear epistemic fruit.
2. It is a habit of fruitful seeing and acting that makes a scientist exceptionally praiseworthy and a model of excellence (and with respect to a whole career, rather than a single isolated success).
3. It allows us to distinguish theoretical and experimental excellence of the *scientist* from excellent *theories* or *experiments* (the excellence of which must in certain cases be credited more to luck than to any distinctively meritorious scientific character of the theoretician or experimenter).

Nor is responsiveness merely an empirical indicator of success. It is epistemically normative with respect to the truth-goal of scientific inquiry; as Kockelmans puts it with respect to physics:

> ...a physical statement is true if it is concerned with nature in the way nature *should* be addressed within the mode of intentionality proper to the physicist. As long as nature continues to give meaningful replies to the questions which the physicist asks, the questions and the statements of the physicist are true; hence a physical statement is true as long as experiments confirm what the physicist suspects on the basis of his theories...Each subsequent experiment may confirm this truth, but it can also apodictically show that nature is 'unfamiliar' with our question, that we question it in a way which does not 'suit' it. (1966, p. 170)

This normative component is essential to distinguishing scientific responsiveness as a genuine virtue rather than a mere skill or talent. If we cannot take seriously the idea that a scientist *should* respond to natural phenomena in some ways, and that other sorts of responses *should not* be made, then arguably the concept of a 'virtue' does not properly apply to the scientific habit of responsiveness. In the following Section I will try to show that excellent scientists often *do* take such responsiveness to be a virtue in the proper normative sense.

4 The Virtue of Perceptual Responsiveness in Historical Scientific Praxis

> What we observe is not nature itself but nature exposed to our method of questioning. Our scientific work in physics consists in asking questions about nature in the language that we possess and trying to get an answer from experiment by the means at our disposal. In this way quantum theory reminds us, as Bohr has put it, of the old wisdom that when searching for harmony in life one must never forget that in the drama of existence we are ourselves both players and spectators. It is understandable that in our scientific relation to nature our own activity becomes very important when we have to deal with parts of nature into which we can penetrate only by using the most elaborate tools. (Heisenberg 1958, p. 58)

Werner Heisenberg's quote, establishing as it does the familiar Baconian trope that science is a questioning of nature, may seem at first to tell us very little about scientific *virtue*. Yet a closer inspection of the quote shows that Heisenberg is doing more than just perpetuating an especially lovely metaphor for scientific inquiry, or personalizing nature as the coy guardian of all the mysteries of existence. Rather, Heisenberg is trying to tell us something important about the *scientist* – namely, that in her role as a questioner, her own pattern of activity conditions the outcome of scientific research far more than she, or we, might realize. Heisenberg is challenging the modern notion of the ideal scientist as a passive witness to truth, a pure recipient of nature's 'objective' answers. Consider these words of Max Planck:

> An experiment is a question which science poses to Nature, and a measurement is the recording of Nature's answer. (1949, p. 110)

The matter is not nearly so straightforward as this statement implies. The posing of a question by experimental means involves a vast array of choices on the part of the experimenter. By means of the mathematical-theoretical language in which she chooses to *form* her question and the 'elaborate tools' of experimental technique that she must devise to *direct* the question to nature, the scientist *shapes* both the form and the content of nature's answers. When a measurement is taken, an answer may be 'recorded', *but the answer did not antedate the question*, for it is entirely the product of the experimental interaction. The experimental conversation with nature, then, like all conversations, is only as illuminating as permitted by the scientist's 'conversational' abilities.

It becomes imperative to ask, then, what makes a scientist a *good* conversationalist with nature? It is here that the parallels with moral and social virtue strike us once more. We know that certain communicative virtues enable a conversationalist who has cultivated them to interact with others in ways that tend to lead to greater consensus, clarity, understanding, and future profitable exchange than is the case for conversationalists who lack those virtues, or worse, who possess those communicative vices that tend to lead interlocutors into conflict, ambiguity, confusion, misdirection, misinformation and premature termination of discourse. Among the social and moral virtues key to excellent communication we might list patience, charity, diplomacy, honesty, and empathy. There is also an intellectual component to communicative excellence; being a good conversationalist requires a kind of discernment, sensitivity to context, and a creative ability to choose from the array of invited responses those most likely to produce fruitful replies.

Of course, a virtuous 'converser' with natural phenomena will not manifest habits of practice *identical* to those of an excellent social conversant, given the different natures of their respective interlocutors. Unlike humans, most scientific phenomena do not become petulant or irritable when questioned too aggressively, do not dissemble out of pride or shame at an impudent question, and do not respond to flattery or benevolence of spirit. Still, there is reason to think that there *does* exist a virtue, or a complex of virtues, which predispose scientific inquirers to be better questioners of, and listeners to, experimental phenomena. If Heraclitus was right that "eyes and ears are bad witnesses for men who have barbarian souls," then there must be a sort of well-disposed 'soul' or *character* that makes *excellent* witnesses of eyes and ears. This is something that Charles Darwin seems to have suspected, and while he described it as a *virtuous instinct* rather than a cultivated *habit*, he clearly understood it as an enduring part of his character, as opposed to an externalized methodology or technique:

> ...I believe there exists, & I feel within me, an instinct for the truth, or knowledge or discovery, *of something of the same nature as the instinct of virtue*, & that our having such an instinct is reason enough for scientific researches without any practical results ever ensuing from them. (1903, p. 61)

In order to flesh out this possibility, let us consider how this virtuous character might be manifest in the research philosophies and histories of two extraordinary scientists. Given the nature of *virtues* as enduring features of a person's character demonstrated in a "complete life" of excellent activity (Aristotle 1999, p. 1098a20), it is important for our considerations that I have chosen two scientists each lionized not for a single experimental or theoretical discovery, but for a lifetime of scientific excellence: seventeenth century experimenter Robert Hooke and twentieth century cytogeneticist Barbara McClintock.

4.1 Robert Hooke

Hooke's reputation as a scientific investigator, however unavoidably dimmed by comparison with his contemporary and fierce rival Isaac Newton, is well established in the annals of science. He made or played a key role in remarkable breakthroughs in physics, chemistry, geology, biology and applied mechanics, to name just some of his experimental interests. His extraordinary skills as a designer and implementer of scientific experiments were widely acknowledged by his contemporaries, and his mentor Robert Boyle ensured his appointment in 1662 as the Royal Society's Curator of Experiments. As a consequence Hooke was not only responsible for his own scientific investigations, but for consulting on and recording the details of the greater share of British experimental science performed in the late seventeenth century.

Hooke's character as a scientific investigator has been described as extraordinary in several respects, most notably the astonishing breadth of his interests; the creativity and range of his mechanical skills in designing instruments and experiments; and his uncommonly good theoretical instincts. Thus despite his broad scientific interests and mechanical and practical leanings, he was neither a scientific dilettante nor merely a master technician. As biographer Margaret 'Espinasse notes, "Hooke did work of first-rate importance" in virtually every subject he pursued, and intuitively grasped the underlying theoretical connections between apparently unrelated branches of study (1956, p. 81). Hooke arrived at a correct theory of combustion decades before his contemporaries, made critical contributions to our understanding of respiration, recognized the biological significance of fossils and related geological phenomena long before Darwin provided an explanatory context for such discoveries; was the first to recognize and describe the function of the cellular structure of plant matter, and anticipated by two centuries the potential manufacture of synthetic fibers (Gunther 1930a, x).

While Hooke was not a *perfect* embodiment of modern scientific excellence (his chief lack being the mathematical genius of a Newton), his intellectual dispositions and talents taken as a whole arguably made him as well-suited for a lifetime of extraordinary scientific achievement as any single human investigator could aspire to be; as 'Espinasse notes, among his peers he most perfectly united and constantly displayed the virtues of both the pure and the applied scientist in a wholly integrated mind (1956, p. 41, 49). He therefore serves as an appropriate model for our inquiry into the nature of scientific virtue.

Let us begin by noting what is *distinctive* about Hooke's scientific character and praxis. Historian R.T. Gunther notes that:

> Hooke would not rest content until he had jotted down *all* the possible solutions and their variants, whether practicable or not, that presented themselves to his extraordinarily active brain. It was evidently a habit he acquired early, for, while still a boy he invented thirty several ways of flying. No one has been more fertile in the devising of experiments, or more systematic in tabulating possible procedures....He believed this method of systematizing ideas to be peculiarly his own, and he is said to have frequently spoken of other researchers, even the most eminent, as 'childishly contenting themselves with partial views of the corners of things'. (1930a, xii–xiii)

I earlier described the virtuous experimenter as "one who, among other things, properly reads or 'decodes' all of the salient invitations to measurement" offered by a phenomenon or theoretical model, and who "creatively finds a way to take up just those invitations whose answer may shed the most light". Hooke seems to have been an exceptionally good 'decoder' and 'answerer' of such invitations. And like his mentor Robert Boyle who defended the scientific value of anyone, even a lowly tradesman or mechanic, in a position to be "very conversant" with Nature ('Espinasse 1956, p. 27), Hooke's descriptions of his own scientific researches consistently suggest an inclination to see himself as involved in a lifelong conversation with nature that evolves according to a systematic praxis of experimental questioning. When faced with the task of defining for the Royal Society the proper scope of experimental observations of air, instead of offering a succinct schematic enclosure of the subject matter, Hooke proceeds to list no less than *95* separate questions, organized under 3 general headings: questions of air's substance, its quantity or scope, and its motions (Gunther 1930a, pp. 113–115). One imagines the patience of some of Hooke's fellows being tested by his exhaustive surveys of the question-spaces opened by experimental phenomena, yet we have described scientific virtue as entailing an extraordinary sensitivity and responsiveness to these very spaces.

Though Hooke occasionally characterizes the experimenter's interactions with nature in the Baconian image of an ardent seducer, manipulator or even violent aggressor (Gunther 1930b, p. 459), his descriptions of his own observations have been noted to reveal a far more sensitive, even "tender-hearted" attitude to natural phenomena ('Espinasse 1956, p. 58). Historian Lisa Jardine describes Hooke's perceptual abilities as sharpened by the opportunity to "affectionately engage" with his natural surroundings, when occasionally freed from the pressures of Oxford to be "entirely in his element" as an 'exuberant' observer of natural phenomena (Jardine 2004, p. 123). This is consonant with a virtue of *perceptual responsiveness* as a parallel of social or *moral responsiveness*; just as the latter entails an appropriate affective attunement to another human being, the former too entails a proper affective relation to natural phenomena. These need not be exactly the *same* sort of affect, and responsiveness of both sorts is compatible with a range of affective styles; but scientific and moral responsiveness are each arguably incompatible with hostile affect, or affective *indifference* to the other.

Hooke's intellectual character is characterized by 'Espinasse as consonant with the Baconian model of dynamic empirical questioning, as opposed to the tendency of eighteenth century thinkers inspired by Newton to idealize a purely inductive procedure. Describing the mindset of experimenters like Boyle and Hooke, 'Espinasse points out that a scientist asks a question of nature "because he expects a certain answer," and yet recognizes that the answer may well invite him to reform the question (1956, p. 31). She quotes early biographer John Ward as describing Hooke as having an unusual talent for responding to such invitations, leading his thought into "an endless round from hypothesis to experiment and back to hypothesis again." (*ibid.*) This hermeneutic structure of experimental motivation is precisely what we described in Sect. 3 as following from the responsive scientist's tendency to regard the experimental interaction not as a one-way or definitive test of nature but as an ongoing conversation, the outcome of which must be continually evaluated according to the totality of motivating evidence given at that particular stage of the conversation.

Among Hooke's greatest achievements is his detailed recording of his own experimental observations in microscopy, *Micrographia* (1665). In the Preface to that work we find evidence of Hooke's own attitudes and convictions with respect to the proper character of the scientist and his manner of relating to nature. Though Hooke aimed to do no more than describe "the true nature of the things themselves," (1961 [1665], unpaginated) his own labors taught him that even the bare description of an experimental phenomenon demanded of the experimenter a range of perceptual virtues beyond mere attentiveness to detail.

It is noteworthy that in his own account of these virtues, Hooke chooses to emphasize not only the intellectual but also the moral, or we may wish to say, 'paramoral', dimensions of experimental excellence. For example, he tells us that "a sincere Hand, and a faithful Eye" are even more important than methodical precision or intellectual rigor. (*ibid.*) One might be tempted to simply dismiss this claim, or to take 'sincerity' and 'faithfulness' as mere rhetorical flourishes on experimenter's mechanical dexterity and acuity, as opposed to genuine virtues of character. But this would be a mistake. Hooke describes the experimenter's task as one of developing an *honest familiarity* with his subject, learning the subject's "manner of walking" (*ibid.*) so that the experimenter may come to recognize the subject's distinctive style under other, more novel circumstances, and also to see how to adapt his experimental methods creatively and appropriately to such circumstances. He thus describes experimental perception as learning to "receive [the thing] in a right manner," (*ibid.*) and I suggest that it would not, for Hooke, be a stretch to compare errors of scientific perception to the blunders of an insensitive friend who manages to both misunderstand and offend his companion by failing to notice and adequately respond to her distinctive style of communication, or by hearing only what he wants or expects to hear.

He describes scientific failures as due not to errors of method so much as to defects either in sensorial or memorial faculty (which are to be ameliorated by the creative invention of instruments), or defects in the "temper and dispositions" of men (*ibid.*), e.g., *scientific vices*. Such vices are characterized by Hooke largely in terms of 'dogmatic confidence' on matters *unmotivated* or *undermotivated* by the evidence, that is, a tendency to regard what can only be provisional conjectures as "unquestionable Conclusions," and a failure to respond appropriately to evidence given. He tells us that "scrupulous choice, and a strict examination, of the reality, constancy and certainty of the Particulars" must be mediated by a well-tuned attention, which while not neglecting even the "most vulgar Instances," is nevertheless intuitively oriented to the "most instructive" evidences. (*ibid.*)

Experimenters without these virtues, on the other hand, may "pretend to be so sharp-sighted as to see what a preconceiv'd Hypothesis tells them should be there, where another man, though perhaps as seeing…can discover no such matter" (Hooke 1961 [1665], p. 158). For Hooke such disagreements over what is 'seen' are evidence *not* of the inherent epistemic weakness of sight, but of the need to cultivate *the virtue of right seeing* as a proper reception of, and response to, natural phenomena. Discussing the controversy over whether nautilus fossils were of living origin, he tells us that:

> Anyone that will diligently and impartially examine both the (fossilized) Stones and the Shells…will, I can assure him, find greater reason to perswade him of the Truth of my position [that Fossils were once alive] than any I have yet urged or can well produce in Words; no Perswasions being more prevalent than those *which these dumb Witnesses do insinuate.* (Gunther 1930b, p. 712, emphasis added)

The virtuous experimenter must learn to attend to Nature and her insinuations "not only in her ordinary course," but in her "doublings and turnings" (Hooke 1961 [1665], unpaginated Preface). This, of course, involves learning to be not simply a passive witness to phenomena as they commonly occur but a responsive questioner, one who knows how to formulate the experimental inquiry that will reveal what is not already obvious. Here too instruments are the powerful aids of perceptual virtue, for by them "the Earth itself, which lyes so neer us, under our feet, shews quite a new thing to us." (*ibid.*) Such instrumental questioning must be patient and yet searching, exploring as many angles of a phenomenon as possible "in several lights, and in several positions to these lights," before drawing conclusions as to the "true form." (*ibid.*) Nor must we become so presumptive as to believe that nature will always confine itself to the scope of our theoretical imagination, however broad: "For who would ever have imagined such a configuration or fabrick, as that of the ring of Saturn?" (Gunther 1930b, p. 739) Even so, there are patterns and hints of patterns in nature to which we must remain perceptually receptive: "there is a real beauty and allurement in truth, that will produce some votaries in the worst of times; and that will in time prevail, and shine out" (Gunther 1930b, p. 741). In a striking parallel with Aristotle's doctrine of the virtuous mean, Hooke describes scientific virtue as consistently following "middle wayes," balancing careful deliberation and severe examination with openness and receptivity, constraining enlargement of knowledge with scrupulous and exacting standards of evidence, and practicing "much slowness in debating, and shyness in determining." (*ibid.*)

Among the most historically significant discoveries discussed in the *Micrographia* is Hooke's account of observing the microstructure of cork, which was the first experimental detection and characterization of the cellular structure of plant matter and the basis for the future of histology (Gunther 1930a, x). Hooke's report of this experimental discovery reflects well our phenomenological model of scientific virtue as motivated responsiveness. Hooke describes his initial visual discovery of the cork's microscopic pores as accompanied by a co-given sense of something subtly communicated to him, namely, a general truth-structure pertaining to this particular phenomenon that invited and promised answers to a range of further theoretical and experimental inquiries:

> I no sooner discerned these (which were indeed the first *microscopical* pores I ever saw, and perhaps that were ever seen, for I had not met with any Writer or Person, that had made any mention of them before this) but me thought I had with the discovery of them, presently hinted to me the true and intelligible reason of all the *Phenomena* of cork... (1961 [1665], p. 113)

Hooke goes on to enumerate all of the further questions about cork to which this initial observation of its cellular structure hinted at answers (Why is cork so light? Why does it not take up water? Why does it not permit air to pass through?), as well as more general theoretical suggestions of a unified structure common to all plants. Yet Hooke describes these not as settled theoretical conclusions, but as invitations that now motivated for him a corresponding range of cognitive and experimental responses. For example, Hooke followed up on this initial discovery by looking for and verifying the presence of a comparable cellular structure in a host of other trees, vegetables and grasses (1961 [1665], p. 115). Other motivated enquiries produced negative results, such as his attempt to discover evidence that these pores served the function of channeling nutritive liquids within the plant (though he acknowledged those results as inconclusive, pending the invention of greater microscopic power). Still other motivated possibilities are described by Hooke as simply beyond the present power of his instrument to attempt to take up, such as the possibility of learning from this structure the general mechanical causes of different degrees of elasticity within all bodies (1961 [1665], p. 114). The encounter with cork exemplifies Hooke's perceptual responsiveness, his greatest virtue as an experimenter: that is to say, his ability to read from a single novel observation of a particular phenomenon a vast scope of invitations to theoretical and experimental response, joined with an extraordinary inventiveness in finding ways to implement the most productive of those responses possible within his experimental range, and the concern to document the remaining invitations as rational motivations for future experimental work, or as was often the case, for making instrumental improvements to his perceptual capacities for questioning.

4.2 Barbara McClintock

Deliberate cultivation of the virtue of responsiveness can also be discerned in the research philosophy and practices of cytogeneticist Barbara McClintock, whose research on the genetic structure of maize earned her the 1983 Nobel Prize in Medicine and Physiology. This completed a chorus of scientific acclaim honoring her lifetime of extraordinary achievement, including the National Medal of Science, the first Macarthur Laureate award, 15 honorary degrees, Brandeis' Rosenstiel Award, Columbia's Horwitz Prize, the Wolf Prize in Medicine and the Lasker Award for Basic Medical Research. The Genetics Society of America lauded her, its former President "for her brilliance, originality, ingenuity and complete dedication to research" (Fox Keller 1983, p. 13).

What made McClintock characteristically suited for scientific excellence? Let us begin by examining her greatest experimental successes. The experiment that made her career, done with her student Harriet Creighton and described as "one of the truly great experiments in modern biology," (Bertsch McGrayne 1993, p. 156) demonstrated that new combinations of physical traits in living organisms were made possible by the 'crossover' of chromosomal parts. McClintock had already succeeded in mapping the ten maize chromosomes by mastering a new technique for staining; this allowed her to discern distinct chromosomal structures that her mathematical analysis showed were correlated with specific inherited traits. Based on this knowledge, she devised an experiment that posed and answered a question of great scientific import at the time: does the crossover of chromosomal parts occur, and is it correlated with the crossover of genetic information responsible for producing new combinations of physical traits (Creighton and McClintock 1931)?

McClintock was *uniquely* able to direct this question *to* the corn plant because she had already developed such an extraordinary experimental acquaintance with it, one which had allowed her to detect two distinctive structures of the ninth chromosome of a particular maize strain that seemed to be responsible for two unique physical traits – waxy, purple kernels. In order to demonstrate that these structures could be exchanged in a manner correlated with the genetic crossover of the associated traits, she carefully fertilized the strain of corn with waxy, purple kernels with a strain without either of those traits; the result was a mix of offspring like the parent with both waxy and purple kernels, offspring like the fertilizing strain with neither trait, and offspring with just *one* trait of the first strain (waxy *or* purple kernels, but not both). A combination of complex mathematical analysis and microscopic inspection of the chromosomal structures showed that the offspring in the third group were the result of distinct parts of the ninth maize chromosome exchanging places in the subsequent generation. This experimental evidence of chromosomal crossover essentially completed the theoretical basis of classical genetics (Fox Keller 1983, p. 58).

A host of other important contributions to genetics are credited to McClintock, including advances in the understanding of chromosomal breakage relevant to cancer and aging research. But her crowning scientific achievement was her discovery in the 1940s of transposons or 'jumping genes,' a discovery that challenged the dominant orthodoxy of genetic information as transmissible only through a fixed and irreversible process. The great significance of her discovery, which led eventually to her Nobel Prize, was not understood or acknowledged until decades later when evidence of transposable elements was also found in bacteria. Transposition turned out to be of universal significance for genetic transmission, and key to understanding the processes of cell division, mutation and cellular repair in all living organisms (Fox Keller 1983, p. 191).

To what did McClintock attribute her record of extraordinary experimental success? One part of the answer is her disposition to become completely absorbed in the perceptual encounter with the cell structure: "You are so absorbed that even small things get big…Nothing else matters. You're noticing more and more things that most people couldn't see because they didn't intently go over each part, slowly but with great intensity. It's the intensity of your absorption" (Bertsch-McGrayne 1993, p. 156). For her such habits of perceptual absorption in minute details, which biographer Fox Keller describes as seemingly dictated by "internal forces" rather than external methodological constraints, were the key to understanding the phenomenon as a whole (1983, p. 101). Of course, we are speaking here of the visual inspection of a cell; how does a such an inspection constitute a 'conversation' with the cell, and what does perceptual absorption have to do with perceptual responsiveness?

We must remember that perceptual inspection of a cell is neither static nor passive; but an extended praxis in which one must be capable of discriminating what, in the vast array of microphenomena the cell presents to the eye, is relevant or new. Some parts of the cell will invite the attentive viewer to look again, to magnify, to probe further; others invite passing over. Let us return to the parallel with conversational or social responsiveness. We often find that we are understood best by those who have the tendency to allow themselves to be absorbed in what we have to say, without impatiently pressuring us to 'get to the point' or self-interestedly directing the subject back to themselves and their own desires. Likewise, the experimenter who is excessively concerned with her own immediate theoretical or practical goals will not be able to fully allow the perceptual encounter its own chance to speak. Yet absorption in the encounter of the other *also* admits of excess – for absorption without discernment and appropriate response is hardly a model of conversational excellence. If I converse with someone who hangs on my every word but never stops and inquires more deeply about something I have said, never asks me to clarify anything, never highlights the significance of any part of what I've said, then the exchange is of relatively little value to either of us. While the person who is incapable of absorption in my words will never truly understand me, the person who seems passively and equally fascinated with every single word that escapes my mouth, offering no discerning response, is either an insincere sycophant or a conversational idiot.

Likewise, a virtuous conversant with nature knows how to truly observe, but is also actively integrating what she observes with a meaningful informational structure that invites specific theoretical and experimental responses:

> You let the material tell you where to go, and it tells you at every step what the next has to be because you're integrating with an overall brand new pattern in mind. (Barbara McClintock in Fox Keller 1983, p. 125)

McClintock also described to her biographer the scientific vices that she believed kept many of her peers from the same insights:

> "...if the material tells you, "It may be this," allow that. Don't turn it aside and call it an exception, an aberration, a contaminant...That's what happened all the way along the line with so many good clues."... "I feel that so much of the work is done because one wants to impose an answer on it," McClintock says. "They have the answer ready, and they [know what they] want the material to tell them." Anything else it tells them, "they don't really recognize as there...*If you'd only just let the material tell you.*" (Fox Keller 1983, p. 179)

Near the end of her career, in her 1983 press release in response to winning the Nobel Prize, McClintock remarked on the pleasure she had taken over the years in "asking the maize plant to solve specific problems and then watching its responses" (Bertsch-McGrayne 1993, p. 172). Yet it is worth noting that McClintock's success has also been credited to her ability to conceive genetic processes as themselves "responsive to signals"; an ability that allowed her to conceive of the chromosomal capacity for self-repair long before her peers (Bertsch-McGrayne 1993, p. 158; Fox Keller 1983, p. 200). As McClintock noted, "the ability of a cell to sense [their] broken ends, to direct them toward each other, and then to unite them so that the union of two DNA strands is correctly oriented is a particularly revealing example of the sensitivity of cells to all that is going on within them" (Bertsch-McGrayne 1993, p. 158). Thus it was at least in part her tendency to relate even to a humble stalk of corn as a potential conversant that allowed her to conceive of its genetic structure as itself in conversation with and responding to its environment, facilitating her groundbreaking discovery that genetic mutations were not always permanent and can be reversed by environmental influences (Bertsch-McGrayne 1993, p. 167). Those peers of McClintock's who long resisted her conclusions might have recognized the value of her discoveries sooner had they themselves better cultivated scientific habits and attitudes of perceptual responsiveness.

What is interesting to consider about Hooke and McClintock is that while each describes their own experimental practices and virtues in terms associated with perceptual responsiveness to nature, there is a vast difference in their scope and style of scientific activity; Hooke is regarded as an exemplar of breadth in scientific research ('Espinasse 1956, p. 45), while McClintock's lifetime focus on the genetic structure of the maize plant is an exemplar of depth. Hooke's was more mechanically than mathematically gifted, while McClintock was a master of complex statistical analysis. This suggests that responsiveness, if it indeed was a common contributor to Hooke and McClintock's enduring excellence, is a scientific virtue compatible with a diversity of other personal styles, interests and talents for scientific work. This should not be surprising: virtuous communicators in human to human contexts may also embody a broad range of personal styles and talents; some put their virtues to work cultivating excellent communication with a few close companions; others put the same virtues to work within a larger social context of education or civic leadership.

It might be objected that Hooke and McClintock represent an older model of scientific practice, one in sharp contrast with the contemporary era of 'Big Science'. Even if we grant that the distinctive virtue of perceptual responsiveness conditioned their successes, why should we think that this same virtue is equally relevant today? Indeed the experimental situation in modern particle physics, to use just one example, with its instrumental complexity, reliance on high-level theory, and intensely social and distributed nature of experimental practice, bears little resemblance to that of a seventeenth century experimenter such as Hooke or even a field geneticist such as McClintock, both of whom could work in conjunction with others but could also produce groundbreaking research in relatively simple and solitary experimental settings. How can the scientific virtues of responsiveness, modeled as they are on the conversational virtues of simple, direct exchange between two subjects, be applied to contemporary experimental life?

While such a question is a reasonable one, I suggest that it misunderstands the scope of the virtue in question. First, the increasing reliance on complex instrumentation in many scientific disciplines is no obstacle *in principle* to the practice of perceptual responsiveness. We saw that evidenced in the research practices and philosophy of Hooke, whose perceptual abilities were seamlessly and even *necessarily* integrated with his advancement of instrumental praxis. That said, there are questions of great epistemic import concerning scientists' increasing reliance on computers to not only *detect* significant phenomena but also to *model* and represent their experimental presence; arguably at some stage we are relying on computers to practice perceptual responsiveness *for* us. While this is an important matter for philosophers of science to investigate, there is no *prima facie* reason to believe that perceptual responsiveness has been rendered moot by contemporary advances in scientific instrumentation.[7]

Nor is the increasingly social and distributed nature of experimental practice a problem for a theory of perceptual responsiveness as a scientific virtue. Habits of perceptual responsiveness do not merely allow researchers to better 'hear' the responses given by nature to their experiments; they allow researchers to more effectively design productive experimental questions, and also to grasp which future questions may be invited by a given experimental result. These products of perceptual responsiveness are not restricted in their value to the research life of the virtuous individuals from whose work they emerge, for they establish the signposts that other researchers may use to inform and enhance their own theoretical and experimental questioning. Ultimately, even Hooke's and McClintock's virtues are far more praiseworthy for the theoretical and experimental possibilities they opened for others to pursue than for the possibilities they were able to pursue themselves; experimental science has always been most successful when integrated within a community of inquirers who communicate their perceptual experiences to one another. The virtue of perceptual responsiveness, then, is no less critical to scientific success in the age of Big Science; rather it is even *more* critical.

5 Conclusions

If one accepts the foregoing as evidence that perceptual responsiveness belongs on the list of the scientific virtues, three further questions follow. First, how is perceptual responsiveness as a virtue of excellent scientists related to the *moral* character

[7] See Vallor (2010) for a related argument.

of those persons? Second, how is perceptual responsiveness related to other, more conventionally familiar scientific virtues? Finally, what light, if any, does the account of perceptual responsiveness shed on the question of whether intellectual or scientific virtues in general are conceptually distinct from the moral virtues?

Let us start with the first question. There is reason to think that the virtue of perceptual responsiveness does *not* entail that its holder be a paragon of moral virtue. Hooke's moral character has been a matter of controversy among his biographers, but there is a general consensus that his social virtues declined from their peak, moving from a "generous, gregarious and good-natured" spirit in his middle years (Jardine 2004, p. 305) to one "Melancholy, Mistrustful and Jealous" as the controversies with Newton, Oldenburg and others weighed heavier upon him (Gunther 1930a, p. 65). It is of course possible that his scientific virtue declined in tandem with his moral qualities, but Hooke does not seem to ever have been judged as a moral *exemplar* either by his contemporaries, or by posterity. Thus on the face of the matter, scientific excellence of the sort Hooke possessed does not also entail moral excellence. Yet he was certainly not without morally positive qualities; as we have noted, he was known early on as a generous and good-natured sort. He was capable of forming and maintaining lifelong friendships with highly respected fellows such as Robert Boyle, perhaps similar to those Aristotle called 'virtue friendship', which imply the possession of fine moral character by both individuals. Hooke was also described by early biographer Richard Waller as having "a piercing Judgment into the Dispositions of others," (*ibid.*) a quality which usually implies notable powers of moral discernment and a strong capacity for empathy.

We might speculate that Hooke's habits of perceptual responsiveness contributed in some significant way to this capacity for understanding his fellows; but as we noted earlier, persons and natural phenomena do not invite identical patterns of human response, even if they require *parallel* virtues of *responsiveness*. It may even be the case that heightened attunement to the patterns of scientific response invited by nature distorts or dulls one's attunement to those different patterns of moral response invited by human interactions, such that persons of great scientific virtue could be *less* likely to excel at moral praxis. We could speculate on this with regard to Barbara McClintock as well; like Hooke, she was capable of forming enduring friendships and strong collegial bonds, but outside her close circle of friends she could be perceived as rigid, impatient and occasionally uncharitable (Fox Keller 1983, p. 50; Bertsch-McGrayne 1993, p. 162, 170). Of course, such perceptions may have been skewed by gender bias, or perhaps her years of relative isolation working in Cold Spring Harbor dulled her social sensitivities. Her lifelong struggle against institutionalized sexism could have made her less open or trusting (Bertsch McGrayne 1993, p. 162, 164). Perhaps all, or none, of these things are true. Regardless, there is little evidential basis for asserting a strong correlation between possessing the scientific virtue of perceptual responsiveness to nature and having the parallel virtue of exemplary moral responsiveness to persons.

What should we say then about the relation between the virtue of perceptual responsiveness and the other scientific virtues? I noted in Sect. 2 that a number of plausible scientific virtues appear to be preconditions for, or implied by, perceptual responsiveness. These plausibly include *open-mindedness, perseverance, diligence, adaptability, sensitivity, insight* and *creativity*. Perhaps even other scientific virtues, such as honesty, fairness and courage, could be related to perceptual responsiveness, though these seem initially to pertain more to scientists' moral relations with other scientists or the public than to their perceptual relations with nature. In Sect. 2 I suggested that while that not *all* scientific virtues may be conditions for or implications of perceptual responsiveness, it does seem that responsiveness can serve as a sort of 'umbrella virtue' that expresses the practical union of several others. This might suggest a parallel function between the virtue of perceptual responsiveness in the intellectual domain and *phronesis* or *practical wisdom* in the moral domain. Might we say that these are in fact the same virtue, allowing us to unite these two domains?

For a number of reasons, we should probably reject, or at least suspend, this identification. The first and primary reason has been hinted at already: the patterns of appropriate response invited by natural phenomena are simply different from those invited by moral agents, and it is not evident that the same habits of mind can produce ideal patterns of response to both. A second reason to hesitate at such an identification is suggested by historical examples; if the same habits of mind or virtue *did* produce both scientific and moral excellence, then we should expect that characteristically excellent scientists would all be morally exemplary persons, and vice versa. History simply does not allow us to draw this conclusion. Finally, we should perhaps resist this identification because it undercuts the important difference between the kinds of normative force at work in the intellectual and moral domains; making a fatal and avoidable error of experimental perception is certainly blameworthy, but the blame that attaches to the agent simply does not carry anything like the gravity of the blame attaching to an agent who commits a moral error of a parallel kind.

Must we then conclude that scientific and moral virtue are unrelated, and that the remarkable parallels we have illustrated throughout this chapter between scientific-perceptual responsiveness and communicative-moral responsiveness are purely accidental? The answer is no. While it is the task of a further inquiry to make a compelling argument to this effect, such striking parallels suggest a far more plausible possibility: that the scientific virtue of perceptual responsiveness is a second-order adaptation and modification of the primary communicative and relational structure of moral virtue. We know that the human brain can recruit, repurpose and adapt one set of skills or habits for a different purpose – consider the way in which the deep brain structures that enabled oral literacy were adapted and modified over a relatively short period of human existence to give rise to a new, but related praxis of the written word. I will pose as a speculative conclusion, then, that the cultivation of an attentive and responsive praxis of epistemic interaction with nature (what I have called 'perceptual responsiveness') adapted a pre-existing praxis of attentive and responsive moral interaction with other human beings, what virtue ethicists call *phronesis* or practical wisdom. While the significant differences in these praxes of virtue confound their identification with each other, there remains the possibility that there is another level at which they may still be united. Linda Zagzebski has suggested that "It may be that at the deepest level the moral and intellectual virtues arise from the same motivation, perhaps a love of being in general" (1996, p. 167). Reading the works of Hooke and of McClintock, it is hard not to feel the pull of this suggestion.

References

Aristotle. 1999. *Nicomachean Ethics*. Trans. Terence Irwin. Indianapolis: Hackett Publishing.

Bertsch McGrayne, Sharon. 1993. *Nobel Prize women in science: Their lives, struggles and momentous discoveries*. New York: Birch Lane Press.

Creighton, Harriet, and Barbara McClintock. 1931. A correlation of cytological and genetical crossing-over in Zea Mays. *Proceedings of the National Academy of Sciences* 17: 492–497.

Darwin, Charles. 1903. *More letters of Charles Darwin: A record of his work in a series of Hitherto unpublished letters, Part One*, ed. Francis Darwin and A.C. Seward. Accessed 1 Feb 2012 from Project Gutenberg at http://www.gutenberg.org/ebooks/2739.

'Espinasse, Margaret. 1956. *Robert Hooke*. Berkeley: University of California Press.

Fairweather, Abrol, and Linda Zagzebski (eds.). 2001. *Virtue epistemology: Essays on epistemic virtue and responsibility*. Oxford: Oxford University Press.

Fox Keller, Evelyn. 1983. *A feeling for the organism: The life and work of Barbara McClintock*. New York: W.H. Freeman and Company.

Gunther, R.T. 1930a. *Early science in Oxford*, vol. VI.

Gunther, R.T. 1930b. *Early science in Oxford*, vol. VII.

Heisenberg, Werner. 1958. *Physics and philosophy: The revolution in modern science*. New York: Harper Perennial.

Hooke, Robert. 1961 [1665]. *Micrographia: Or some physiological descriptions of minute bodies made by magnifying glasses with observations and inquiries thereupon*. London: Dover Publications.

Husserl, Edmund. 1982 [1913]. *Ideas Pertaining to a Pure Phenomenology and to a Phenomenological Philosophy: First Book*. Trans. F. Kersten. Dordrecht: Kluwer.

Husserl, Edmund. 1989 [1952]. *Ideas Pertaining to a Pure Phenomenology and to a Phenomenological Philosophy: Second Book*. Trans. R. Rojcewicz and A. Schuwer. Dordrecht: Kluwer.

Jardine, Lisa. 2004. *The curious life of Robert Hooke: The man who measured London*. New York: Harper Collins.

Kockelmans, Joseph. 1966. *Phenomenology and physical science*. Pittsburgh: Duquesne University Press.

Merleau-Ponty, Maurice. 1968. *The visible and the invisible*. Ed. C. Lefort, trans. A. Lingis. Evanston: Northwestern University Press.

Planck, Max. 1949. *Scientific Autobiography*. Trans. F. Gaynor. New York: Philosophical Library.

Vallor, Shannon. 2010. The pregnancy of the real: A phenomenological defense of experimental realism. *Inquiry* 52(1): 1–25.

Zagzebski, Linda T. 1996. *Virtues of the mind: An inquiry into the nature of virtue and the ethical foundations of knowledge*. Cambridge: Cambridge University Press.

A Matter of Phronesis: Experiment and Virtue in Physics, A Case Study

Marilena Di Bucchianico

In this paper I tell the story of the Balkanization of the theory community in Condensed Matter Physics (CMP) and superconductivity research, and focus on the way experiments actively contribute to the formulation of theories. I claim that there is a tension between the different methods and aims of two scientific traditions, which I call *descriptive* and *principled,* as they implement the contribution from experiments. The interplay of the two scientific traditions evident in the hundred-year long quest for the understanding of superconductivity will be presented to explore the meeting of theoretical and experimental driving forces and their impact on the evaluation of theories and research programmes.

I will begin by introducing the debate. Then I will characterize the traditions as embodying preferences and characterizing practical wisdom, or phronesis. Presenting selected episodes from the history of superconductivity, I will offer my analysis and argue that the set of preferences expressed by the two traditions underlie different conceptions of what it means to formulate a theory and to succeed in problem solving, informing the crisis of consensus found in current superconductivity research.

Conventional or Low Temperature Superconductors (LTS) were first observed a century ago and later explained through the Bardeen-Cooper-Schrieffer (BCS) theory. BCS theory was formulated in the 1950s and won its proponents the Nobel Prize in 1972, becoming not only the leading theory in Condensed Matter Physics but also one of the most compelling successes of theoretical physics. BCS derived a formula for the temperature at which a superconducting material makes a transition to the superconducting state (Tc) and from that an approximate upper limit to the transition temperature of 30 K (Anderson and Morel 1962). Materials at a temperature higher than that limit were not supposed to display superconductivity.

M. Di Bucchianico (✉)
San Francisco State University, 1600, Holloway avenue, Philosophy, HUM,
94132 San Francisco, CA, USA
e-mail: dibucchi@gmail.com

A. Fairweather (ed.), *Virtue Epistemology Naturalized: Bridges Between Virtue Epistemology and Philosophy of Science*, Synthese Library 366, DOI 10.1007/978-3-319-04672-3_17,

In 1986, though, new types of materials broke through that roof and showed unusually high transition temperatures (up to a recently patent pending superconductor which exhibits an extraordinary Tc of 200 K!). Ever since (Bednorz and Müller 1986), the CMP community has been struggling with the problem of how superconductivity arises in certain materials at higher temperatures than that of conventional superconductors. Most scientists still agree that the electrons pair in the new materials, as they do in BCS theory, but disagree on just how that mechanism is mediated. Despite much intensive research in the 25 years that followed the discovery of high temperature superconductors and many promising leads, an answer to this question has so far eluded scientists.

"The field of HTS", said theorist Chandra Varma in my interview,[1] "is full of dissent. There is an enormous amount of confusion and diversity of views, partly because this is a field with an enormous number of experiments. Every known technique in science has been brought to bear on these materials and lots of different anomalies have been discovered". Tens of theories have been offered so far and the community is rife with dissent and disagreement.

There is often a tension between representing the world with a reasonably complete and accurate description, and the principled formulation of highly consistent theories. This is naturally the case in the HTS field, which, explains Varma in my interview, is stuck between the unprecedented complexity of the phenomena observed experimentally in the new materials, and the sea of theoretical accounts which attempt to derive the phenomena from first principles, but can only tackle a small portion of the evidence. For decades, first-principled theories have advanced, but their inability to address the phenomena at large has led to considerable debate.

Given the complexity of the experimental landscape of HTS, compromising the descriptive and predictive power of theories for the sake of logical coordination can be a justifiable step in the progression to a more satisfactory and general theory, but I stress that the *selection* of crucial experimental features of the materials (for theory-formulation purposes), and the assessment of the *relevance* of these features, is largely open to debate and discussion (Curty and Beck 2003). Aiming at a theoretical representation of a given phenomenon, scientists normally have to identify and characterize the target phenomenon; in this task exploratory experiments play an important epistemic role, shaping the language of future theories of the phenomenon and linking, in many ways, observations to conceptualizations (e.g. recognition of patterns, formulation of empirical laws and regularities, and creation of novel concepts to fit and explain new taxonomies).

In the case of HTS, the problem is borne out of a pre-existent theoretical framework for superconductivity, developed after decades of cumulative work by brilliant scientists, which had reached the stage of consensus and international recognition in the community. The problem of superconductivity – or LTS – was considered solved with BCS theory; the discovery of HTS was welcomed initially with the belief that, even though BCS had been shown to be insufficient to

[1] Chandra Varma interviewed first in Dresden on July 12, 2006 and a second time at UC Irvine on April 19, 2008.

fully understand the new materials, such pre-existent successful framework would have at least greatly aided and guided theoretical efforts in that direction, so the solution ought to be at bay.

A dozen theories emerged, but the experiments largely retained a 'life of their own', fuelled by the growing interest in applications and HTS technology. Experimentalists still take a pragmatic approach most of the times. Bob Cava, an eminent experimentalist, said in our interview[2] that one should, "I say, believe all religions, for one may be right! So if I hear an interesting idea from a theoretician about how something works, I try to put that into my chemistry searches, somehow"; otherwise, one keeps working independently and with limited help from theory. Many of these experiments can be seen as resembling the exploratory experiments that Bernd Matthias pioneered in the Fifties, aimed at identifying regularities and empirical rules among different materials and properties (Matthias and Hulm 1952; Matthias 1955). But they are different from the exploratory experiments that Steinle discusses,[3] which occur in contexts in which theories are genuinely lacking and cannot fulfill any guiding role. Rather, HTS experiments are pursued every day with the clear knowledge that the results are going to meet a whole zoo of preconceived theoretical interpretations. Furthermore, pre-existing concepts such as Cooper pairs, Bose-Einstein condensation, d-wave and s-wave states, and so on, are taken as background knowledge for those experiments. A metaphor for experimenting in HTS can remain 'exploration'; however, it is no longer that of the adventurer exploring a mysterious jungle, but rather that of a puzzle enthusiast tackling a riddle he has extensively studied before. The search for regularities and patterns, typical of the exploratory experiments in standard cases, is maintained even in the presence of pre-existent theoretical backgrounds (backgrounds that experimentalists use as a toolbox rather than as a manual, as expressed by Cava). This exploratory attitude *re*defines the problem, and *re*assesses our understanding of the phenomena.

The last two decades have brought elucidation of what is to be modeled and explained, but this process has been variously controversial for different theorists. No single theory does (or can) address all the data or features that have been associated with the superconducting phenomenon (Scalapino 2007). For each proponent, the selection of experimental results and of certain measurable parameters brings about a (retrospective) definition of the target phenomenon. The debate on theories then becomes a complex mixture of controversies involving related factors. *The relevance of different experimental results* – the definition of the target problem – *is not only disputed in itself, but it is also assessed differently along different preferences,* and it is for this reason that I have developed an analysis based on different "traditions" which mirror preferences in these choices.

[2] Robert Cava interviewed in Dresden on July 13, 2006.

[3] Among the types of systematic experimentation that are not guided by theory, Friederich Steinle (1997, 2006) introduces one called 'exploratory'. In his view, it typically occurs in those periods in which scientists are entering new research fields. In my view, and maybe compatibly with Steinle's, exploratory experimentation practices are found even in later, more advanced, stages of theory, particularly when, as in the case of superconductivity, the phenomena are very complex.

In fact, it seems to me that a central relevant factor is found in the clash between different views on theory formulation, and consequently different criteria for consensus, as I intend to show with this case study. Behind different criteria for consensus lie different preferences over the methodologies employed to achieve the desired solution, and, ultimately, behind that, different views on what the desired solution is supposed to look like. The majority of theorists in HTS assume without hesitation that what is missing, and needed, is an explanation of the phenomena from first principles; this is rarely disputed. In my view, the bone of contention is the different 'senses of duty' towards the *accurate description of the phenomena.* Since experiments have created a very complex picture of the phenomena that contains even more features and anomalous properties than in the phenomenological description of low temperature superconductors (LTS), different scientists can create all sorts of different descriptions of the high Tc superconductors' phenomena by selecting different combinations of experimentally observed features which they deem relevant. From these descriptions, a theorist can then build hypotheses and explanations to derive the high Tc superconductors' behavior from first principles. Then, when the theory is sufficiently developed, a theorist may be offered different pieces of evidence which may or may not fit into her description, and which may or may not be accountable for by her theory. *At this point, the importance given to the aim of providing the most accurate description, relative to the importance given to the aim of the logical coordination of the theory, becomes crucial.*

Nobel laureate P.W. Anderson, a key figure in the HTS landscape, maintains that "there is enough irrelevant complexity that an unwitting theorist may never reach the neighborhood of the actual problem" (Anderson 1994). This has to be remedied by cutting out the irrelevant bits (a somewhat dangerous and potentially arbitrary practice). Once the irrelevant complexity has been excluded, and a 'proper' description that contains only relevant features is set, the maze of alternative paths, says Anderson, can be reduced by simple logic ("When one has found a way through the maze of conflicting requirements, that is certain to be the way, no matter how many deep-seated prejudices it may violate and no matter how unlikely it may seem to those trained in the conventional wisdom", (p. 638)). But what are the criteria for relevance here?

Anderson's claim that relevance is obvious and implicitly determined through the simple use of logic is surely an oversimplification, and probably only rhetorical. It is impossible to specify criteria for relevance, and the subjective skills and intuition of different scientists guide different decisions. But this complexity does not exclude the possibility of philosophical insight. This is exactly what I claim is provided by thinking of practical knowledge, or Phronesis, in terms of the two traditions. Behind the methodological difference between physicists like Anderson, Bernd Matthias, Richard Feynman and John Bardeen among many others, lies an assumption that was clearly expressed by Feynman when he claimed that achieving a first-principles derivation of *any* property meant *grabbing the tiger by the tail* – obtaining at least a true *aspect* of the underlying mechanism for superconductivity. While for Matthias an explanation that failed to include the whole set of observed regularities was inadequate, for Anderson the inclusion of regularities not initially fitting the principled theory would be achieved by further development of the theory; this final step was for him *warranted* by the plausibility of the theory in terms of first principles. This latter attitude allows a theorist to ignore evidence not initially concordant with/included in the theory, since the logical coordination of the theory, says Anderson, is the needed warrant – and experimental inconsistencies will most likely prove to be "apparent more than real" (therefore they should not be too worrying for the theorist) (Anderson 1994).

Under this lens, the high level of dissent in the HTS debates can be interpreted in a way that goes beyond controversies on the mere interpretation of data. Dissent becomes a bitter war among theorists armed with different preferences on how to define the problem, how to solve it, and how to reach consensus on the solution. The concept of Phronesis in Virtue Epistemology becomes then the mediating virtue in a complex situation where different virtues and desiderata meet in conflict and demand a weighed assessment.

Let me then introduce explicitly the two traditions that I see at play here, before showing them in action.

Take first these two aims, which a virtuous theory of a complex phenomenon should ideally reach:

(a) To give the most accurate and complete description of the phenomena.
(b) To explain the phenomena in a way that is consistent with accepted/acceptable principles and theories.

It is clear that ideally scientists would like to achieve both aims. Some philosophers have argued that this is impossible. It is often the case that abstraction, and consistency with principles, conflict with accuracy in describing the many features of a phenomenon.[4] In this paper it is sufficient to recognize that in general an actual theory in a non-ideal case will only be able to achieve these aims *partially*, and that leaves room for future improvements.

This being the case, we will find that when we need to evaluate theories or models (or practices) we can try to assess how well they fit the two different desiderata; we would then evaluate their merits differently according to the different weights that we may assign to the implementation of each of the two aims.

As Duhem famously remarked in "The Aim and Structure of Physical Theory" (1914), requiring that a theory be logically coordinated cannot be justified or proven as necessary or reasonable. He maintains that the requirement is legitimate "because it results from an innate feeling of ours which we cannot justify by purely logical considerations" (such as the principle of contradiction or the law of economy of thought) "but which we cannot stifle completely either". In his view, "Reason" naturally aspires to the logical unity of physical theory, even if we have nothing more than common sense to back-up this aspiration. On the other hand, "Imagination", faced with a conceptual framework and what we could call the yet-unrelated parts of theory, desires to embody these diverse parts in concrete representations (p. 102–103). Different scientists have different mixtures of these two tendencies, as if differently dominated by reason and imagination.

[4] Cartwright for example argues that in the strongest sense this is actually always the case (1999).

In my view, these remarks inform deeply the two previously stated desiderata for theories, and define the two traditions.

For the ***descriptive*** *tradition a larger preference is given to practices, experimentation, conceptualizations, and theorizations that give the most accurate and complete description of the phenomenon we want to tackle.* Representing the world (doing justice to its manifest complexities) is under this light the primary goal in our efforts towards understanding it.

For the ***principled*** *tradition, by contrast, a larger preference is given to practices, experimentation, conceptualizations, and theorizations that are able to explain the phenomenon in a way that is logically coordinated and consistent with accepted/acceptable principles and theories.* Logically coordinating the abstract components of theories makes the world intelligible to us. This takes priority.

Notice that the two definitions share the kinds of "material content" (they both comprise *practices, experimentation, conceptualizations, and theorizations)* while *selecting* differently (upon) within that *same* set. It is not the case that the descriptive tradition will select preferentially experimentations and practices, nor is the case that the principled tradition selects preferentially conceptualizations and theorizations. The two traditions do not separate experimentalists and theorists in camps.

At first sight, this may resemble the bottom-up/top-down distinction in methodology, but it does not map precisely onto it. When taken to an extreme, descriptive preferences may lead a scientist to ignore or reject the principles of our best and most harmonious theories with little remorse, when this appears necessary to accommodate facts, much to the horror of the principled-minded. Conversely, principled preferences may lead a scientist to shamelessly devise a series of ad hoc modifications and tangled-up repairs of the "worm-eaten columns of a building tottering in every part" (Duhem 1914, p. 217), which would seem unacceptably childish from the point of view of the descriptive-minded.

The aims and methods of what Steinle calls *exploratory* experimentation seem to fit more closely the descriptive extreme of the spectrum, and the top-down approach fits loosely the principled end. Phenomenological theories are therefore no longer in contraposition with principled theories, but in a somewhat mediating relationship between the sets of preferences.

In fact, while on the one hand phenomenological approaches fully acknowledge the basic autonomy of observed regularities, and build up parcels of theory starting from there, on the other hand they accommodate the logically ordering desire, by providing conceptualizations that can even be, in the best cases, subsequently mapped directly onto a first-principle account of the same phenomenon.

In the metaphor used by Aristotle, we acquire (through practice) an "eye" that enables us to identify and assess the worthiness of appropriate actions according to the contingencies of specific situations (Aristotle 1941, p. 1033; Dunne 1993). In my use of it, *phronesis* is embodied in character and concerns the ways scientists match knowledge to the specific circumstances of the problem at hand. In this Aristotelian fashion, it is the practical wisdom that guides the scientist towards the desired solution, informing goals and the methodology that best suits the reaching of the goals. When coupled with the traditions presented here, phronesis becomes the capacity to select the appropriate mixture of virtues among those presented by the two traditions.

In the two traditions we find different preferences among the same epistemic virtues. The importance of one virtue is, at most, denied "locally" in relation to the other. Scientists with preferences aligned with the principled tradition such as Anderson, believe that a theory which suffers to some extent under the weight of experimental evidence is still saved by its consistency with first principles; this suggests not only how important internal logical coherence is for Anderson but also the extent to which empirical adequacy and predictive power are seen by him as secondary. More precisely, it shows beliefs about some epistemic virtues as *consequent* to others.

Anderson, for example, may still maintain that empirical adequacy is a clear desideratum for theories, but he would suggest that this desideratum should only be the focus of the last phase of development of a theory (Anderson and Matthias 1964). This is not a trivial matter of prioritising, but is justified on the basis that finding the (usually single) way to account for the phenomena consistently with first principles represents "usually the end of the story", after which quantitative agreement will inevitably follow.[5]

For Bernd Matthias, by contrast, a model that lacks internal consistency and a first-principle derivation has a promising chance of leading to the true mechanism as long as it seems to be empirically adequate; it is then from there that a more principled account can be built, if we need or want one (though for purely practical purposes we may not) (Bromberg 1994). In an ideal situation both camps aim at a theory that ultimately equally satisfies the epistemic virtues of empirical adequacy, and of logical consistency (with other accepted principles and within itself/self-consistency). It is also a possibility that, in ideal cases, approaches starting from the different ends of the spectrum would reach their final stages and find that at that point they are equivalent or compatible. A final theory may even claim robustness on the basis of its predictive power, when this was a virtue that was not *initially* considered, by its principled advocates, to be of primary importance. Alternative accounts which initially performed better in the light of that epistemic virtue may have actually been discarded as unpromising, sometimes even as unreasonable. In general, then, a characterization simply in terms of different, static, epistemic virtues, would fail to account for the complexities of the disagreements found in the superconductivity case.

The difference seems to lie in phronesis and in the methodological considerations concerning the best path to success for theories. This though is not only an issue of interest for methodology. In fact, adding the *diachronic* dimension and stressing the historical evolution of the theories and the historical evolution of their acceptance in relation with experiments, I claim we learn something that goes beyond a history of methodology. Not only do we gain new insights into both the nature of scientific theories and theoretical and experimental practices, through the eyes of physicists; we also gain a new perspective on the issue of dissent in science. I agree with Lakatos

[5] This point is also re-stated in Anderson's "central dogmas", which I discuss elsewhere.

that when we ask a question about Nature, how we find an answer is part of the answer. Although much work remains to be done to clarify these issues, by shifting the discussion to the different criteria for consensus we can instead make sense of 'dissent' in HTS as the clash of evolving preferences over virtues and of different traditions. Highlighting the role played by implicit epistemic preferences in scientific practice is not meant to criticize or praise the philosophical ideas of non-philosophers (scientists engaged in this debate). What matters is that from the point of view of a rational *a posteriori* reconstruction of the contentious issues between theories, this issue of contention on methodology, and criteria for consensus, may be invisible. By looking at the dialogue between the two different traditions in problem solving in physics, the issue becomes visible, open to interpretation, and able to contribute to our models of scientific progress and to the discussion of dissent in science as informed by virtue epistemology.

As scientists try to answer a scientific question or solve a problem, such preferences (can) remain silent if no controversy or disagreement arises, even if different methods and aims are being used in practice. The case of superconductivity is exemplary in terms of the extent of controversy; and I claim that it shows the meeting and the clash of such preferences in practice. The history of superconductivity is in fact incredibly rich. It spans over 100 years, during which the field has undergone impressive challenges, upheavals, successes and shocking discoveries. This is one of the reasons why I believe this to be a golden opportunity for philosophers keen on investigating scientific practice: the debate's longevity, filled with controversies and dissent, has magnified the conceptual components which are proper of a research field in its crucial forming stages. Accessing the abundant historical and sociological data on this case is akin to observing in *slow-motion* an object following sophisticated trajectories.

Given this richness, and wanting to present more than one episode from this history, I will need to limit the depth of the historical analysis. I have selected a few of the many episodes which are seminal for this analysis, firstly to show how these preferences *met*, and subsequently, by analyzing some more specific views held by some of the most prominent figures in the debate, to show how they *clashed*. In other words, I will explore these episodes to show how behind different criteria for consensus lie different phronetic stances: different preferences over the methodologies employed to achieve the desired solution.

[For this historical survey I have relied heavily on the wonderful book my Matricon and Waysand "The Cold Wars – A History of Superconductivity" (2003). For convenience, references to this source will simply state page numbers-after the first citation]

1 Door Meten tot Weten: From Measurement to Knowledge

Low Temperature physics is a relatively young field in the history of physics. From its early days at the end of the nineteenth century until the end of the Second World War, the main centre for low temperature research was undisputedly Kamerling

Onnes' laboratory in Leiden. Holland had been extremely active in physical theory since the seventeenth century, hosting scientific figures like Huygens, Spinoza and Descartes, and fostering pioneering work in the applied sciences. The Leiden laboratory, though, started a small revolution in the actual practices of physicists involved in what we today call 'condensed matter physics', and arguably for the whole of science. In fact, as I will show shortly, Kamerling Onnes quickly realised that the discipline needed a different foundation, and required theorists, mechanics, experimentalists, and glass blowers to actively work together (Matricon and Waysand 2003). He not only created the model for many subsequent laboratories for low temperature research, but he also initiated developments that characterize modern scientific activity.

One of the most important quests in the field has been the liquefaction of gases, and of Helium in particular. It had firstly captured the interest of Faraday in the early nineteenth century, and turned out to be one of the greatest driving forces for technology.

The first results, and then the work of eminent scientists such as Joule and Kelvin, brought a new qualitative understanding of the complex problem, i.e. that it was not just pressure but also lower temperature that was needed for gases' liquefaction. This started the 'Age of Low Temperatures' (p. 3). The techniques developed by researchers such as Louis Cailletet and Raoul Pictet marked the start of cryogenics.[6]

The link between cryogenic successes and artifacts such as the 'Dewar' made it quite clear that a neat separation of tasks and specializations would hinder the quest for liquefaction. Onnes quickly realized, for example, that scientific labs needed the art of glassblowing, to design and produce the technology needed for lower temperatures, so that a lab could be independent, flexible and systematic in its exploration of the properties of matter. He advocated this and much more in a scientific programme that he pushed very strongly at Leiden.

Inside the history of superconductivity, his scientific programme is the first appearance of the descriptive tradition's influence, standing at the base of the very discovery of the superconducting phenomenon.

In 1882, in honor of his appointment in Leiden, Kamerlingh Onnes gave his first public lecture, outlining a manifesto for research:

> "Physics owes its fruitfulness in creating the elements of our material well-being, and its enormous influence on our metaphysics, to the pure spirit of experimental philosophy. Its primary role in the thoughts and actions of contemporary society can be maintained only if, through observation and experimentation, physics continues to wrest from the unknown ever-new territory. […]
>
> Perhaps, like a poet, his work and all his activities are motivated solely by a thirst for truth; to penetrate the nature of matter might be his principal goal in life. Nevertheless, the courage to accept a position that makes it possible to realize these goals must come from the conviction that his activities will be useful only if he follows certain well-defined principles.
>
> What I believe is that quantitative research, establishing relationships between measured phenomena, must be the primary activity in experimental physics. FROM MEASUREMENT TO KNOWLEDGE (Door meten tot weten) is a motto that I want to see engraved on the door of every physics laboratory" (p. 17)

[6]A term coined in 1878.

Onnes strongly imposed this agenda in Leiden. Contrary to the practice of his most notable colleagues (for example Dewar) who had established a monopoly on their apparatus, machinery, and even technicians, Onnes allowed visits to Leiden to anybody interested. In fact he managed to 'steal' some of the best glassblowers in Germany, and then started a school for scientific instrument makers and glassblowers at the side of his laboratory. This school used to send his graduates to work at physics laboratories spread all over Europe and it is still graduating students nowadays. What is more, as a graduate student in physics in Leiden, the typical pupil would be assigned 50 hours of compulsory work and training in glassblowing and metalworking.[7]

Onnes' focus on 'well-ordered' experimental practice and his eagerness for autonomy and efficiency were evident to the rest of the scientific community, which did not spare him their criticisms. The Dutch physicist Hendrik Brugt Gerhard Casimir, a student and assistant of Ehrenfest and Pauli, who was at Leiden and later would become famous for the Casimir effect, criticized Onnes' epistemological stance in his memoirs, pointing out that we can start to measure only when we know what to measure (Casimir 1983). Casimir objected to "Door meten tot weten" writing that qualitative observation has to precede quantitative measurement. "By making experimental arrangements for quantitative measurement", he warned his colleagues, "we may even eliminate the possibility of new phenomena appearing".[8] His remarks express a well-known discussion in philosophy of science and exemplify the popularity of the principled tradition in fundamental scientific research.

I want to stress again that it is sterile to contrapose the two traditions as incompatible scientific paradigms. Even though Onnes' manifesto may sound dogmatic, it did not imply that theoretical speculation prior to experiment was of no use, nor that we shouldn't let theory tell us what is there to be measured (when theory manages to express that guidance). Nonetheless, Onnes emphasizes the need to recognize the role of "establishing relationships between measured phenomena through quantitative measurement" as a fundamental one inside experimental physics, without dealing directly with concerns of coherence among different sets of relationships and larger explanatory goals (at least initially). His manifesto exemplifies the importance of meticulous and open-minded exploratory research, and the great epistemic

[7] Emilio Segrè, who was a particle physicist and a student of Fermi, pointed out that Onnes' lab represented the forerunner of the institutions of *Big Science*. He noticed that usually scientists or scholars associate the passage of physics to the large scales with the introduction of particle accelerators. While the seeds of big science are certainly visible there, several features that characterize the large scale model had already emerged in Leiden. "The association of science with engineering, the collective character of the work, the international status of the laboratory, the specialization of laboratories centred on one technique, the division of the personnel into permanent staff and visitors. A laboratory with all these characteristics had been formed by Heike Kamerlingh Onnes at the end of the nineteenth century for the study of low-temperature phenomena" (p. 18).

[8] He supports this with the example of the discovery of X-rays, interpreted in his own way: the physicist Lenard "had an experimental set-up which was better for certain quantitative measurement than Rontgen's, so he did not discover X-rays" (p. 161).

value he appointed to it; it is, in my view, an expression of preference that aligns him to the descriptive tradition. His work affirmed the strength of the descriptive tradition as a valuable asset in scientific progress, and his manifesto aimed at developing this strength to its full potential.

With the liquefaction of Helium, Onnes had all the ingredients he needed for his scientific programme. Helium was just the starting point. He decided that "the entire laboratory would embark on a systematic programme of measurements of the properties of matter at low temperatures" (p. 24). As in the manifesto of his first lecture at Leiden, he assumed physics' job was *to wrest from the unknown ever-new territory, through observation and experimentation*, without having a particular framework or theoretical goal in mind. Once again, even though his description may seem dogmatically empiricist in nature, Onnes did not hide that he had his own theoretical hypotheses derived from previously observed experimental patterns, for example the famous decrease of resistance (R) with decreasing temperature (T). His intuitions and hypotheses, though, were not the consequence of a theory he maintained as valid. As is common in the descriptive tradition, he would focus on regularities and observed patterns to search for general empirical rules, and then try to formulate adequate representations of those regularities. Theoretical hypotheses were used as tools, and he was open to alternative incompatible ones as long as they could provide a better understanding of the observed patterns, without worrying too much about coherence. And then Superconductivity happened.

Even today, a typical course on thermodynamics teaches us that as we get closer and closer to absolute zero we approach a state of atomic immobility. As we cool Nature down, from cold to colder, we get the picture of its atoms slowing down to a stall. Helium appears in this story as the first serious challenge to this incorrect view. Through a race to ever-lower temperatures, and a stress on techniques devised to make sure that not a trace of air was left in the system, Leiden's laboratory achieved Helium's liquid state. They observed a metal with no detectable electrical resistance, a bizarre form of 'eternal movement' appearing as temperatures dropped: something that nobody had expected. Immobility was a concept that would naturally fit solids but was problematic for liquids. Yet it seemed impossible to solidify Helium. It is no wonder that the simple observation of this phenomenon gained Onnes a Nobel Prize.

How could one make sense of the bizarre phenomenon of superconductivity? The real difficulty, as physicists would later find out, was that no one actually understood *simple* conductivity, even in ordinary metals (p. 29). The theories of matter and electricity were clearly inadequate. This is one of the merits of revolutionary discoveries: They not only create new questions to be answered, and display new phenomena for theory to explain, but also often force the theoretical community to wake up to the notion that they had not actually understood sufficiently, or even understood at all, what they thought they had understood.

In the Thirties and Forties, while more and more features of metals were being explained, and the creation of a larger coherent picture of solid state physics (nowadays called 'condensed matter physics') was advancing remarkably, superconductivity managed to resist the most determined efforts of the best physicists around.

Old and new generations met in failure; from Bohr to Einstein, from Bethe to Brillouin, from Heisenberg to Lorentz, all had tried and failed. Bloch expressed the exasperating nature of the tricky phenomenon by formulating a theorem; it seemed to him evident at this point that [Bloch theorem:] "*Every theory of Superconductivity is a false theory*" (p. 44).

At the beginning, superconductivity meant simply absence of resistance. Then another discovery changed forever the way superconductivity was understood: The Meissner effect.

Meissner' experiments (Meissner and Ochsenfeld 1933) showed that one could introduce a new and very unexpected feature as characteristic of superconductivity: B=0; the magnetic field inside the superconductor is always zero. The fact that inside a superconductor E=0 was expected and uncontroversial, since there is no other way to have a voltage difference inside the superconductor, which by definition has R=0.[9] If B is always zero at every point in the superconductor, *independently of the path followed* (whether B is applied before or after the sample is cooled through the superconducting transition), then the transition from the normal to the superconducting state is *reversible* (London 1937). Yet Maxwell's equations predicted frozen flux and not the expulsion of the flux. The "bible" had spoken against the alleged observed facts.

It was Fritz London who then had the nerve to suggest that maybe something was missing in the bible. He and his brother Heinz decided to proceed in a revolutionary way, adding a fifth equation to the pillars of Maxwell.

The London brothers (1935) started by making E=0 and B=0 an *initial unjustified assumption*. They imagined electrons as if they were freely moving under the influence of a uniform external electric field. These electrons, according to Lorentz's law, would encounter a uniform force, and thus they would accelerate uniformly. This simple observation, starting from taking E=0, is contained in the first London equation.[10]

London's ideas were initially very unpopular. Gorter and Casimir came up with an alternative model that had immediate appeal to the theoretical community.

---

[9] If V=RI, and R=0 then V=0.

[10] (1) (First London equation, for E)

Applying Faraday's law to (1), one obtains a differential equation for B. *This equation permits both constant and exponentially decaying solutions but London recognized that constant non-zero solutions were non-physical, because they would disagree with the Meissner effect.* The resulting simplification led to the second London equation, which was postulated to complement Maxwell's:

(2) (Second London equation, for B)

The equation for B states that the curl of the current, j_s, is proportional to the magnetic field, B. The terms e and m are the charge and mass of the electron, but n_s was a new phenomenological constant loosely associated with the number density of superconducting carriers.

The proportionality factor turned out to have the dimensions of a length, and has since been called 'the London penetration depth', designated l_L.

This suggested a more sophisticated point. The Meissner effect did not mean that the permeability of superconductors was zero; it is just that the magnetic field cannot penetrate the surface layer beyond the London penetration depth. This startling prediction has been confirmed by many experiments, but the first ones only appeared in 1940.

Theirs was the first thermodynamic approach to superconductivity, called the *"two fluid" model* (1934). This was simpler than London's, and interpreted the idea of a phase transition in the most conservative way, by postulating a second fluid that appeared at the transition temperature. Thus all thermodynamic quantities were linear combinations of the contributions from the normal and the superconducting electrons belonging each to a fluid. While it accepted the conclusion $B = 0$, the model said nothing about how the circulation of two fluids was supposed to make it happen. Nevertheless the two fluid model remained for a long time the favored approach (p. 65).

The attention of the community of theorists was in fact almost completely caught by models that proposed simple mechanisms but had no predictive power. Gorter and Casimir's was the first, and shortly afterwards Mendelssohn added another one. London's equation, by contrast, had no theoretical intuitive justification, yet its *solutions* were easily describable and were appealingly intuitive.

In accordance with the central epistemic goals of exploratory experimentation, London's approach was that of *finding appropriate representations by means of which general empirical rules or equations can be formulated.* In fact, he started from recognizing the Meissner effect, observed in experiments, as crucial. Then he recognized Maxwell's equation as incomplete if we are to account for this effect. Contrary to the precept of starting from [accepted] first principles, he boldly suggested changing one of the pillars of theoretical physics by adding an equation for the simple purpose of accommodating this new evidence.

Having derived his phenomenological equations, he proposed that these could be consequences of the coherence of a quantum state, borrowing the intuition he had explored in his molecular research. In this way, he introduced a completely new concept in the theoretical landscape, that of a macroscopic quantum coherent state.[11]

As I said, his equations had no theoretical motivation; but with the conceptual development of a macroscopic state, they satisfied the desire to fill the gap between a theoretical representation of Nature and the complexities of the superconducting materials exposed by the new evidence. In this sense London expressed quite clearly a preference for conceptualizations and theorizations that, even if they had to go beyond the structure of prior established principles, gave the most accurate and complete description of the phenomenon – in this case, the magnetic behavior of superconductors. This is characteristic of the descriptive tradition.

The negative reaction that London received – when he was not merely ignored – was a treatment with which he was familiar. Although now considered unequivocally the true father of superconductivity, he spent almost all of his life as an outsider. He started his studies with both physics and philosophy, and at the age of 21 had published a paper in logic on "the formal conditions of purely theoretical perception" in

[11] Most physicists seem to think that it is not until after the Second World War that the idea of macroscopic quantum order appeared in the scientific literature. At a time when quantum mechanics was applied only to microscopic phenomena, London's ambition of applying it to molecules to explain macroscopic effects was a true novelty, and one that sounded almost too exotic for the average theorist (p. 71–72).

Husserl's Journal, the *Jahrbuch fuer Philosophie und Phaenomenologishe Forschung*. From his early days his scientific work displayed an unusual conceptual basis that made him radically different from the majority of his colleagues.

His reputation as a theoretical *chemist* kept his voice at the very back of the choir. He struggled to find a stable position; fleeing from the Nazis, he even had to submit another thesis in order to convince the establishment of his fitness for a professorship in France (p. 72).

There was only one physicist whose independent line of thought on superconductivity was similar to London's, even though the two had never met, separated by an "iron" wall of ideology. That was Lev Landau. Another physicist, John Bardeen, could be counted among London's few sympathetic readers. He had received the London brothers' article from his Harvard teacher, John C. Slater, and discussed it with him with great fascination. Landau and Bardeen went on to become the two pillars of Superconductivity theory.

1.1 Bardeen's Phronesis

BCS theory has been widely discussed, in both its formal and technical content, in the literature (French and Ladyman 1997) (Cartwright et al. 1995). Set against London's theory, and Landau's, BCS has been shown to be different in important ways. For example, contrary to their phenomenological approach, BCS provides us with both a Hamiltonian, and a description of the superconducting materials which arguably justifies it. It describes a microscopic mechanism for the emergence of the superconducting quantum state. Nevertheless, while these features of BCS theory achieved the highest recognition in the theoretical development of superconductivity, I wish to offer a less common perspective, following Bardeen's methodological ideas (Hoddeson 2001). This will allow me to discuss Bardeen's *Phronesis*, his stance on the dialogue between theory and experiment, and hence on the two traditions.

John Bardeen was originally an electrical engineer. He had been a child prodigy in mathematics and developed very early a keen interest in physics. He nevertheless landed in graduate school, seduced by the news that Einstein would be going to Princeton, only after years of working in an oil company.[12] By the time he had returned to his initial interest in attacking the problem of Superconductivity, 20 years later, he had already been awarded a Nobel Prize for the invention of the transistor, giving him a place in the list of "100 Most Influential Americans of the Century"

[12] Einstein, unfortunately, started his position at the Institute for Advanced Studies, which is adjacent but not part of Princeton. Furthermore he did not intend to take any graduate students. Bardeen's supervision, however, did not end up in poor hands. Eugene Wigner mentored him, and led him to the publication of an important calculation from first principles of electron–phonon scattering in a metal, a calculation which turned out to be very useful to his theory of superconductivity (p. 146).

(Barnes 1991). Awarded a second time for his work in Superconductivity and BCS theory, he still is the only physicist to have received two Nobel prizes.

Bardeen made his methodology quite explicit to all his collaborators, as they have recollected later. David Pines, a major player in superconductivity research and Bardeen's early assistant (then his office neighbor at Bell laboratories for 32 years), reports, in a biographical article on his mentor, Bardeen's agenda for solving scientific problems:

- "*Focus first on the experimental results, by careful reading of the literature and personal contact with members of leading experimental groups.*
- *Develop a phenomenological description that ties the key experimental facts together.*
- *Avoid bringing along prior theoretical baggage and do not insist that a phenomenological description map onto a particular theoretical model. Explore alternative physical pictures and mathematical descriptions without becoming wedded to a specific theoretical approach.*
- *Use (thermodynamic and) macroscopic arguments before proceeding to microscopic calculations.*
- *Focus on physical understanding, not mathematical elegance. Use the simplest possible mathematical descriptions.*
- *Keep up with new developments and techniques in theory, for one of these could prove useful for the problem at hand.*
- *Don't give up! Stay with the problem until it's solved.*" (Pines 1992)

Let me offer three remarks on my own reading of them.

Firstly, by 'phenomenological description' I believe Bardeen means something close to what Cartwright calls *representative* models. These are "models that are intended to be reasonably accurate accounts of target phenomena and their sources" (1999). Specifically, he had in mind models like that of London, with *ad hoc* assumptions based on experimentally observed regularities. He wanted to develop them to account for more of the features and regularities that were emerging from further experiments.

Secondly, to "avoid bringing along prior theoretical baggage" should not be read as a literal claim. Taken literally, we would nowadays consider it naïve. It is impossible to start from scratch, without assuming *any* theoretical *datum*. He better explains what he means when he specifies that what is important is not to insist that a phenomenological description maps onto a particular theoretical model. The focus then is not on avoiding theoretical baggage entirely, but on avoiding becoming wedded to any specific theoretical approach. His key guideline is to engage in open exploration of the alternative physical pictures and mathematical descriptions, for the sake of better representing the phenomena.

Lastly, his remark on mathematical elegance should clearly not be read as a statement of preference for solutions that are not elegant. Read in the context of his methodological ideas, where he puts it in contrast with 'physical description', mathematical elegance refers to the activity of building sophisticated mathematical structures for the sake of bringing logical coordination to a description – without necessarily adding content to it. It is in fact a common explicit assumption of physicists that the process of unifying some otherwise independent aspects of a theory, or of independent theories, requires mathematical sophistication. The mathematics must be able, for example, to transform, or map, separate parameters (or concepts or laws) onto special instances or expressions of a single or more general, more fundamental, parameter. Physicists usually refer to the many cases of successful unification in science as examples of this. So with his remark on mathematical elegance, Bardeen is stressing the importance of maintaining focus on "physical understanding", recommending an instrumental use of theories and theoretical tools for the purpose of the description of the phenomena ("Keep up with new developments and techniques in theory, for one of these could prove useful for the problem at hand").

I will return to the above points shortly.

Following these guidelines and focusing initially on exploratory research, Bardeen had accumulated almost all that he needed to formulate a theory of superconductivity. Cooper and Schrieffer provided the missing bits.

Bardeen, after the departure of Pines as his postdoctoral collaborator, was looking for a bright young theorist, and in 1955 the young Leo Cooper was recommended to him. At the same time, Bardeen encouraged his new PhD student, Robert Schrieffer, to join him on the quest for an account of superconductivity. The starting point was inspired by London's phenomenological theory and in agreement with Landau's theory, that superconductivity had to come from the long-range order of some quantity.

In 1956, Cooper found out that even arbitrarily weak attraction could lead to pair formation in the presence of a Fermi sea. These pairs would behave as bosons, therefore explaining the emergence of the condensate that had been suggested by analogy with superfluidity and hinted at by experiments. In fact, positive ions are attracted to negative electrons and this, he suggested, polarizes the ions towards the electron. As the first electron escapes, a second electron sees this positive cloud and is attracted to its location, leading to the formation of what have been called *Cooper pairs*.

Cooper pairs were exactly the kind of thing that London would have liked. They extended in all directions, like a sphere: their large spatial dimensions conferred to them a clear macroscopic character; there was no sense in talking of them as conventional particles, or simply as two electrons together, moving in classical trajectories. They were a new quantum state unlike any other known.

Bardeen's painstaking implementation of his maxims led to the seminal publication, in 1957, of the BCS paper.

The set of ideas used by Bardeen and his colleagues in the BCS paper are not completely theoretically justified. For the Hamiltonian, they make assumptions about what states are significantly interacting, e.g. that for pairs of electrons with equal and opposite momenta the scattering interactions would be much more significant. These assumptions are ultimately left to be tested by the success of the theory they constitute – by judging the theory's capacity to account for the many features of superconductivity. The BCS Hamiltonian has been argued to be, in this sense, both theoretically principled and phenomenological (*ad hoc*) (Cartwright 1999). This leaves us in a somewhat ambivalent position.

According to Pines, Bardeen "believed in a bottom-up, experimentally based approach to doing physics, as distinguished from a top-down" approach. "To put it another way", he explains, "deciding on an appropriate model Hamiltonian was John's penultimate step in solving a problem, not his first". This, however, is also indecisive and does not mean that we can automatically enroll Bardeen in the same phenomenological camp as Landau and London, for the following reasons.

Landau-Ginzburg's and London's approaches shared with Bardeen's a commitment towards the experimental grounding of the initial description, and, more importantly, a solid base in phenomenological descriptions as the starting point. From that common starting point, though, the aims of the two camps separated. While achieving a similar macroscopic description of superconductivity, BCS provided a microscopic mechanism and a Hamiltonian for it. Most importantly, BCS achieved the final Hamiltonian for the superconductor's quantum state through a process of abstraction, neglecting some effects, such as anisotropy, and inserting several simplifying assumptions to avoid disturbances that may undermine the model.

The concepts usually employed in philosophical discussions both of the relationship between theory and experiment and of the role of models in superconductivity – discussions in which the theories of London and BCS are set against each other – are unsatisfactory for use in my integrated historical/philosophical analysis. This is one of the reasons why I suggested that adopting an interpretation of the practices of science in terms of the two traditions could bring a better understanding of the issues at stake in the debate on superconductivity theory. In fact, Bardeen's approach to the problem of superconductivity is emblematic of a *cooperative phase of interaction* between the two traditions, a very successful phronesis. While we cannot locate him in the phenomenological or bottom-up camp without incurring into ambiguities, we can locate him along the descriptive tradition.

In fact, Bardeen was naturally aiming at achieving both goals, namely to give the most accurate and complete description of the phenomena and to explain the phenomena in a way that is consistent with accepted principles and theories. Nevertheless, the previous description of Bardeen's ideas for the formulation of theories is to show that *he clearly conceived of the two goals as significantly related and most importantly believed that the road to the second is via the first.* While this observation may seem trivial at first sight, let me show why I think it is not.

An objection to the claim that the road to a logically coherent explanation of the phenomena is best pursued via the accurate (phenomenological) description of those phenomena may be that this claim trivially reduces to the discovery/justification distinction. One could say that it is not at all surprising that even the most logically based theoretical account needs to be discovered via all sorts of considerations, including phenomenological ones. However, granting the fact that logically coordinated theories originate from a discovery phase that naturally includes (in some loose way) experimental observations, Bardeen's belief in the need for description to precede and guide first principle derivation is not generally shared. For example, I mentioned earlier that, for Bardeen, deciding on an appropriate 'model Hamiltonian' was the penultimate step in solving a problem, not the first (Pines 1992). This is in contraposition, as we will now see, with Richard Feynman's line of attack on the superconductivity problem – which was expressed (with exceptionally bad timing) just as Bardeen, Cooper, and Schrieffer were completing the famous BCS paper.

So let me show that Bardeen's is not just a trivial phronetic stance by showing an opposite paradigm in physics, exemplified by Feynman and embraced by many.

2 Don't Ask What the Theory Can Do for You. Ask What You Can Do for the Theory

Just a few months before the publication of the BCS paper, an important theoretical conference took place in Seattle. Richard Feynman, who, while not yet a Nobel Laureate, was already widely considered one of the greatest theorists alive, gave the main lecture on the status of theory in superconductivity, published soon after in Reviews of Modern Physics (1957). He had tried his diagrammatic techniques on the electron–phonon system but had not found the expected change from the normal metal behavior. Such was his reputation that many physicists have confessed to being worried at the time that if he had tried extensively to solve the problems mathematically and had still failed, there was more likely to be something wrong with the formulation of the problem than with his math.

Feynman suggested a radical approach, the opposite of London's. In his opinion, the highest ambition that a theorist could have was to deduce an explanation from first principles, which, for superconductivity, Feynman thought meant Schrödinger's equation. Having thus set out his starting point, he suggested an unusual approach. His own words (Feynman 1957) clearly express the solution he devised for the problem that the theoretical community faced:

> It does not make any difference what we explain, as long as we explain some property correctly from first principles. If we start honestly from first principles and make a deduction that such and such a property exists—some property that is different for superconductors than for normal conductors, of course—then undoubtedly we have our hand on the tail of the tiger because we have got the mechanism of at least one of the properties (p. 209)

The goal was getting a single success in one bit of the problem, as opposed to the problem as a whole. In this, he clearly assumes that succeeding in the first-principle derivation of a single, isolated, aspect of the problem would most likely provide clues for explaining the other properties, grabbing the tiger by the tail. Anything would do; it was not very important which property was explained:

> If we have it correct we have the clue to the other properties, so it isn't very important which property we explain. Therefore, in making this attempt, the first thing to do is to choose the easiest property to handle with the kind of mathematics that is involved in the Schrödinger equation. I decided it would be easiest to explain the specific heat rather than the electrical properties. [..] But we do not have to explain the entire specific heat curve; we only have to explain any feature of it, like the existence of a transition, or that the specific heat near

> absolute zero is less than proportional to T. I chose the latter because being near absolute zero is a much simpler situation than being at any finite temperature. Thus the property we should study is this: Why does a superconductor have a specific heat less than T? (p. 210).

In the end, Feynman reported Casimir's conclusion that there was only one way to tackle the problem. It had to be to simply *guess* the quality of the answer. "The only reason that we cannot do this problem of superconductivity", he concluded, "is that we haven't got enough imagination" (p. 212).

Feynman's talk came at the end of a long phase of acute frustration in physical theory for researchers in the field of superconductivity. While this was soon to be relieved by the BCS solution, Feynman's attitude was neither accidental nor temporary. It was a clear expression of the long-standing theoretical tradition based on first principles, an approach that had found in the complexities of superconducting materials 'the toughest crowd for their show', since the very early days of cryogenic research with liquid helium. Indeed, this approach was not damaged as a result of the success of Bardeen – which had in any case achieved (at least partially) a principled derivation. It is common to many fundamental theorists.

After the conference at which Feynman had given his speech, BCS published their paper and quickly achieved success. This does not mean that their results were immediately clear. As Matricon & Waysand remark, "Bardeen had followed all [of his] precepts in his decisive work on superconductivity, the lack of mathematical elegance included" (p. 159). Their paper was a real craft of ambition. Filled with complex mathematics, and containing radical ideas interconnected in complex ways, combined with experimentally derived intuitions, it was from a technical point of view hard to swallow, and difficult to digest. Anderson observed that it was very lucky for his colleagues that the BCS article was so poorly written, since this opened the way for many publications to flourish in the following years, giving order, simplification, and a more robust derivation, to the BCS ideas, to the satisfaction of the more radically principled scientists.

What is now in fact meant when one mentions BCS theory is often its most important re-statement, due to Bogoliubov and his collaborators (Bogoliubov et al. 1958). The main virtue of Bogoliubov's version is the translation of BCS theory into the more sophisticated language of self-consistent fields (Hartree-Fock theory). This put BCS' disordered results into the desired order, guided by first principles.

I am now going to make few comments on Feynman's talk, relating it to Bardeen's view and to the traditions as I conceive of them.

It is interesting to note that Feynman ends his talk with a remark by Casimir, and this provides an amusing parallel. Casimir was the critic of Onnes' empiricist agenda for science. Casimir had objected to Onnes that qualitative observation has to precede quantitative measurement; we cannot go on to properly measure anything if theory does not first tell us what to measure. Feynman's remarks can be seen as analogous to Casimir's. He suggests that we cannot know what observations are relevant, and that it is theory that shall guide the choice as we evaluate the fitness of that observation for a principled derivation.

One of the most striking features of Feynman's approach is its pragmatic spirit. Among the criticisms that theorists usually direct at phenomenological theories is the charge of being too 'utilitarian' and pragmatic. Chemists and experimental physicists in particular, attempting to solve quantitatively bits of a problem for concrete returns in terms of applications, are usually negatively labeled as 'pragmatist'. Putting aside a discussion of the misguided normative implication of this, I think that Feynman's approach can be seen as similarly pragmatic, though in the following different sense. The choice of the feature to be explained by theory is not justified in any way, in his view, except by the simple fact that theory 'worked', explaining that feature. Given the apparent stalemate – the baffling behavior that Nature seemed to display whenever physicists looked deeper into the phenomena – theory had to compromise. Not wanting to give up the aim of a 'royal' derivation from first principles, yet also not accepting a full description of the phenomena without first-principle justification, Feynman had resolved to work out some small percentage of the phenomena, a part that seemed most likely to be soluble by known techniques and accepted principles. Hard times call for hard measures, and the best heroic attitude was for Feynman to be heroically pragmatic.

Most importantly, in contrast to Feynman's "anything goes" attitude, London had maintained that explaining diamagnetism was not just *a* path, but an *essential* path – the true face of superconductivity. In arguing for the importance of starting from a phenomenological description of the phenomena, Bardeen had accepted London's view of the superconductor *as a whole.* The Meissner effect was crucial. This view was not justified by a theory, and was motivated by qualitative arguments based on both the assessment of the general experimental features, and analogy with intuitively similar experimental situations.

It is in this sense, then, that I claim that while London was 'instrumentalist' in the use of theory, Feynman was 'instrumentalist' in the use of experiment.

To some extent, I see Bardeen's approach as embodying a part of both views, and it is in this sense that I identify within his research the cooperative phase of interaction between the traditions. The line of cooperation between the traditions, though, has a clear orientation (as I have anticipated). This is a crucial point, and I will now explain it further.

On one side, the epistemic role of the descriptive tradition in its exploratory nature, not guided by theory, which is evident in London's approach, becomes a significant part of Bardeen's approach and constitutes one difference from Feynman's. Bardeen's stress on the importance of starting from an accurate phenomenological description, not from first principles, led him to build the theory using concepts that were able to account for Meissner's effect. Looking for a microscopic mechanism, he was guided by considerations arising from that phenomenological description. Bardeen had to find something that explained the formation of a phase with higher order, which led him to the restriction to a macroscopic entity. Since the coupling between electrons (fermions) was unable to account for Meissner's effect, his discussion shifted to bosons, by analogy with superfluidity.

This, though, was not the full story. Having restricted the space of possibilities, he could start looking for a specific microscopic mechanism; in doing so, he returned to phonons, which he had studied extensively since his graduate years. Maintaining that this choice, like any other, had to be firmly grounded in experimental considerations, he extensively studied the emerging experimental literature, and noticed that the isotope effect could be accounted for via a specific electron phonon coupling. This displays similarities with Feynman's approach. In fact, while it was certainly a striking experimental feature, the selection of the isotope effect as relevant was somewhat arbitrary. The hidden strength of the choice was the fact that phonons seemed to be able to account for it. So in this sense Bardeen had singled out a feature mostly because of its fitness to be derived from first principles, in line with Feynman's approach. Nevertheless, Bardeen did not believe that phonons *had* to be the answer. They would be so only insofar they provided a mechanism that was helpful to the full description of the phenomena. The knowledge that phonons led to the isotope effect actually meant little until he could see the whole macroscopic picture coming together and accounting for the features that constituted the phenomenological model which he kept with him at all times.

The difference, then, between scientists is in their degree of confidence that starting from, on the one hand, first principles or, on the other hand, the phenomena and their description, will lead them to the desired solution. This is then not merely a difference of goals, but a difference in conceiving the connection between them. It is not just a matter of preferring a complete description to a principled explanation. It is a matter of the scientist's judgment as to beliefs and confidence over which one is the (better) starting point from which to achieve the other.

These different methodologies do not exhaust their importance in methodology: They reflect the set of preferences expressed by the two traditions, and underlie different conceptions of what it means to formulate a theory and to succeed in problem solving. The set of criteria for the evaluation of theories employed by scientists underlie the reaching of a consensus on theories, and thus contribute to define what 'scientific success' means. In claiming that these criteria are deeply dependent on the preferences exemplified in the two traditions, I suggest that this dependence is thus transmitted to the issue of consensus, which is found in a crisis in current superconductivity research as the task of achieving theories that give a reasonably full description of the phenomena suffers in favor of using criteria for theories based on first principles. Many HTS players denounce this crisis; in this climate, balkanization of the theoretical HTS community is encouraged, and theories become like toothbrushes: We each have our own, and we do not share.

Acknowledgements I am indebted above all to Nancy Cartwright and Hasok Chang for comments and invaluable mentorship. Special thanks to Tom Ryckman and Micheal Friedman, to Conrad Heilmann, Alice Obrecht, and Sally Riordan for the comments and encouragement received. Thanks also go to Abrol Fairweather, the wonderful SFSU philosophy department and the lands of Big Sur. And finally, I express deep gratitude to my parents, Duccio and Fernanda.

References

Anderson, P. 1994. *A career in theoretical physics*, World scientific series in 20th century physics, vol. 7. Singapore: World Scientific Publishing.

Anderson, P.W., and B.T. Matthias. 1964. Superconductivity. *Science* 144(3617): 373–381.

Anderson, P.W., and P. Morel. 1962. Calculation of the superconducting state parameters with retarded electron–phonon interaction. *Physical Review* 125: 1263–1271.

Aristotle. 1941. In *The basic works of Aristotle*, ed. Richard McKeon. New York: Random House.
Barnes, B. 1991, January 31. John Bardeen, nobelist, inventor of transistor, dies. http://www.highbeam.com/doc/1P2-1047095.html. Washington Post.
Bednorz, J.G., and K.A. Müller. 1986. Possible high Tc superconductivity in the Ba – La – Cu – O system. *Condensed matter, Zeitschrift für Physik B* 64(1): 189–193.
Bogoliubov, N., V. Tolmachev, and D.V. Shirkov. 1958. *A new method in the theory of superconductivity*. Moscow: Academy of Sciences Press.
Bromberg, J.L. 1994. Experiment Vis-a-Vis theory in superconductivity research. The case of Bernd Matthias. In *Physics, philosophy, and the scientific community: Essays in the philosophy and history of the natural sciences and mathematics in honor of Robert S. Cohen*, ed. K. Gavroglu, 1–10. Dordrecht: Springer.
Cartwright, N. 1999. *The dappled world: A study of the boundaries of science*. Cambridge, UK: Cambridge University Press.
Cartwright, N., T. Shomar, and M. Suárez. 1995. The tool box of science: Tools for the building of models with a superconductivity example. In *Poznan studies in the philosophy of the sciences and the humanities*. Amsterdam: Rodopi.
Casimir, H.B. 1983. *Haphazard reality, half a century of science*. New York: Harper and Row.
Casimir, H.B., and C.J. Gorter. 1934. The thermodynamics of the superconducting state. *Physik Z* 35: 963.
Curty, P., and H. Beck. 2003. Thermodynamics and phase diagram of high temperature superconductors. *Physical Review Letters* 91: 257002.
Duhem, P. 1914. *The aim and structure of physical theory*. Princeton: Princeton University Press.
Dunne, J. 1993. *Back to the rough ground: "Phronesis" and "techne". modern philosophy*. South Bend: University of Notre Dame Press.
Feynman, R.P. 1957. Superfluidity and superconductivity. *Reviews of Modern Physics* 29: 205–212.
French, S., and J. Ladyman. 1997. Superconductivity and structures: Revisiting the London account. *Studies in History and Philosophy of Modern Physics* 28B(3): 363–393.
Hoddeson, L. 2001. John Bardeen and the theory of superconductivity. *Journal of Statistical Physics* 103(3/4): 625–640.
London, F. 1937. *Une conception nouvelle de la supraconductibilite*. Paris: Hermann.
London, F., and H. London. 1935. The electromagnetic equations of the supraconductor. *Proceedings of the Royal Society A* 149(866): 71.
Matricon, J., and G. Waysand. 2003. *The cold wars – A history of superconductivity*. Trans. C. Glashausser. New Brunswick: Rutgers University Press.
Matthias, B.T. 1955. Empirical relation between superconductivity and the number of valence electrons per atom. *Physical Review* 97: 74–76.
Matthias, B.T., and J.K. Hulm. 1952. A search for new superconducting compounds. *Physical Review* 87: 799–806.
Meissner, W., and R. Ochsenfeld. 1933. Ein neuer Effekt bei Eintritt der Supraleitfähigkeit. *Naturwissenschaften* 21: 787.
Pines, D. 1992, April. An extraordinary man: Reflections on John Bardeen. *Physics Today*, 64–70.
Scalapino, D. 2007. Novel aspects of superconductivity. Retrieved 2009, from blogspot: http://novelsc.blogspot.com/2007/07/tuesday-31st-july.html
Steinle, F. 1997. Entering new fields: Exploratory uses of experimentation. *Philosophy of Science* 64: 65–74. Chicago: The University of Chicago Press.
Steinle, F. 2006. Concept formation and the limits of justification: "Discovering" the two electricities. In *Revisiting discovery and justification*, ed. J. Schickore and F. Steinle, 183–195. Dordrecht: Springer.

Part IV
Understanding, Explanation and Epistemic Virtue

Knowledge and Understanding

Duncan Pritchard

1 The Knowledge Account of Understanding

What is the relationship between knowledge and understanding? On a very popular picture, one that has a lot of intuitive appeal and which is especially prevalent in the philosophy of science, the relationship between these two epistemic standings is fairly straightforward. In a nutshell, the idea is that understanding is essentially a type of knowledge—*viz.*, knowledge of causes. More specifically, to have understanding of why X is the case is to know why X is the case, where to know why X is the case is to know that X is the case is because of Y. Call this the *knowledge account* of understanding.[1] One finds defences of the knowledge account, in varying levels of explicitness, in the work of such authors as Peter Achinstein (1983), Wesley Salmon (1989), Philip Kitcher (2002), James Woodward (2003) and Peter Lipton (2004).[2]

---

[1] Note that we are here implicitly focussing on a particular kind of understanding, one which concerns understanding why something quite specific is the case. Typically, this will involve knowing why a specific event occurred. This kind of understanding is sometimes contrasted with a more general kind of understanding which concerns, say, an entire subject matter (e.g., '*S* understands quantum physics'). While there are connections between the two kinds of understanding, and while one would clearly desire an account of understanding which could accommodate both types, it would take us too far from the main thread to consider this more general notion of understanding here. For further discussion of these two types of understanding, see Brogaard (2007) and Kvanvig (2009).

[2] Consider the following remark made by Lipton, for example:

> Understanding is not some sort of super-knowledge, but simply more knowledge: knowledge of causes. (Lipton 2004, 30)

Grimm (2014) also attributes the knowledge account, broadly conceived anyway, to Aristotle, Lewis (1986), Miller (1987), Strevens (2008) and Greco (2014).

D. Pritchard (✉)
University of Edinburgh, Edinburgh, UK
e-mail: duncan.pritchard@ed.ac.uk

A. Fairweather (ed.), *Virtue Epistemology Naturalized: Bridges Between Virtue Epistemology and Philosophy of Science*, Synthese Library 366, DOI 10.1007/978-3-319-04672-3_18,

So, for example, imagine a scientist—let's call her 'Kate'—in a lab observing a certain chemical reaction. Kate, let us stipulate, understands why this chemical reaction took place. On the knowledge account, Kate's understanding of why this chemical reaction took place is constituted by her knowing why it took place. And her knowing why this chemical reaction took place is constituted by her knowing that it took place because, say, the substances in question were reacting to the oxygen that she introduced. Kate's understanding of why the chemical reaction took place is thus constituted by her knowing what caused it to occur.

The simplicity of this picture makes it very attractive, and I think that for a wide range of cases it does generate the correct result. But, as I've argued elsewhere—e.g., Pritchard (2009) and Pritchard et al. (2010, ch. 4)—ultimately it is not quite right, and it is important to recognise why. The problem is that understanding and the corresponding knowledge of causes comes apart in both directions—*viz.*, you can have understanding while lacking the relevant knowledge of causes, and you can have knowledge of causes while lacking the relevant understanding.

Consider first how an agent might have knowledge of causes while lacking the corresponding understanding. The point here is that there are ways in which one might gain knowledge of causes which wouldn't suffice for understanding. So, for example, consider a counterpart of Kate, Kate*. Kate* comes to know that it was the introduction of the oxygen which caused the chemical reaction not because she figured this out for herself, but because a fellow scientist, who has specialised expertise in this regard which our hero lacks, informs her that this is the cause of the reaction. Furthermore, let us stipulate that Kate*, while generally proficient in chemistry, does not have any sound epistemic grip on why the introduction of oxygen should have this effect on the substances in question.

Given that Kate* has gained this (true) information about the cause of the chemical reaction from someone she recognises to be an expert in the field, she surely counts as knowing what the cause of the chemical reaction was. Moreover, it is also surely right that Kate* knows why the chemical reaction took place, given that she has this knowledge of the cause of the reaction. So Kate* knows why the chemical reaction took place, and she knows that it took place because of the introduction of oxygen. Crucially, however, Kate* does not understand why the chemical reaction took place, because in order to possess understanding in such a case it is surely required that she should have a sound epistemic grip on why cause and effect are related in this way. Since Kate* lacks this, she lacks understanding. One can thus have the relevant knowledge of causes (along with the relevant knowledge why) and yet lack understanding.

Consider now how one can possess understanding even while lacking the relevant knowledge of causes. Imagine a second counterpart of Kate, Kate**. Kate** is like Kate except that her belief regarding the cause of the chemical reaction, while true, is only luckily true, in the sense that she could so very easily have formed a false belief. So, for example, suppose that Kate** forms her belief regarding the cause of the chemical reaction by employing an instrument. We can now imagine two ways in which the use of this instrument could introduce luck into Kate**'s acquisition of the target true belief.

The first way would be via the instrument malfunctioning (unbeknownst to Kate**), but happening to produce the right result nonetheless. This would be akin to a standard Gettier-style case, in that Kate** would be forming a justified true belief in the target proposition (*viz.*, that oxygen is the cause of the chemical reaction), but failing to gain knowledge of this proposition because her belief is only luckily true (i.e., had the instrument not happened by chance to produce the right result, then Kate** would have believed falsely in this case).

The second way in which we could introduce epistemic luck into Kate**'s acquisition of the target true belief is more subtle. Suppose, for example, that the instrument which Kate** uses is not malfunctioning, and so delivers her the correct result. But suppose further that the instrument very nearly did malfunction, such that in most near-by possible worlds where Kate** employs this instrument it would be malfunctioning and so at best only delivering the correct result by chance. Let us also stipulate that had the instrument malfunctioned Kate** would not have noticed. Again, Kate** would form a justified true belief in the target proposition by using this instrument, but her belief would not amount to knowledge in virtue of how it is only luckily true (i.e., she could so very easily have formed a false belief in the target proposition).

Although our hero ends up with a belief which is only luckily true in both cases, the type of epistemic luck at issue in the second case is very different from that in play in the first. After all, in the second case Kate** really is using a properly functioning instrument in order to gain her true belief, unlike in the first case. Elsewhere I've called the particular kind of epistemic luck in play in the second case *environmental epistemic luck*—see, for example, Pritchard (2009)—on the grounds that the luck in question specifically concerns the epistemic environment that the agent is in (in this case, that the properly functioning instrument available to Kate** could so very easily have undetectably been malfunctioning).[3]

The reason why it is important to mark this specific kind of epistemic luck is that while a luckily true belief can never amount to knowledge, even when the kind of epistemic luck in play is environmental, understanding is compatible with at least environmental epistemic luck.[4] In order to see this, consider again the two variants of the Kate** case that we just described. In both cases our hero doesn't gain knowledge of what caused the chemical reaction, since the belief that she forms is only luckily true. In both cases, then, she doesn't know that the chemical reaction was because of the introduction of the oxygen, even though she truly believes this in both cases (and, what is more, believes this with justification). But while it is also true in the first case that Kate** lacks an understanding of why caused the chemical reaction, she does seem to acquire this understanding in the second case. For while Kate** surely can't acquire an understanding of what caused the chemical reaction by using a malfunctioning instrument which only happens to produce the right

[3] It is specifically environmental epistemic luck which is at issue in the famous 'barn façade' case, or so I have argued anyway. See, for example, Pritchard (2009) for more on this point.

[4] For further discussion of the claim that a luckily true belief cannot amount to knowledge, see Pritchard (2013). For a critical response to this claim, see Hetherington (2013).

result, remember that in the second case the instrument being used is not in fact malfunctioning at all, but rather working just as it is supposed to. So what barrier would there be to Kate** gaining understanding in this case (unless, of course, one is already convinced that understanding requires the corresponding knowledge)?[5]

There are thus problems with the idea that understanding is to be conceived of in terms of knowledge of causes, and this means that the knowledge account is under threat. In the next Section I will suggest that the situation isn't just that the knowledge account is problematic, but moreover that there is an alternative account available which offers a far superior picture of how these two epistemic standings relate. This is the *cognitive achievement* account.

2 The Cognitive Achievement Account of Understanding

As I have argued elsewhere—e.g., Pritchard (2009) and Pritchard et al. (2010, ch. 4)—I think there is a good reason why understanding and knowledge come apart in the particular ways just specified, one which further supports the reading offered of the cases in question. This is that understanding, unlike knowledge, is a specific kind of achievement. Achievements are, roughly, successes that are because of ability; that is, where the success in question is primarily creditable to the agent's exercise of the relevant ability. So, for example, an archer's success at hitting the bull's eye counts as her achievement so long as this success is primarily creditable to her exercise of those abilities relevant to archery and not to other factors (such as a lucky gust of wind).

Interestingly, achievements, while not generally compatible with lucky successes, are compatible with lucky successes where the luck in question is entirely environmental. For consider an analogous case to the Kate** example involving environmental epistemic luck which concerns our archer. Suppose that the archer skilfully fires an arrow at a target, thereby successfully hitting the target as intended. Unbeknownst to our archer, however, this success could very easily have been failure. Imagine, for example, that it was pure chance that the archer fired in normal environmental conditions, such that in most near-by possible worlds there would be very high winds which would have prevented the archer from hitting the target.

Our archer's success is thus lucky, in that it is a success that could so very easily have been a failure. But nonetheless, isn't it correct to say of our archer that her success was primarily creditable to her archery abilities, and thus that this was no less of an achievement as a result? After all, although it is true that she could so very easily have been unsuccessful, in fact nothing did get in the way of her displaying

[5] Interestingly, Kvanvig (2003) argues that understanding, unlike knowledge, is compatible with the kind of epistemic luck at issue in standard Gettier-style cases too. As I claim in Pritchard (2009), however, I think he reaches this conclusion because he fails to make the distinction between standard Gettier-style epistemic luck and environmental epistemic luck.

her archery abilities. Hence, there seems no reason to deny that her success constituted a genuine achievement, even despite the environmental luck in play.

Compare this with a case where a standard, non-environmental, type of luck is in play. Suppose, for example, that our archer skilfully fired the arrow but that it was knocked off course by an unlucky gust of wind, and then knocked back on course again by a second lucky gust of wind, so that the arrow does indeed hit the bull's eye after all. In such a case we would have a display of archery ability and we would have the relevant success to go with it, but I take it that we wouldn't say that this success constituted a genuine achievement. After all, the success was not primary creditable to the agent's display of archery ability, but was rather down to the fortuitous second gust of wind. So while achievements are consistent with mere environmental luck, they are not in general consistent with lucky success.[6]

What goes for achievements goes for cognitive achievements—*viz.*, cognitive successes (i.e., true beliefs) which are because of the exercise of cognitive ability (where this means that the cognitive success in question is primary creditable to the agent's cognitive ability). In a standard Gettier-style case, one's cognitive success does not constitute a cognitive achievement since it is not primarily creditable to the agent's cognitive ability (even though the agent does exercise the relevant cognitive ability) but rather down to the epistemic luck in play. But in cases that involve mere environmental epistemic luck, however, a cognitive achievement is nonetheless exhibited, since it remains that the cognitive success, while lucky, is even so primarily creditable to the cognitive agency of the subject.

This is one reason why we shouldn't equate cognitive achievement with knowledge, as some have been tempted to do.[7] For while knowledge is incompatible with lucky cognitive success, cognitive achievement is compatible with lucky cognitive success so long as the epistemic luck in question is purely environmental. In this sense, knowledge can be *more* demanding than cognitive achievement. But there is also a sense in which knowledge can be *less* demanding than cognitive achievement, and this is brought out quite nicely by the case of Kate* offered above where she gains her knowledge of causes by trusting the word of an expert. For while, as we noted there, this is a perfectly respectable route to knowledge, it is not a route to understanding, since this requires the agent to be able to carry the relevant cognitive load by herself (enough of it, anyway).

What goes for understanding also goes for cognitive achievement, and this is no coincidence. Think about Kate*'s cognitive success in coming to know, purely via this testimonial route, that the chemical reaction was caused by the introduction of oxygen. Is this cognitive success primarily creditable to *Kate*'s* cognitive abilities? Surely not. Indeed, if anything, it is primarily creditable to the cognitive abilities of her expert informant. This case thus doesn't just demonstrate that one can have

[6] For further discussion of the nature of achievements, see Pritchard et al. (2010, chs. 2 & 4) and Pritchard (2010).

[7] For the main proponents of a view of this general form (though often not expressed in quite these terms), see Sosa (1988, 1991, 2007, 2009), Zagzebski (1996, 1999) and Greco (2003, 2007, 2008, 2009a, b, c).

understanding without the corresponding knowledge demanded by the knowledge account, but also demonstrates that one can acquire knowledge without thereby exhibiting a cognitive achievement.[8] As noted above, I don't think this is a coincidence.

This is because the kind of epistemic standing involved in understanding is in its nature a cognitive achievement. That is, it is in its nature the kind of cognitive success which is primarily creditable to the exercise of the subject's cognitive ability. As such, there should be no surprise (i) that it is compatible, along with achievements more generally, with environmental (epistemic) luck; and (ii) that it is not the kind of epistemic standing that one can acquire by for the most part trusting the word of another (no matter how authoritative one's informant is).

I am thus suggesting an alternative picture of the relationship between knowledge and understanding to that offered by the knowledge account. Rather than think that understanding is constituted by a particular kind of knowledge (i.e., knowledge of causes), as the knowledge account demands, we should instead recognise that understanding is a kind of cognitive achievement and as such differs in some respects from knowledge, to such an extent that the knowledge account is untenable. Call this the *cognitive achievement account* of understanding.[9]

3 Grimm *Contra* the Cognitive Achievement Account

One of the principal defenders of the knowledge account of understanding has been Stephen Grimm (2006; cf. 2010). In an important new article, Grimm (2014) has challenged the cognitive achievement view of understanding and in the process has offered a new defence of (a version of) the knowledge account. As we will see, what is key to his defence of the knowledge account is to offer a very different construal of how this account should be understood, to the extent that I think it is best to label Grimm's new rendering of this view as a novel third position in the debate, albeit one that is in the spirit of the knowledge account. For reasons that will become apparent, we will refer to this new account of understanding as the *grasping account*. Before we consider the grasping account of understanding, however, it will be useful to review a particular critical line that Grimm makes regarding the cognitive achievement account, in that this critical line is completely independent of the grasping account of understanding that he offers.

Grimm proposes to deal with the case offered above of an agent (Kate*, in our example) who has knowledge of the cause of an event and yet fails to have the

[8] It is an interesting question why knowledge should be such that it marks out an epistemic standing distinct from cognitive achievements, though it is not one that I can usefully engage with her. For further discussion of this issue, see Pritchard et al. (2010, ch. 3) and Pritchard (2011).

[9] Drawing on my work on this topic, Hills (2009, 2010) has developed her own variant of the view that I am here calling the cognitive achievement account, though her focus is specifically on moral knowledge and understanding.

corresponding understanding by, in effect, forcing a dilemma. On the one hand, Grimm argues that if the agent really does have no real conception of how cause and effect can be related, then it isn't even plausible to suppose that this agent has the relevant knowledge. For instance, he writes that in such a case "it is not clear that [*the agent*] understands the content of that proposition [*i.e., the proposition concerning cause and effect*] well enough to actually believe it." (Grimm 2014, §2). On the other hand, Grimm (2014, §5) argues that insofar as we do credit the agent with having at least some conception of how cause and effect can be related, then both the ascription of knowledge and the corresponding understanding will be apt. Grimm defends this point by arguing that understanding comes in degrees, and thus that we shouldn't be so quick to conclude from the fact that an agent has very little understanding that she has no understanding.

I think we can safely ignore the first horn of the dilemma being posed here, since it was never part of the argument against the knowledge account to suppose that the agent concerned had no conception at all of how cause and effect might be related, to the extent that we could seriously doubt whether the agent even had the conceptual resources to believe the target proposition. Consider how we described this case above. Since Kate* is a scientist she surely has *some* conception of how the introduction of oxygen might cause the chemical reaction in question. It is not then as if this relationship between cause and effect is something that is completely opaque to her, as might be the case if one, say, told a medieval alchemist that a certain chemical reaction was caused by the introduction of oxygen. As far as the medieval alchemist goes, Grimm might well be right that we wouldn't even credit such a person with a belief about what caused the chemical reaction (much less knowledge), no matter who their informant is. But certainly we would credit Kate* with both the relevant belief and the corresponding knowledge too. Nonetheless, the point remains that she does not have a sufficient conception of how cause and effect are related to count as having understanding.

The crux of the matter is that there is more to understanding why an event took place than simply having some conception of how cause and effect *might* be related. In particular, what is required is some sort of grip on how this cause generated this effect, a grip of the kind that could be offered as an explanation were someone to ask why the event occurred. Significantly, if Kate* were asked the question of why the introduction of the oxygen caused the chemical reaction, she would be unable to respond. Indeed, one would expect her to instead direct the questioner towards her more knowledgeable informant.[10]

So, *contra* the first horn of Grimm's dilemma, the claim being made is that our hero does have a sufficient conception of how cause and effect might be related to

[10] I think that what is muddying the waters here is that the example that Grimm focuses on—found in earlier work by myself (see, e.g., Pritchard 2009)—concerns a child who gains knowledge of the cause of an event while nonetheless lacking the corresponding understanding. As such, it is perhaps natural to wonder, as Grimm does, whether this child really does have knowledge of the cause that is being credited to her. As the case of 'Kate*' here illustrates, however, it is entirely incidental to this objection to the knowledge account that the agent concerned is a child.

genuinely count as having the relevant causal knowledge. What then of the second horn of the dilemma? In treating Kate* as having some conception of how cause and effect might be related, are we therefore committed to supposing that she has some limited degree of understanding of the event in question? The foregoing remarks suggest not, in that there is a distinction to be drawn between, on the one hand, having a sufficient conception of how cause and effect might be related to enable the agent to have the relevant causal knowledge, and, on the other hand, having a sufficient explanatory grip on how this particular cause generated this particular effect in order to possess the corresponding understanding. If we are to be speared on the second horn of Grimm's dilemma, then he needs to make a case for thinking that this distinction is illusory.

Let's look again at how Grimm motivates this second horn of the dilemma:

> [...] when I start chopping onions and my eyes begin to water, I think I understand why my eyes are beginning to water, namely, because I am chopping the onions. I don't think it is because of the time of day, or the colour of the shirt I am wearing, or anything like that; it's because of the onions. But obviously someone with a greater understanding of onion (and eyeball) chemistry would be able not just to identify the onions as the cause but would be able to say what it was about the onions that was bringing this about—in this case, the particular sulphur compounds that were being broken down and released into the air when I did the chopping. [...] what these facts seem to illustrate is not that the person who appeals to the compound understands while I fail to understand, but that understanding comes in degrees; I have less of it, and he has more. (Grimm 2014, §5)

To begin with, we need to consider whether it is enough to be credited with understanding that one can merely identify the cause, even if one lacks a conception of how cause and effect are related. This is, after all, what Grimm is suggesting here, in that he says that while he knows that it is the onions which are causing his eyes to water, he doesn't know much more than that, still less does he know how onions cause eyes to water. And yet the claim is that he has some limited understanding of why his eyes are watering purely in virtue of knowing this cause.

If Grimm were right about this case, then it would follow that someone like Kate* can gain not just knowledge of the cause of the target chemical reaction by receiving the testimony of her authoritative colleague, but can also thereby gain an understanding of this event too. The case would then be neutralised in terms of the challenge it raises for the knowledge account. But how plausible is it that mere knowledge of a cause can suffice for understanding? I think that on closer inspection it isn't plausible at all, even if we factor in the point that the kind of understanding in play is quite minimal.

We noted above that what we are looking for when we credit someone with understanding is more than just a general conception of how cause and effect might be related. What is required is rather a grip on how this cause generated this effect, a grip of a kind that could be offered as an explanation were someone to ask why the event occurred. Imagine, for example, that Grimm were to represent himself to others as understanding why his eyes are watering, but when asked for further information merely pointed to the cause of this event (i.e., the chopping of the onion). Wouldn't Grimm's audience regard him as having misled them? The point is that in representing oneself as being in possession of an understanding of some event, no matter how limited, one is representing oneself as not merely being able to identify the cause of that event, but also as being able to offer a sound explanatory story regarding how cause and effect are related. If one cannot offer such an explanatory story, then one doesn't count as having understanding, not even a limited understanding.

Note that this is not to deny Grimm's point that understanding comes in degrees, and thus that sometimes one can genuinely possess an understanding which is quite minimal. So, for example, one could imagine someone having a rudimentary grasp of how chopping onions can cause one's eyes to water which suffices for a limited kind of understanding of the target event, albeit one which is very deficient when compared with that possessed by someone who has the additional relevant chemical knowledge that Grimm mentions in the cited passage above. But the point is that the kind of example that we have posed as a problem for the knowledge account are precisely *not* of this sort. That is, they are not cases where the agent is in possession of a sound, if rudimentary, explanatory story relating cause and effect, but rather cases where the agent, while knowing full well what the cause of the event is, *lacks* the further explanatory story.

Think of Kate*, for example. As just noted, she will have some conception of why introducing oxygen might cause the target chemical reaction, but crucially she wouldn't be able to offer even a rudimentary explanatory story about how the oxygen caused this effect. As noted above, merely having a conception of how something *might* cause something else is not the same thing as having even a rudimentary explanatory account of how a particular cause and effect are related. Grimm's point about degrees of understanding is thus by-the-by, since once we properly understand the case at issue there is no temptation to ascribe even a limited degree of understanding to the agent.

The dilemma that Grimm poses for the cognitive achievement account of understanding is thus illusory. But even if this account of understanding can survive this critique, it still might be the case that the alternative account of understanding that Grimm offers should be preferred. Accordingly, let us now turn to examining this proposal.

4 The Grasping Account of Understanding

Grimm holds is that the knowledge account is essentially right, to the extent that understanding *is* constituted by the possession of the relevant causal knowledge. Where the knowledge account goes wrong, according to Grimm, is to fail to notice that the causal knowledge in question is not propositional in nature. Instead, it is a kind of non-propositional *grasping*, where this means grasping the modal relationship between cause and effect.

Before we explore what this proposal amounts to, there is an odd feature of the dialectic of Grimm's argument that deserves note, which is that it isn't entirely clear on his view what is meant to be wrong with the knowledge account. The only type of counterexample that he offers to this proposal is that which is put forward by the cognitive achievement account, and yet, as we saw in the last section, Grimm's own view is that this counterexample doesn't work. As such it isn't clear what problem with the knowledge account Grimm's alternative proposal is meant to solve. Still, if Grimm's alternative proposal is better than both the knowledge account and the cognitive achievement account, then that would be reason enough to adopt it as the best proposal available, and so we will explore it in that spirit.

In developing the grasping account of understanding, Grimm appeals to the recent literature on *a priori* knowledge. As he notes, there are some long-standing problems with the idea that *a priori* knowledge is to be understood in terms of knowledge of necessary truths. Knowledge of a necessary truth is clearly not sufficient for *a priori* knowledge, since this knowledge could be gained in an empirical fashion (e.g., via testimony), but the problem is that it is hard to see what would need to be added to this knowledge to ensure that it is *a priori* knowledge.[11] For example, knowing a necessary truth and knowing that it is necessary won't suffice, since knowledge of both propositions could be gained in an empirical fashion.

One solution to this problem that has been proposed has been to think of *a priori* knowledge not in the usual propositional terms, but rather as a kind of grasping which is due to rational insight.[12] That is, we should not think of *a priori* knowledge in terms of knowing a particular proposition (or set of propositions), but rather as directly grasping, via rational insight, some feature of modal reality. The thought is that once we conceive of *a priori* knowledge along these lines, then the kind of problem just indicated cannot arise.

Grimm's idea is that when it comes to understanding we should think of our knowledge of causes along the same lines. That is, having knowledge of the cause is not to be understood merely in terms of propositional knowledge, but should instead be regarded as a kind of non-propositional grasping. As he puts it, it is "seeing, or grasping, of the terms of the causal relata, their modal relatedness." (Grimm 2014, §4) That is, "what would be seen or grasped would be how changes in the value of one of the terms of the causal relata would lead (or fail to lead) to a change in the other." (*Ibid*) The kind of counterexample that I have posed to the knowledge account is thus meant to be avoided by the grasping account of understanding that Grimm offers. For while Grimm may grant that Kate* can come to acquire, via testimony, propositional knowledge of the cause, such knowledge falls short of the sort of grasping of modal reality that Grimm has in mind and that explains why she lacks understanding.

While I think Grimm's proposal is ingenious, I also think it is deeply problematic. Let's begin with the putative parallel between the kind of knowledge of causes that is at issue when it comes to understanding and *a priori* knowledge. Now Grimm is quite clear that he wants his account of understanding to be assessed independently

---

[11] Note that I am here setting to one side the potential complication that might be posed by *a priori* knowledge of *contingent* propositions, a possibility defended by Evans (1979) and Kripke (1980), amongst others.

[12] The foremost exponent of this particular way of thinking about *a priori* knowledge is Bonjour (2001, 2005, cf. Bonjour 1998).

of the account of *a priori* knowledge that he appeals to. But even so I think we might legitimately ask how plausible this account of *a priori* knowledge is, particularly given that it is simply one view amongst others in this regard, since if it turns out to be suspect, then we might well be less impressed by any attempt to motivate an analogous view as regards the epistemology of understanding.

For example, one obvious question we might ask about this view is what this 'grasping' amounts to insofar as it is not reducible to knowledge of some suitable set of propositions, since this is far from clear. But I think that even if we grant the cogency of this non-propositional notion of grasping, we still have cause to be sceptical about the coherence of the view. For one thing, a key part of the proposal is that the grasping in question is via a faculty of rational insight. This is essential to the view, since if the grasping were the result of some empirical process, then clearly this wouldn't be playing the required role in a theory of specifically *a priori* knowledge. But if we are able to appeal to such a faculty in our account of grasping, why can't we appeal to this faculty in order to deal with the cases which supposedly create problems for the propositional view of *a priori* knowledge? That is, why can't we say that *a priori* knowledge is propositional knowledge (e.g., knowledge of a necessary truth, and knowledge that it is necessary) which is gained via rational insight rather than through an empirical process?

Even setting aside these concerns about the account of *a priori* knowledge in play, there are worries about the parallels between *a priori* knowledge and the epistemology of understanding. For example, while there might be specific reasons for thinking that an appeal to the obscure notion of a non-propositional grasping via a faculty of rational insight is required in the case of *a priori knowledge* because of the special problems this type of knowledge faces, recall that we haven't been given any reason for thinking that there is a parallel crisis when it comes to the epistemology of understanding. Indeed, as noted above, the only problem that Grimm considers for the knowledge account of understanding—the account of understanding most closely related to Grimm's own proposal—is one that he claims, on independent grounds, is illusory.

Moreover, there are, in any case, important disanalogies between the two domains. As noted above, it is vital to the grasping account of *a priori* knowledge that this grasping be the result of rational insight. In contrast, it is crucial to the grasping account of understanding that the grasping *not* be the result of rational insight. After all, knowledge of causes is not something which is gained in an *a priori* manner. But with that point in mind, why should we be so confident that the account of grasping which is meant to be applicable in the case of *a priori* knowledge is applicable here? In particular, even if we are confident that the grasping, via rational insight, of the necessary relatedness of certain properties that is involved in *a priori* knowledge cannot be cashed-out in terms of propositional knowledge, why should it follow that the very different kind of grasping which is involved when it comes to knowledge of causes (which concerns the 'modal relatedness' of the causal relata) is also not susceptible to being cashed-out in terms of propositional knowledge?

We can bring this point into sharper relief by considering the non-empirical grasping of causes that Grimm has in mind. He says that when someone grasps the cause of the event in the way that he has in mind, then one is "sensitive not just to how things are, but to how things stand modally, and in particular to how things might have been, if certain conditions had been different." (Grimm 2014, §5) But why would the proponent of the cognitive achievement account dispute this? After all, this proposal insists that the agent must be in possession of a solid explanatory account of how the target cause and effect are related, and that will inevitably entail that one has a conception of how things might have been if certain conditions had been different. But there is nothing here which in itself suggests that we can't conceive of this sensitivity to such modal facts in propositional terms, such as in terms of further propositional knowledge. For example, Kate*'s informant, who clearly understands why the oxygen caused the chemical reaction, will surely also be able to answer various further questions about what might have happened if things had been different (e.g., if the oxygen had been mixed with another gas, or if the pressure of the oxygen were increased).

The case that Grimm makes for modifying the knowledge account along the lines that he has suggested is thus very weak, and certainly does not give us a reason for preferring his proposal over the cognitive achievement account. I conclude that the cognitive achievement account of understanding remains the most compelling proposal currently available.

Acknowledgements This paper was written while I was in receipt of a Phillip Leverhulme Prize, and I am grateful to them for their support. Thanks also to Abrol Fairweather, Georgi Gardiner, Emma Gordon, Allan Hazlett, Alison Hills, Andrew Mason and Lani Watson for helpful discussions of topics related to this paper. Special thanks to Stephen Grimm and Dory Scaltsas.

References

Achinstein, P. 1983. *The nature of explanation*. Oxford: Oxford University Press.
BonJour, L. 1998. *In defense of pure reason*. Cambridge: Cambridge University Press.
BonJour, L. 2001. Replies. *Philosophy and Phenomenological Research* 63: 673–698.
BonJour, L. 2005. In defense of the *a priori*. In *Contemporary debates in epistemology*, ed. M. Steup and E. Sosa, 98–105. Oxford: Blackwell.
Brogaard, B. 2007. I know. Therefore, I understand. *typescript*.
Evans, G. 1979. Reference and contingency. *The Monist* 52: 161–189.
Greco, J. 2003. Knowledge as credit for true belief. In *Intellectual virtue: Perspectives from ethics and epistemology*, ed. M. DePaul and L. Zagzebski, 111–134. Oxford: Oxford University Press.
Greco, J. 2007. The nature of ability and the purpose of knowledge. *Philosophical Issues* 17: 57–69.
Greco, J. 2008. What's wrong with contextualism? *Philosophical Quarterly* 58: 416–436.
Greco, J. 2009a. *Achieving knowledge*. Cambridge: Cambridge University Press.
Greco, J. 2009b. The value problem. In *Epistemic value*, ed. A. Haddock, A. Millar, and D.H. Pritchard, 313–321. Oxford: Oxford University Press.
Greco, J. 2009c. Knowledge and success from ability. *Philosophical Studies* 142: 17–26.

Greco, J. 2014. Chapter 13 – *Episteme*: Knowledge and understanding. In *Virtues and their vices*, ed. Kevin Timpe and Craig Boyd. Oxford: Oxford University Press.
Grimm, S. 2006. Is understanding a species of knowledge? *British Journal for the Philosophy of Science* 57: 515–535.
Grimm, S. 2010. Understanding. In *The routledge companion to epistemology*, ed. S. Bernecker and D.H. Pritchard, 84–94. London: Routledge.
Grimm, S. 2014. Understanding as knowledge of causes. In *Virtue scientia*, ed. A. Fairweather. Dordrecht: Springer.
Hetherington, S. 2013. There can be lucky knowledge. In *Contemporary debates in epistemology*, 2nd ed, ed. M. Steup and J. Turri. Oxford: Blackwell.
Hills, A. 2009. Moral testimony and moral epistemology. *Ethics* 120: 94–127.
Hills, A. 2010. *The beloved self: Morality and the challenge from egoism*. Oxford: Oxford University Press.
Kitcher, P. 2002. Scientific knowledge. In *Oxford handbook of epistemology*, ed. P. Moser. Oxford: Oxford University Press.
Kripke, S. 1980. *Naming and necessity*. Cambridge, MA: Harvard University Press.
Kvanvig, J. 2003. *The value of knowledge and the pursuit of understanding*. Cambridge: Cambridge University Press.
Kvanvig, J. 2009. The value of understanding. In *Epistemic value*, ed. A. Haddock, A. Millar, and D.H. Pritchard, 95–112. Oxford: Oxford University Press.
Lewis, D. 1986. Causal explanation. In his *Philosophical papers* (vol. 2). Oxford: Oxford University Press.
Lipton, P. 2004. *Inference to the best explanation*. London: Routledge.
Miller, R. 1987. *Fact and method*. Princeton: Princeton University Press.
Pritchard, D.H. 2009. Knowledge, understanding and epistemic value. In *Epistemology*, Royal institute of philosophy lectures, ed. A. O'Hear, 19–43. Cambridge: Cambridge University Press.
Pritchard, D.H. 2010. Achievements, luck and value. *Think* 25: 1–12.
Pritchard, D.H. 2012. Anti-luck virtue epistemology. *Journal of Philosophy* 109: 247–279.
Pritchard, D.H. 2013. There cannot be lucky knowledge. In *Contemporary debates in epistemology*, 2nd ed, ed. M. Steup and J. Turri. Oxford: Blackwell.
Pritchard, D.H., A. Millar, and A. Haddock. 2010. *The nature and value of knowledge: Three investigations*. Oxford: Oxford University Press.
Salmon, W. 1989. *Four decades of scientific explanation*. Minneapolis: University of Minnesota Press.
Sosa, E. 1988. Beyond skepticism, to the best of our knowledge. *Mind* 97: 153–189.
Sosa, E. 1991. *Knowledge in perspective: Selected essays in epistemology*. Cambridge: Cambridge University Press.
Sosa, E. 2007. *A virtue epistemology: Apt belief and reflective knowledge*. Oxford: Clarendon.
Sosa, E. 2009. *Reflective knowledge: Apt belief and reflective knowledge*. Oxford: Clarendon.
Strevens, M. 2008. *Depth: An account of scientific explanation*. Cambridge, MA: Harvard University Press.
Woodward, J. 2003. *Making things happen: A theory of causal explanation*. Oxford: Oxford University Press.
Zagzebski, L. 1996. *Virtues of the mind: An inquiry into the nature of virtue and the ethical foundations of knowledge*. Cambridge: Cambridge University Press.
Zagzebski, L. 1999. What is knowledge? In *The Blackwell guide to epistemology*, ed. J. Greco and E. Sosa, 92–116. Oxford: Blackwell.

Understanding as Knowledge of Causes

Stephen R. Grimm

What is the epistemic gain that occurs when we move from knowing that something is the case to understanding why it is the case—for example, from knowing that the eclipse occurred to understanding why it occurred, or from knowing that the coffee spilled to understanding why it spilled? According to one prominent view, with roots at least as far back as Aristotle, the move from knowing that p to understanding why p occurs when we acquire knowledge of the cause of p. As Peter Lipton puts the idea, the transition to understanding why is not accomplished by acquiring "some sort of superknowledge, but simply more knowledge: knowledge of causes" (Lipton 2004, p. 30).[1] Because of its longstanding appeal, we can think of this idea—the idea that understanding derives from knowledge of causes—as the traditional view of understanding.

Over the last several decades, the traditional idea has encountered a number of objections. According to some critics, knowledge of causes is not necessary for understanding—either because some state short of knowledge is enough for understanding,[2] or because understanding can arise from non-causal sources.[3] According to others, the problem is that knowledge of causes is not sufficient; if these critics are right, it is not difficult to produce cases in which one knows the cause of p while nonetheless falling short of understanding why p.[4]

One thing I will argue in this paper is that all of these concerns turn on an inadequate idea of what it means to have knowledge of the cause. Properly understood, I will claim, the traditional view can avoid the objections that have been leveled against it. My main strategy will therefore be to try to respond to these objections

[1] Further contemporary support for the knowledge of causes view can be found in Salmon (1984, pp. 19–20), Miller (1987, p. 60), Woodward (2003), Strevens (2008), and Greco (2010, pp. 8–9, 2013).

[2] See e.g. Kvanvig (2003, 2009) and Elgin (2004, 2009).

[3] See e.g. Hempel and Paul (1965), Railton (1978), Achinstein (1983), Kitcher (1985), and Ruben (1992).

[4] See e.g. Pritchard (2008, 2009, 2010) and Hills (2009, 2010).

S.R. Grimm (✉)
Fordham University, Bronx, USA
e-mail: sgrimm@fordham.edu

A. Fairweather (ed.), *Virtue Epistemology Naturalized: Bridges Between Virtue Epistemology and Philosophy of Science*, Synthese Library 366, DOI 10.1007/978-3-319-04672-3_19,

by first spelling out in more detail how I think the "knowledge of causes" formula should be understood. Or, perhaps better, I will use these objections to try to gain a better sense of how knowledge of causes gives rise to understanding.

1 The Propositional Model

In trying to fill out the traditional view, let us start with the idea, found in David Lewis (1986) among others, that to understand an event is just to "possess" causal information about that event. But what sort of possession is at issue here, and what sort of causal information, exactly, is most relevant to understanding?

There is one way of possessing causal information that clearly does not seem sufficient for understanding. Suppose that your knee bumps the table at your local coffee shop, leading your cup to spill, and that I am a few tables over, taking this all in. I will now possess a good deal of causal information relevant to the spill, but I might nonetheless possess the information in an "unconnected" way. For example, if my mind is now preoccupied with something else (with the engrossing gossip at the next table, say), then even though I will have registered this information at some level I might nonetheless have failed to do the cognitive work necessary to connect or bring together the information in the appropriate way.

Alternatively, even though I might possess the relevant causal information, I might not be in a position to recognize it as such. Thus, and to switch to a different example, I might know that my son is undergoing an anaphylactic reaction—know that he is breaking out in hives, that he finds it difficult to breathe, and so on—and I might know that (among other things) he just ate some peanuts, but I might not be able to identify the eating of the peanuts as the cause of the reaction. That is, I might not be able to connect up my knowledge about the eating of the peanuts and the reaction in the right way.

But what would "the right way" amount to in these cases? How exactly should these different bits of causal information be connected or brought together? One natural suggestion here is that they should be brought together in the form of a *causal proposition*, that is, a proposition that specifies the causal relationship that holds between the explanandum and the explanans. For example, it would be to possess the causal proposition *that the coffee spilled because the table was bumped*, or *that my son is undergoing an anaphylactic reaction because he just ate some peanuts*.[5]

The following general picture has therefore seemed very tempting to many, namely, that:

(a) S has knowledge of the cause of p

[5] On this way of looking at things, moreover, the appropriate way to "possess" a causal proposition of this sort would presumably be, not just by assenting to the proposition, but by assenting to it in a way that amounts to knowledge. Although one might mentally possess a causal proposition in some other way (say, by disbelieving, or by withhold judgment about it), supporters of the "knowledge of causes" view would obviously take this possession to be of the knowing kind. Although later we will return to the question of what this knowing might amount to, and whether something less than knowing might do, we can let this stand for now.

just in case

(b) S knows that p because of q.

To the extent that philosophers have tried to specify what it means to have knowledge of the cause, this model—what we might call "the propositional model"—seems to be the dominant one in the literature. Thus Lewis is quite clear that what is possessed, when one possesses causal information in a way that is relevant to understanding, is a proposition (Lewis 1986, p. 218), and Jaegwon Kim goes so far as to say that the propositional model is *entailed by* the view that causal relations are real metaphysical relations in the world (a position that Kim calls "explanatory realism").[6] As Kim puts it, such a realist view "makes 'having' an explanation a matter of knowing a certain proposition to be true" (Kim 2010 [1988], p. 157). It is no surprise, then, that when contemporary philosophers such as Duncan Pritchard and Alison Hills have turned to evaluate the traditional knowledge of causes view, it is the propositional model that they have had in mind.

2 Some Cases

If Pritchard and Hills are correct, however, the traditional view needs to be abandoned, or at least supplemented, because one can have knowledge of the cause of p while nevertheless not understanding why p.[7]

To start with one of Pritchard's examples, suppose that my house burns down while my family and I are away for a few hours, and that as we return home to the scene my young son asks the fire chief why it burned down.[8] The chief then tells him that it burned down because of faulty wiring, and my son accepts this based on his say-so. How much does my son now know? For one thing, he presumably knows that his house just burned down; he can see that with his own eyes. More importantly for our purposes, however, he also seems to know that his house burned down because of faulty wiring. He did, after all, receive this information from a perfectly reliable source, and in a language he could understand. On the propositional model described above, he would therefore have not just knowledge of the cause of the fire but would, thereby, understand why his house burned down.

According to Pritchard, however, this last step is implausible because it seems wrong to say that my son now understands why his house burned down. As he writes: "He has no conception of how faulty wiring might cause a fire, so we could

[6] See Kim (2010 [1988], esp. pp 156–59). There Kim summarizes his point by claiming "We have also seen that explanatory realism entails the propositional account of explanatory knowledge" (Kim 2010 [1988], p. 158).

[7] Unlike Pritchard's, Kvanvig's discussion of understanding focuses on cases of what he calls "objectual understanding" rather than on cases of "understanding why." For more on his distinction, see Kvanvig (2003, ch. 8, 2009). The discussion below will eventually broaden out to include those cases as well.

[8] I have adapted Pritchard's first-person story to my own.

hardly imagine that knowing this much suffices to afford him understanding of why his house burned down. Nevertheless, he surely does know that his house burned down because of faulty wiring, and thus also knows why his house burned down" (Pritchard 2010, p. 81; cf. Pritchard 2009, p. 38). The thought therefore seems to be that understanding requires not just being able to identify the cause, in the sense of knowing a causal proposition that specifies or picks out the cause, but requires some conception of how the cause might bring about the effect in question.

Appealing to a different example, but drawing explicitly on Pritchard's work, Alison Hills likewise claims that one can know that p is the case, know that p because of q, and yet nonetheless fail to understand why p. In her case, to be clear, she does not appeal specifically to the notion of a *cause* but rather to the notion of a *reason*. The basic thrust of the argument, however, is the same.[9]

Here is her example: suppose that eating meat is wrong, that I come to believe this as a result of trusting a reliable authority, and that I likewise come to believe (by trusting that same authority) that eating meat is wrong because of the suffering of animals under modern farming methods. According to Hills, even though I might well know all these things (supposing they are true), it does not follow that I would thereby understand why eating meat is wrong. Why? Because, Hills claims, even armed with this knowledge I might nonetheless not be able to "draw relevant distinctions," or to "come to correct conclusions about similar cases." If asked "What about fish?," for example, or "What about animals reared under better conditions?," I might well draw a blank (Hills 2010, p. 192; cf. Hills 2009, p. 100).

In short, according to both Pritchard and Hills, knowing the cause of p or the reason why p—in the sense, more generally, of knowing a correct explanation to the effect that p because of q—is not sufficient for understanding why p. In order for genuine understanding to appear on the scene, something else—perhaps some sense of how the cause brings about the effect (Pritchard), or an ability to answer closely related questions (Hills)—is apparently needed.

Of course, one might dispute Pritchard's and Hills's judgments about these cases. In particular, and focusing for simplicity on Pritchard's case, one might disagree with the claim that my son now knows, solely on the basis of the fire chief's testimony, that his house burned down because of the faulty wiring, because (at least as Pritchard tells the tale) it is not clear he understands the content of that proposition well enough to actually believe it. I will return to this thought at the end of Sect. 5. In the next few sections, however, what I will argue instead is that even if we grant Pritchard and Hills that there is a way in which someone might know the cause that is not sufficient for understanding, there is another way in which one might know the cause that is. More exactly, what I will suggest is that the place where their objection goes wrong is in its innocent-looking first step, the one that supposes the best way to understand the "knowledge of causes" formula is according to the propositional model. If this step is mistaken—if there are other viable ways in which the "knowledge of causes" formula might be understood—then it opens up the possibility that there are other ways in which one might have knowledge of the cause that *do* suffice for understanding.

[9] Especially if, as I will recommend in Sect. 7, we adopt an expansive notion of causation.

But where should we look for these other viable models, for these other ways of thinking about the "knowledge of causes" formula? In the following Section I will suggest that an alternative, and more appealing, way of thinking about the traditional view can be gleaned from considering how a parallel debate plays out in the case of *a priori* knowledge. The *a priori* case is particularly worth considering because, just as in discussions of understanding, appeals to the metaphors of "grasping" or "seeing" are nearly ubiquitous in the literature. A look at the *a priori* therefore promises to give us a better sense of how these central metaphors should be understood in the case of understanding as well.

3 "Grasping" or "Seeing" *a Priori*

Although there are dissenters,[10] perhaps the most common point of agreement among writers on the *a priori* is that *a priori* knowledge is essentially knowledge of necessary truths: knowledge of truths such as that 2+3=5, or that no object can be red all over and green all over at the same time.[11] But even this initial characterization of the *a priori* seems to gloss over something important.

For example, and as several philosophers have noted,[12] one basic problem with this way of characterizing the *a priori* is that not all knowledge of necessary truths amounts to *a priori* knowledge. Suppose that I can't be bothered to do a particular calculation, so I trust your judgment (or perhaps, my calculator's) that 207+86=293. Then I will come to have knowledge of a necessary truth, but it will not be an instance of *a priori* knowledge. Instead, it seems best to think of it as an instance of testimonial knowledge (or the like).

But what then is needed to transform my knowledge *into* an instance of *a priori* knowledge? Presumably the main problem here is that even though I now have knowledge of a necessary truth, I nonetheless fail to see or grasp or in some way appreciate its necessity. So suppose we add that I learn from you not just that 207+86=293, but also that this is a necessary truth, or perhaps you tell me that necessarily, 207+86=293. Does this bridge the gap? Once again, it seems not, for once again it seems that I can accept these propositions based on your say-so while nonetheless not acquiring an instance of *a priori* knowledge.[13]

---

[10] For example, Gareth Evans (1979), Saul Kripke (1980), and John Turri (2010) have argued that there can be *a priori* knowledge of contingent propositions. For critical discussion see BonJour (1998) and Casullo (2003).

[11] For defenders of the idea that *a priori* knowledge is knowledge of necessary truths, see (among others) BonJour, Plantinga, Butcharov and others [Chisholm?].

[12] See Chisholm and Plantinga.

[13] Or, if it is hard to imagine that simple sums of this sort might simply be accepted on the word of another, substitute something more challenging: for instance, that the continuum hypothesis is independent of ordinary set theory (from Plantinga 1993, p. 106). And suppose I learn as well from the authority that this is a necessary truth, and I take that at face value. Here again I don't have an instance of *a priori* knowledge, but I do have knowledge of a necessary truth (and knowledge that it is necessary, and so on).

The upshot therefore seems to be that to know some necessary truth *a priori* I need to do more than just assent to a necessary proposition,[14] or assent to a necessary proposition along with the further stipulation *that* it is necessary. What I need instead, it seems, is to "see" or "grasp" the necessity itself. But how do I do that?

If philosophers such as Chisholm, Plantinga, and BonJour are correct, it seems that to manage this seeing or grasping we need to bring a new power of the mind to bear, a power they refer to as *reason* or *rational insight.* When all goes right, moreover, what this power does is "see" or "grasp," of the sum of 207 and 86, that it could not be otherwise than 293—that there is no possible world where the sum of 207 and 86 does not add up to 293. What the metaphor of "seeing" seems to involve, then, is something like an apprehension of how things stand in modal space—an apprehension, that is, that there are no possible worlds in which the sum of 207 and 86 does not equal 293. Just as, in seeing with one's eyes, one takes in or apprehends how things stand in the physical terrain, so too the basic idea here seems to be that in "seeing" with the eye of the mind, one takes in or apprehends how things stand in the modal terrain: one apprehends what cannot be otherwise, or how certain changes will lead, or fail to lead, to other changes.[15]

If this picture is correct, in any case, then for our purposes the important thing to see is that *what* is grasped or seen, when we grasp or see *a priori*, is not in the first instance a proposition but rather a modal relationship between properties (or objects, or entities) in the world. Put another way, what is primarily grasped or seen is not (e.g.) a proposition such as:

(a) that $207+86=293$; or perhaps that
(b) that $207+86=293$ is necessarily true.

Rather, what is grasped or seen is (something along the lines of):

(c) Of the sum of 207 and 86, how it could not be otherwise than 293.

In other words, on this view what *a priori* knowledge amounts to is a kind of *de re* knowledge—a knowledge that comes from grasping or seeing, of certain properties (objects, entities) that they are modally related in a particular way. As Laurence BonJour puts the idea, *a priori* insights, "are thus putative insights into the essential nature of things or situations of the relevant kind, into the way reality in the respect in question *must* be.... [I]t is often and quite possibly always a mistake to construe them as propositional in form" (BonJour 2005, pp. 99–100).[16]

---

[14] Following Bruce Russell, by a necessary proposition I mean a proposition that cannot be false (Russell 2007).

[15] Cf. Panayot Butchvarov: "While both necessary and contingent truths have fundamentally different objects, in both cases such objects are, in a very general sense, *perceived*" (Butchvarov 1970, p. 179).

[16] Given our purposes here, it is also worth pointing out that BonJour's position in this quote represents a conspicuous and deliberate change from his earlier 1998 book on the *a priori*. While in that book he essentially took for granted that propositions were the object of *a priori* knowledge, he subsequently (as in the 2005 article just quoted) came to believe that this was "a serious mistake" (BonJour 2001, p. 673, and again on p. 678). BonJour credits Paul Boghossian (2001) with helping

The basic idea here is therefore not that propositions have no role to play in *a priori* knowledge, but rather that they play a secondary or derivative role. If Bonjour is right—and I think he is—the primary object of *a priori* knowledge is the modal reality itself that is grasped by the mind, and it is on the basis of this grasp that we then (typically) go on to assent to the proposition that describes or depicts these relationships.

4 Parallels

Suppose that these thoughts about the *a priori* are on target. How does this shed light on our question about how the "knowledge of causes" formula might best be understood?

What the parallel debate concerning the *a priori*—with its parallel notions of "grasping" and "seeing"—suggests is that just as the "knowledge of necessary truths" formula can be understood in a variety of ways, a similar ambiguity can be seen in the traditional "knowledge of causes" formula. In other words, just as the notion of "knowledge of necessary truths" can pick out either the state of:

(a) assenting to a necessary proposition on reliable grounds (on the basis, say, of reliable testimony, or reliable memory),[17]

or the state of:

(b) grasping or seeing the necessary relatedness of certain properties (objects, entities).[18]

---

him to see the shortcomings of the propositional view, because it was Boghossian who pointed that no amount of assenting to propositions *about* necessary facts could add up to a single act of *a priori* insight, a single apprehension of how the constituents of these facts were necessarily related. According to Boghossian, this is the lesson we should have learned from Lewis Carrol's famous dialogue between the Tortoise and Achilles, where the Tortoise effectively notes that it is one thing to assent to the premises of an inference, and quite another to see or grasp how the conclusion follows from the premises. But once this fact is appreciated, BonJour came to think, we should give up the idea that *a priori* insight—at least in many cases—is directed at propositions at all. As he writes: "Moreover, once this possibility is appreciated, it becomes clear at once that at least many other *a priori* insights are also of this non-propositional sort. Consider the one involved in the color incompatibility case. What is most fundamentally grasped or apprehended there, I would now suggest, is the actual relation of incompatibility between the two colors, the way in which the presence of one excludes the presence of the other, with the propositional awareness that this is so, that nothing can be red and green all over at the same time, being again secondary and derivative. And something similar seems to me to be true in many, many other cases. Indeed, the question that arises, but which I will not try to answer here, is whether there are any cases where the most basic insight is propositional in form" (BonJour 2001, p. 677).

[17] Or, perhaps, to a necessary proposition, along with the further information that it is necessary, etc.

[18] Alternatively, one might say here "grasping or seeing, of certain properties (objects, entities), how they are necessarily related."

so too the notion of "knowledge of causes" can refer either to:

(a*) assenting to a causal proposition on reliable grounds (on the basis, say, of reliable testimony, or reliable memory),
(b*) seeing or grasping the modal relatedness of the terms of the causal relata.[19]

In both cases of (b), moreover, the object of knowledge seems to be different than it is in the (a) cases, where the objects are propositions. More exactly, with the (b)s, what is grasped or seen is something like the modal relationships that obtain between the properties (objects, entities) at issue. In the case of knowledge of causes in particular, what would be seen or grasped would be how changes in the value of one of the terms of the causal relata would lead (or fail to lead) to a change in the other.

It is worth emphasizing that even though this "b*" way of thinking about the knowledge of causes formula parallels the "b" way of thinking about the *a priori*, the success or failure of the "b*" claim should not be thought to stand or fall on the success of the "b" claim (a good thing, given that debates about the *a priori* do not seem like they will be resolved any time soon). As I will try to show throughout the remainder of the paper, this "b*" way of thinking about the knowledge of causes formula is defensible in its own right, and apart from any connection to the *a priori*. Nevertheless, as I noted earlier, given how pervasive the appeals are to notions such as "grasping" and "seeing" in both the literature on the a priori as well as the literature on understanding, it would be good if our theory of understanding helped to shed some light on this connection, and this is one thing that I take that the "knowledge of causes" approach to understanding (properly construed) promises to do.

5 Back to the Fire

Applied back to Pritchard's case of the house fire, what the previous section suggests is that whether or not we take my son to understand will depend a great deal on which of these ways of "knowing the cause" we have in mind. On the first way of looking at things, the way that Pritchard presumably has in mind, what we would be imagining is that my son "simply assents" to the causal proposition relayed by the fire chief—that is, he accepts the information, he gives a mental "yes" to it, he is ready and willing to repeat it to his friends, and so on—but that he then, as it were, cognitively leaves it at that.

On the second way, however, we would be imagining that my son does not leave it at that, but that he processes the information at a deeper level, so that he sees or grasps, of the terms of the explanation, how they are related. In this case, he would be bringing a new and different power of the mind to bear: one that is sensitive not just to how things are, but to how things stand modally, and in particular to how things might have been, if certain conditions had been different. For example,

[19] Or again, one might say here "grasping or seeing, of certain properties (objects, entities), how they are modally related."

on this way of looking at things my son would now see or grasp that if the condition of the wires had been different—if the wires had not been faulty—then the house would not have burned down (ceteris paribus).

Now, one might object (along with Pritchard, it seems) that even on this second way of looking at things more must be required of my son, if we are to credit him with understanding—that he must not simply grasp that if the wiring had been in order, the fire would not have occurred (ceteris paribus), but he must also be able to identify what it was about the faulty wiring that led to the fire. This, after all, seems to be what the fire chief himself grasps, and thus it might seem that this is what is really essential to understanding. But rather than conclude, on the basis of this difference, that the fire chief understands why the fire occurred while my son does not, it seems better to say that the difference between the chief and my son is not one of kind but of degree; in particular, the idea would be that while my son has *some* understanding of why the fire occurred, the chief has a much deeper or more sophisticated sort of understanding.

Consider the following parallel: when I start chopping onions and my eyes begin to water, I think I understand why my eyes are beginning to water, namely, because I am chopping the onions. I don't think it is because of the time of day, or the color of the shirt I am wearing, or anything like that; it's because of the onions. But obviously someone with a greater understanding of onion (and eyeball) chemistry would be able not just to identify the onions as the cause but would be able to say what it was about the onions that was bringing this about—in this case, the particular sulfur compounds that were being broken down and released into the air when I did the chopping. What such a person would therefore grasp, which I would not, is that I if were to chop some other vegetable with these same compounds, my eyes would likewise begin to water (ceteris paribus); or perhaps, that if were to chop an onion specially designed to lack these compounds, my eyes would not water; and so on. But again, what these facts seem to illustrate is not that the person who appeals to the compounds understands while I fail to understand, but that understanding comes in degrees; I have less of it, and he has more. And similarly, it seems best to say, for the case of my son and the fire chief.[20]

All that said, one might still object that my portrayal of the two ways in which one might have knowledge of the cause of the fire is misleading (or inadequate or in some other way misguided) because it hardly seems possible for my son to know the cause in the first way I suggested—where my son simply assents to the causal proposition and then mentally "leaves it at that." The main concern here, I take it, is that my son's attitude in this imagined case would seem to be so simple—so mentally thin—that it is not even clear that he would be assenting to the proposition at all. Instead, he might simply be accepting the information as a parrot might—ready to repeat it, but

[20] Among other reasons that it seems best to say this is because otherwise a massive scepticism about understanding seems to threaten. Why, for example, think that the chemical story is sufficient for understanding? Why not insist that we go all the way down to basic physical properties? But if this is really required, much of the ordinary understanding we take ourselves to have would disappear.

without really grasping *what* is being said (or being repeated). Or again, it might be thought that what my son is assenting to is not the proposition *that my house burned down because of the faulty wiring* but rather a "nearby" proposition, such as *that whatever the fire chief just said is true* (or, a bit more formally, that *whatever proposition his sentence just expressed is true*). Either way, it seems that my son would not genuinely know the causal proposition at issue for the simple reason that he would not genuinely assent to it.

What these last thoughts suggest is that it is hard—perhaps impossible—to genuinely assent to a causal proposition without doing the sort of extra cognitive work that I claimed was characteristic of the second way in which one might have knowledge of the cause. That is, that it is hard—perhaps impossible—to genuinely assent to a causal proposition without in some way grasping that what it means for these two items to stand in the "because" relation is that a change in the state of the former will lead to a change in the state of the later (ceteris paribus).

For our purposes, however, it is not necessary to try to settle this question. Indeed, if it turns out that it is impossible for someone genuinely to assent to a causal proposition while mentally leaving it at that, then so much the better for our view, because it would imply that genuinely assenting to a causal proposition automatically generates the sort of modal grasping or seeing ability that is (on the view here) characteristic of understanding. Put another way, it would imply that there are not two ways in which have knowledge of the cause—one "thinner," leaving-it-at-that way, which is not sufficient for understanding, and one "thicker," modal-grasping way, which is—but only one, viz., the thicker way. And if that's right, then Pritchard and Hills would not have identified a genuine way in which one might have knowledge of the cause that does not suffice for understanding in the first place.

6 Another Model

So far I have claimed a few things. First, that the propositional way of thinking about the "knowledge of causes" formula is not mandatory, and that it instead seems possible to know the cause in such a way that the object of one's knowledge—of one's grasp—is the modal relationship that obtains between the terms of the explanation.[21] Second, that knowing the cause in this second way appears to be sufficient for understanding—at least, that it appears to be sufficient for *some* degree of understanding.

[21] This is not to deny the fallibility of whatever power of the mind it is that evaluates how things stand modally. One might therefore "see" or "grasp" (or seem to see or grasp?) modal relationships that do not obtain. It might therefore be better to say that the object of understanding in these cases, and perhaps in all cases, is our abstract representation *of* the relationships that we take to obtain in the world—as it were, our "mental model" of these relationships—so long as one recognizes a difference between abstract representations which take the form of models and those which take the form of propositions. For more on different ways of construing the object of understanding, see Grimm (2010) and Greco (2013).

For convenience, let us think of this second way of thinking about the "knowledge of causes" formula as the modal-model—that is, a model on which what is grasped when one has knowledge of the cause is the modal relationship that obtains between the terms of the explanation.

The modal model therefore seems not just to be not a possible way of construing the "knowledge of causes" formula, but a more charitable way, because it frees the view from counterexamples (at least, of the sort described above). Another strength of the view is that it ties in naturally with claims about the nature of understanding which have been independently popular among philosophers. According to Linda Zagzebski, for example, "understanding is not directed towards a discrete proposition, but involves grasping relations of parts to other parts and perhaps the relations of part to wholes" (2009, p. 142; cf. 2001, p. 242). And according to Julius Moravcsik, "What we understand are systems of various sorts; in a world in which elements do not constitute the relevant structures there can be no understanding" (1979, p. 56). Although put in different ways, the common thought here seems to be that the primary objects of understanding are the relationships (or structures) that hold among the various elements of reality—that it is in grasping how things are related in this way that we grasp how the world is structured.[22] But then this way of thinking about the object of understanding naturally accords with the way we construed the knowledge of causes formula above, on which the objects of that knowledge were the modal relationship that obtained between the terms of the explanation, rather than the propositions that described those relationships.

Our proposal also accords with the common idea that to have understanding is to have a kind of *ability* or *know how*.[23] On our proposal, "seeing" or "grasping" would count as a kind of ability, because the person who sees or grasps how certain properties (objects, entities) are modally related will characteristically have the ability to answer a variety of what James Woodward (2003) has called "What if things were different?" questions. That is, the person will be able to see or grasp how changes in some of these items will lead (or fail to lead) to changes in the others. Of course, as we noted earlier, some of us will be able to answer many more of these "What if things had been different?" questions than others. But again, what this illustrates is simply the truism that understanding comes in degrees; and indeed, it is a further virtue of our proposal that it naturally accommodates this fact.

---

[22] For further defenses of the idea that the object of understanding is non-propositional, see Riggs (2003), Kvanvig (2003, p. 192), and Brewer (2009, pp. 298–99). Roberts and Wood (2007, pp. 46–47) claim that the object of understanding could be either propositional or nonpropositional, though when it comes to propositional understanding they seem to have in mind the sort of "semantic grasping"—the grasping of words or concepts—that is not our main focus here.

[23] See, for example, Zagzebski (2001, pp. 241–43), Hills (2009, 2010, ch. 9), and Grimm (2010, 2012). De Regt and Dieks (2005), as well as de Regt (2009), argue for a more idiosyncratic version of the understanding as know how view, according to which someone who understands a theory "can recognize qualitatively characteristic consequences of T without performing exact calculations" (de Regt 2009, p. 33).

7 Causation and Dependence

We therefore have good grounds for thinking that knowledge of causes, suitably understood, is sufficient for understanding. But is it necessary? Recall again the beginning of the paper, where we noted two problems for this claim. On the one hand, there was the objection that understanding can come from non-causal means. If this is right, it would be the "of causes" part of the "knowledge of causes" formula that spells trouble, because one could acquire understanding by means of knowing something other than causes. On the other hand, and more recently, there is the objection from Kvanvig that one can have understanding in the absence of knowledge. If this is right, it would be the "knowledge" part of the "knowledge of causes" formula that spells trouble.

The question of whether understanding can come from non-causal means (or, as it is sometimes put in the literature, whether non-causal explanations are legitimate) is a large one, and I will not attempt to settle the debate here.[24] In this Section I will instead piggyback on what I take to be the most promising response to this concern, in order to show I think how the "knowledge of causes" formula needs to be refined (or, perhaps better, clarified) to address this issue.

I take it that the basic concern behind the first objection is that the appeal to causation is too limited, because causes are most naturally understood as the pushers and pullers of the world, and yet some of our understanding does not appeal to pushers or pullers at all but rather to other sorts of relationships that seem to obtain between the explanans and the explanandum. To appeal to some of David-Hillel Ruben's examples: Why was St. Francis a good man? Because he was benevolent. Or again: Why is that painting beautiful? Because of its color composition. And so on.[25] As Ruben notes, it is through grasping these relationships that we come to understand the thing in question, but the relationship grasped does not seem to be a causal one. It is not as if, for example, St. Francis's benevolence *caused* him to be a good man; it seems more natural instead to say that his goodness was in some sense constituted by his benevolence.

What should we make of examples of this sort? One attractive way for the causal theorist to respond is by thinking of our notion of causation more expansively. In particular, the strategy would be to expand the notion of causation so that, as Woodward claims, "any explanation that proceeds by showing how an outcome depends… on other variables counts as causal" (2003, p. 6).[26] So understood, Ruben's examples would properly count as "causal" because they capture how the one property (goodness or beauty) metaphysically depends on the other property (benevolence or color composition).

---

[24] As noted earlier, for examples of those who favor non-causal explanations, see Hempel and Paul (1965), Railton (1978), Achinstein (1983), Kitcher (1985), and Ruben (1992). For some responses on behalf of the causal view, see Salmon (1984), Lewis (1986, pp. 221–24), and Woodward (2003, pp. 5–7).

[25] For further discussion and examples, see Ruben (1992, ch. 7).

[26] Greco likewise argues for a broad reading of the causal relation: "Understanding involves 'grasping,' 'appreciating,' or knowing causal relations taken in the broad sense: i.e., the sort of relations that ground explanation" (Greco 2010, p. 9; cf. Greco 2002).

Alternatively, and given how closely our notion of causation is tied to pushing and pulling[27]—to exerting causal force—a perhaps more attractive strategy would be to demote the notion of causation from its central role and instead to appeal more generally to the notion of dependence. On this view, dependence would be the genus category, with different kinds of dependence—causal dependence being but one—playing the role of species. As Jaegwon Kim puts the idea:

> [M]y claim will be that dependence relations of various kinds serve as objective correlates of explanations. Dependence, as I will use the notion here, is a relation between individual states and events; however, it can also relate facts, properties, regularities between events, and even entities. We speak of the "causal dependence" of one event or state on another; that is one type of dependence, obviously of central importance. Another dependence relation, orthogonal to causal dependence and equally central to our scheme of things, is *mereological dependence* (or "mereological supervenience," as it has been called): the properties of a whole, or the fact that the whole instantiates a certain property, may depend on the properties had by its parts. (Kim 2010 [1994], p. 183)

As Kim goes on to note, moreover, there seem to be a variety of further dependence relations beyond the causal and the mereological: thus the widowing of Xanthippe seems to depend on the death of Socrates, evaluative facts (such as considered just above) seem to depend on the non-evaluative facts on which they supervene, and so on (Kim 2010 [1994], pp. 183–84).[28]

Since understanding seems to arise from a grasp of all these different types of dependence, it might therefore be better, or perhaps less misleading, to adapt our original way of putting things and claim that understanding consists not of "knowledge of causes" but rather of something like "knowledge of dependency relations" or perhaps just "knowledge of dependencies."[29] Although I can see why one would be tempted by this description, the point worth emphasizing here is that the difference between the "knowledge of dependencies" formula and the "knowledge of causes (broadly understood)" formula is not a substantive philosophical one but rather simply comes down to a difference in terminology, or perhaps in marketing. Since the traditional use of the word "cause" has a certain elegance and simplicity about it, however, I see no great danger in continuing to support the "knowledge of causes" formula, provided the term "causes" is thought of in the broader or more expansive way just noted. For those who would prefer to use the term "dependencies" to pick out this grounding relation, I have no complaint.

---

[27] Our modern notion of causation at least; Aristotle's notion of causation was more expansive, along the lines developed here.

[28] For another advocate of this approach, see Strevens: "I suggest that while the causal influence relation is one kind of raw metaphysical dependence relation that can serve as the basis of the difference-making relation, there are others as well, and that any of the difference-making relations so based is explanatory" (2008, pp. 178–79).

[29] In addition to avoiding misleading connotations, such an approach would also make it easier to see cases such as Hills's from Sect. 2—in which she claimed that the wrongness of eating meat supervened on, or depended upon, the suffering of the animals involved—as falling into the same category as more "overtly" causal cases such as Pritchard's.

8 Kvanvig on Understanding

A second way in which one might think that knowledge of causes is not necessary for understanding would be if one thinks, along with Kvanvig, that something less than knowledge of the cause is sufficient for understanding.

To illustrate why one might think this, suppose that, in a room full of elaborately falsified history books, you randomly pick out the sole accurate book in the room and come to believe all of its claims about the past.[30] To focus on one claim in particular, suppose that you start to read the part of the book describing the Comanche dominance of the southern plains of North America during the eighteenth century, and that you come to believe that the Comanches dominated because of their superior horsemanship.

Suppose that you grasp this explanation, it makes sense to you, and so on. Would you then understand why the Comanches dominated the southern plains during this period? According to Kvanvig, it seems that you would. After all, he notes, you can now correctly answer a wide range of questions about the Comanche dominance, pointing (let's say) to some particular aspects of their horsemanship that brought about this result, and so on.

But now suppose we ask not whether you understand these various things about the Comanches but whether you *know* them. As Kvanvig points out, there is considerable pressure here to say that you do not, for one of the standard lessons of the Gettier literature seems to be that if your beliefs might easily have been mistaken, then even if they are both justified and true they will nonetheless not amount to knowledge. To have knowledge, it looks like a more secure connection to the truth is required; acquiring the truth by chance, through history books or otherwise, is not enough.

If Kvanvig is right, understanding is therefore in a way less demanding than knowledge and in another way more. It is less demanding, because it seems that one can have understanding of some subject even though one might easily have been mistaken about that subject. But it is more demanding in that it requires that the internal connections among one's beliefs actually be "seen" or "grasped" by the person doing the understanding. When it comes to knowledge, by contrast, especially knowledge of propositions, no such internal grasp seems required.

Kvanvig's objection is therefore directed at a particular kind of knowledge—propositional knowledge—for which the elements of grasping, seeing, and the like do not seem obligatory.[31] How then does it bear on our way of thinking about the

---

[30] For concreteness, we can imagine that you are living in some sort of Orwellian regime, intent on falsifying the past, and that by chance you pick up the one accurate book not destroyed by the regime. In this and the following paragraph I am adapting the example developed by Kvanvig in his (2003, pp. 197–98).

[31] As Kvanvig notes in his recent comment on his 2003 book: "In my book on the value of knowledge, I argued in favor of a conception of epistemology that gives strong place to what I termed "objectual understanding"…. I argued that such understanding was not explicable in terms of propositional knowledge, and thus does better than propositional knowledge in addressing a certain value problem about various epistemic states. It is this type of understanding that I want to argue here has special value" (Kvanvig 2013, pp. 1–2, typescript).

"knowledge of the cause" formula, on which grasping or seeing how the explanandum is connected to, or depends upon, the explanans is crucial?

I think that the genuine answer here is that it is not easy to say. Consider, for example, Kvanvig's oft-repeated claim that what we "focus on" when we are considering whether someone understands is different from what we "focus on" when we are considering whether someone knows, and now ask: What do we focus on, when we are considering whether someone has knowledge of the cause in the sense defended above?[32] Again, I find it hard to say. Or rather, insofar as I have a good idea of what this "focus" test amounts to, I think that what I focus on is the "internal" element of grasping or seeing how the different causal elements depend upon one another in our representation of the world, rather than on "external" facts about the etiology of the grasping or the like.

If this is right,[33] then it looks like the standard claim that knowledge is incompatible with luck is mistaken, at least in its unrestricted form. I say "in its unrestricted form" because what these points suggest is that while there might be some forms of knowledge—perhaps all examples of propositional knowledge fall into this category—that are incompatible with luck, there are other forms which are not. For example, according to Kvanvig's "focus" test, knowing the cause in the way that we have construed it above would seem to be compatible with luck, and if Ted Poston is right, cases of "know how" are compatible with luck as well (see Poston 2009). Indeed, it is perhaps not surprising that philosophers have regularly been tempted by the thought that epistemic states which emphasize the notion of an ability—in the way that understanding, know-how, and knowledge of the cause (in our sense) all seem to—are compatible with luck in a way that states which do not emphasize this ability are not. Why? Perhaps because Kvanvig is right that our main concern, in evaluating these states, is whether the ability in question is actually present, and we are less concerned with whether the ability came to exist in a chancy or haphazard way.[34]

In short, for the very same reasons that it seems that there can be lucky understanding it seems that there can be lucky knowledge of causes as well. But then what Kvanvig's examples seem to show is not that understanding is not a species of knowledge, but rather that it is particular kind of knowledge. On our view, a knowledge of causes.[35]

---

[32] For Kvanvig's "focus" test, see his (2003, p. 198) and (2009, pp. 98–99).

[33] And again, I'm not sure it is, because Kvanvig's "focus" test is not so clear to me.

[34] Compare the view of Greco (2010) and Sosa (2011), on which what matters is not whether the ability was acquired by chance (think of the Swampman case), but rather whether the ability, however acquired, is reliably employed.

[35] Thanks to Anne Baril, John Greco, Allan Hazlett, David Henderson, Mikael Janvid, Kareem Khalifa, Soazig Le Bihan, Bob Roberts, and Josh Thurow for helpful comments on an earlier version of this paper.

References

Achinstein, Peter. 1983. *The nature of explanation*. New York: Oxford University Press.

Boghossian, Paul. 2001. Inference and insight. *Philosophy and Phenomenological Research* 63: 633–640.

BonJour, Laurence. 1998. *In defense of pure reason*. New York: Cambridge University Press.

BonJour, Laurence. 2001. Replies. *Philosophy and Phenomenological Research* 63: 673–698.

BonJour, Laurence. 2005. In defense of the *a priori*. In *Contemporary debates in epistemology*, ed. Matthias Steup and Ernest Sosa. Malden: Blackwell.

Brewer, Talbot. 2009. *The retrieval of ethics*. New York: Oxford University Press.

Butchvarov, Panayot. 1970. *The concept of knowledge*. Evanston: Northwestern University Press.

Casullo, Albert. 2003. *A priori justification*. New York: Oxford University Press.

de Regt, Henk. 2009. Intelligibility and scientific understanding. In *Scientific understanding: Philosophical perspectives*, ed. Henk de Regt, Sabina Leonelli, and Kai Enger, 21–42. Pittsburgh: Pittsburgh University Press.

Elgin, Catherine. 2004. True enough. *Philosophical Issues* 14: 113–131.

Elgin, Catherine. 2009. Is understanding factive? In *Epistemic value*, ed. Adrian Haddock, Allan Millar, and Duncan Pritchard. New York: Oxford University Press.

Evans, Gareth. 1979. Reference and contingency. *Monist* 62: 161–189.

Greco, John. 2002. Virtues in epistemology. In *The oxford handbook to epistemology*, ed. Paul Moser. New York: Oxford University Press.

Greco, John. 2010. *Achieving knowledge*. New York: Cambridge University Press.

Greco, John. 2013. Episteme: Knowledge and understanding. In *Virtues and their vices*, ed. Timpe Kevin and Boyd Craig. Oxford: Oxford University Press.

Grimm, Stephen. 2010. Understanding. In *The Routledge companion to epistemology*, ed. Sven Berneker and Duncan Pritchard. New York: Routledge.

Grimm, Stephen. 2012. The value of understanding. *Philosophy Compass* 7: 2.

Hempel, Carl, and Paul Oppenheim. 1965. Studies in the logic of explanation. In *Aspects of scientific explanation and other essays in the philosophy of science*, ed. Carl Hempel. New York: Free Press.

Hills, Alison. 2009. Moral testimony and moral epistemology. *Ethics* 120: 94–127.

Hills, Alison. 2010. *The beloved self: Morality and the challenge from egoism*. New York: Oxford University Press.

Kim, Jaegwon. 2010 [1988]. Explanatory realism, causal realism, and explanatory exclusion. In *Essays in the metaphysics of mind*. New York: Oxford University Press.

Kim, Jaegwon. 2010 [1994]. Explanatory knowledge and metaphysical dependence. In Essays in the metaphysics of mind. New York: Oxford University Press.

Kitcher, Philip. 1985. Salmon on explanation and causality: Two approaches to explanation. *Journal of Philosophy* 82: 632–639.

Kripke, Saul. 1980. *Naming and necessity*. Cambridge, MA: Harvard University Press.

Kvanvig, Jonathan. 2003. *The value of knowledge and the pursuit of understanding*. New York: Cambridge University Press.

Kvanvig, Jonathan. 2009. The value of understanding. In *Epistemic value*, ed. Adrian Haddock, Allan Millar, and Duncan Pritchard. New York: Oxford University Press.

Kvanvig, Jonathan. 2013. Curiosity and the response-dependent special value of understanding. In *Knowledge, virtue and action: Essays on putting epistemic virtues to work*, ed. Henning Tim and P. David. New York: Routledge.

Lewis, David. 1986. Causal explanation. In his *Philosophical papers* (vol. 2). New York: Oxford University Press.

Lipton, Peter. 2004. *Inference to the best explanation*, 2nd ed. New York: Routledge.

Miller, Richard. 1987. *Fact and method*. Princeton: Princeton University Press.

Moravscik, Julius. 1979. Understanding and knowledge in Plato's philosophy. *Neue Hefte für Philosophe* 15: 53–69.

Plantinga, Alvin. 1993. *Warrant: the current debate*. New York: Oxford University Press.
Poston, Ted. 2009. Know how to be gettiered. *Philosophy and Phenomenological Research* 79: 743–747.
Pritchard, Duncan. 2008. Knowing the answer, understanding, and epistemic value. *Grazer Philosophische Studien* 77: 325–339.
Pritchard, Duncan. 2009. Knowledge, understanding, and epistemic value. In *Epistemology*, Royal institute of philosophy lectures, ed. Anthony O'Hear. New York: Cambridge University Press.
Pritchard, Duncan. 2010. Knowledge and understanding. In *The nature and value of knowledge: Three investigations*. Co-authored with Alan Millar and Adrian Haddock. New York: Oxford University Press.
Railton, Peter. 1978. A deductive nomological model of probabilistic explanation. *Philosophy of Science* 45: 206–226.
Riggs, Wayne. 2003. Understanding 'virtue' and the virtue of understanding. In *Intellectual virtue: Perspectives from ethics and epistemology*, ed. Michael DePaul and Linda Zagzebski, 203–226. New York: Oxford University Press.
Roberts, Robert, and Jay Wood. 2007. *Intellectual virtues: An essay in regulative epistemology*. New York: Oxford University Press.
Ruben, David-Hillel. 1992. *Explaining explanation*. New York: Routledge.
Russell, Bruce. 2007. *A priori* justification and knowledge. In *Stanford encyclopedia of philosophy*, ed. Ed Zalta. Accessed 7 Apr 2014.
Salmon, Wesley. 1984. *Scientific explanation and the causal structure of the world*. Princeton: Princeton University Press.
Sosa, Ernest. 2011. *Knowing full well*. Princeton: Princeton University Press.
Strevens, Michael. 2008. *Depth: An account of scientific explanation*. Cambridge, MA: Harvard University Press.
Turri, John. 2010. Contingent a priori knowledge. *Philosophy and Phenomenological Research* 83: 327–344.
Woodward, James. 2003. *Making things happen: A theory of causal explanation*. New York: Oxford University Press.
Zagzebski, Linda. 2001. Recovering understanding. In *Knowledge, truth, and duty: Essays on epistemic justification, responsibility, and virtue*, ed. Matthias Steup. New York: Oxford University Press.
Zagzebski, Linda. 2009. *On epistemology*. Belmont: Wadsworth.

Knowledge, Understanding and Virtue

Christoph Kelp

1 Introduction

According to a thesis that has enjoyed a high degree of popularity in the philosophy of science:

> UK. Understanding is a species of knowledge.

While there may be some disagreement over how to unpack this thesis in more detail, it seems fair to say that the received view, apparently dating back as far as Aristotle (see Greco 2010: 9), is that understanding is knowledge of causes. Peter Lipton states the view nicely in the following passage:

> Understanding is not some sort of super-knowledge, but simply more knowledge: knowledge of causes. (Lipton 2004: 30)

Other proponents of and sympathisers with UK include Peter Achinstein (1983), Wesley Salmon (1989), James Woodward (2003) and Philip Kitcher (2002). One of the obvious selling points of UK is its simplicity and elegance. Another one concerns considerations about the aim of inquiry. As Alan Millar (in Pritchard et al. (2010: 98)) has aptly pointed out, a natural way of expressing the goal of our ordinary everyday inquiries is in terms of knowledge. In inquiring into things like whether the bank will be open on Saturday, where the meeting will take place or who took the car keys, we are trying to come to know the answers to these questions. At the same time, a natural way of expressing the goal of scientific inquiries, and one that a number of philosopher's of science have been attracted to (see e.g. Salmon 1998; Lipton 2004; DeRegt and Dieks 2005; Strevens 2006), is in terms of understanding. Astronomy aims to understand celestial objects, biology aims to understand various aspects of living organisms etc. UK promises to unify these two plausible conceptions of the aim of ordinary and scientific inquiry.

C. Kelp (✉)
Centre for Logic and Analytic Philosophy, KU Leuven, Belgium
e-mail: Christoph.Kelp@hiw.kuleuven.be

A. Fairweather (ed.), *Virtue Epistemology Naturalized: Bridges Between Virtue Epistemology and Philosophy of Science*, Synthese Library 366, DOI 10.1007/978-3-319-04672-3_20,

At the same time, virtue theories of knowledge have been on the rise in recent epistemology. According to virtue theories:

VK. One knows that *p* if and only if one's believing *p* truly is due to the exercise of cognitive competence.

If one accepts that successes due to the exercise of competence are achievements, VK is equivalent to the thesis that knowledge is a sort of achievement. Accordingly, the view is also sometimes stated as follows:

VK*. Knowledge is a cognitive achievement.

Contemporary proponents of versions of VK include Ernest Sosa (e.g. 2007, 2010), John Greco (e.g. 2010) and Wayne Riggs (e.g. 2002, 2009). I have also defended a version of the view in Kelp 2011. Among the obvious advantages of VK are its simplicity and elegance. Furthermore, champions of VK have claimed that VK offers a solution to the Gettier problem. Most notably for present purposes, champions of virtue theories have argued that VK yields an account of the value of knowledge according to which knowledge is valuable for its own sake, or *finally* valuable. (Greco 2010: 99)

Combining UK with VK gives us a virtue theory of both knowledge and understanding. This seems desirable in view of the fact that a "basic commitment [of virtue epistemology] is that intellectual agents and communities are the primary source of epistemic value and the primary focus of epistemic evaluation." (Greco and Turri 2011: §1) The thought here is that properties of agents rather than properties of beliefs are the primary source of epistemic evaluation. In view of this commitment, it is desirable that one have a virtue theory of all epistemic standings if one has a virtue theory of any one such standing. Another benefit for proponents of UK is that they get a plausible account of the value of understanding, according to which understanding is finally valuable, for free.

These considerations make UK and VK an appealing package deal. However, a number of epistemologists have objected to both VK and UK. The most prominent foes of VK are Jennifer Lackey (2007, 2009) and Duncan Pritchard (e.g. in his contribution to Pritchard et al. (2010)), while UK has been challenged by Jonathan Kvanvig (2003, 2009), Catherine Elgin (1996, 2006, 2009), Linda Zagzebski (2001) and Pritchard (2009, and his contribution to Pritchard et al. (2010)). The various attacks on UK can be distinguished in terms of the conception of understanding they are directed towards. It is by now fairly standard in epistemology to distinguish between "objectual" understanding, such as understanding phenomena, people and theories one the one hand, and "propositional" understanding, such as understanding why something is the case or how to do something on the other. I would like to suggest that the objections due to Kvanvig, Elgin and Zagzebski are best understood as objections to knowledge based accounts of objectual understanding, while Pritchard's objections concern propositional understanding and, more specifically, understanding why.

In a different paper (Kelp 2013), I have developed a novel knowledge based account of objectual understanding and argue (a) that it avoids Kvanvig, Elgin and

Zagzebski's objections and (b) that there is reason to prefer it to the non-knowledge based competitors Kvanvig, Elgin and Zagzebski offer. Once we have a version of UK for objectual understanding in play, we should of course be especially keen to have a version of UK for propositional understanding, if only for unity's sake. For that reason, in this paper, I will turn to Pritchard's objections and his alternative accounts of understanding and knowledge. More specifically, what I will try to do is to defend the VK-UK package deal against Pritchard.

2 Pritchard's Arguments and Alternative

2.1 Pritchard's Argument Against UK

Pritchard interprets UK as a thesis about propositional understanding. More specifically, according to the thesis Pritchard attacks:

> UK_P: [U]nderstanding why *X* is the case is equivalent to knowing why *X* is the case, where this is in turn equivalent to knowing that *X* is the case because of *Y*. (Pritchard et al. 2010: 74)

Against UK_P, Pritchard argues that knowing that *X* is the case because *Y* is neither necessary nor sufficient for understanding why *X* is the case. I will start with the argument against the sufficiency thesis. Here Pritchard offers the following case:

> *Young Son.* Ernie arrives back home and discovers to his horror that his house it on fire. The firefighter in charge tells Ernie that faulty wiring caused the house to be on fire. Ernie's young son asks him why his house is on fire and Ernie tells him that it is on fire because of faulty wiring.

According to Pritchard, Ernie's son's belief that the house is on fire because of faulty wiring qualifies as knowledge. At the same time, Ernie's son may have "no conception of how faulty wiring might cause a fire" (Pritchard et al. 2010: 81) and as a result Ernie's son does not understand why his house burned down.

I don't find Pritchard's case convincing essentially for the reasons given by Stephen Grimm (2014) in his contribution to this volume, which is why I will not discuss the case any further here. Instead I would like to turn to Pritchard's second case, which is intended to show that knowledge of causes is not necessary for understanding why. Here goes:

> *Fake Firefighters.* Ernie arrives back home and discovers to his horror that his house is on fire. He approaches a firefighter who is standing in front of the house and asks him what happened. The firefighter tells Ernie that his house burned down due to faulty wiring. Unbeknownst to Ernie, he is talking to the only real firefighter among a group of loiterers in firefighter outfits who would have given him a false answer. (Pritchard et al. 2010: 79)

Pritchard points out that the case is structurally analogous to the infamous fake barn case (see also below) and so Ernie doesn't know that his house is on fire because of faulty wiring. At the same time, the thought is that intuitively he does understand why the house is on fire. We are thus said to have a case in which someone understands why *X* but does not know that *X* because of *Y*.

2.2 Pritchard's Argument Against VK

Pritchard's objections to the VK-UK bundle do not stop with his worries about UK. On the contrary he also offers a number of arguments against VK. Pritchard also thinks that VK doesn't state a necessary condition on knowledge:

> *Landmark.* Rosita arrives at the train station in an unknown city and asks the first passerby she encounters for directions to a famous landmark. Her informant is a knowledgeable resident of the city who tells her that the landmark is straight ahead on Greenwich Street and Rosita forms the corresponding belief.

Intuitively, Rosita knows that the landmark is on Greenwich Street. However, argues Pritchard, her belief is not true due to Rosita's competence. Rather, if anything, it is true due to the competences of her informant. Again, if Pritchard is right about this, Rosita does know but does not satisfy the right-hand side of VK and we have a further problem for VK. (Pritchard et al. 2010: ch. 2.6)

I agree with Pritchard that Landmark poses a problem for VK if the due to relation is unpacked in terms of explanatory salience; in particular, if a success is due to the exercise of competence only if the success is primarily creditable to the exercise of competence. After all, it seems right that, in Landmark and similar cases, the testifier's cognitive competences are more salient in the explanation of the testifiee's cognitive success than the testifiee's own competences. However, there is excellent independent reason to believe that champions of VK had better not construe the due to relation in this way (see e.g. Sosa 2007: 86). A more promising alternative is to construe the due to relation in terms of competence manifestation (see e.g. Sosa 2010). As I argue elsewhere (Kelp 2009a, 2011), this account avoids the problems posed for VK by cases like Landmark. Thus cases like Landmark do not pose a decisive problem for VK.

Finally Pritchard also argues that VK's competence condition is not sufficient for knowledge. More specifically, he takes the infamous fake barn case to establish this:

> *Fake Barns.* Grover, a reliable barn spotter, drives through the countryside, sees a barn in the field on the right and comes to believe that he is facing a barn. Unbeknownst to Grover, the barn he is looking at is the only real barn in a field otherwise populated with barn façades that are so cleverly constructed that Grover could not distinguish them from real barns from his position on the road.

Intuitively, Grover doesn't know that he is facing a barn. At the same time, it looks as though Grover truly believes that he is facing a barn due to his reliable barn spotting competence. If this is correct, then Grover lacks knowledge whilst satisfying the right-hand side of VK. Fake Barns thus constitutes a problem for VK. (Pritchard et al. 2010: ch. 2.5)

2.3 Pritchard's Alternative

Pritchard does not stop with pointing out some problems for VK and UK. On the contrary he offers alternative accounts of both knowledge and understanding which he considers superior to VK and UK because they accommodate the intuitions in all the cases he thinks pose a problem for VK and UK. More specifically, Pritchard proposes to deal with cases like Fake Barns by placing an additional safety condition on knowledge. At the same time, Pritchard acknowledges that safety alone isn't sufficient for knowledge. A further competence condition on knowledge is needed. However, since Pritchard thinks that cases like Landmark show that the competence condition at issue in VK is too strong, he offers a weaker version of the competence condition that, he claims, can accommodate the intuitions in these cases. More specifically, the account of knowledge Pritchard ends up with takes the following shape:

> PK. S knows that p if and only if S's safe true belief that p is the product of her relevant cognitive abilities (such that her safe cognitive success is to a significant degree creditable to her cognitive agency). (Pritchard 2012: 273)

Moving on to understanding, Pritchard takes Fake Firefighters to show that there is no safety condition on understanding. At the same time, he takes understanding to be a genuine cognitive achievement and so endorses (roughly) the following account of understanding:

> PU. Understanding why *p* is true belief that *p* because *q* that is due to the exercise of cognitive competence.[1]

Pritchard thus offers his PK-PU bundle as alternative to the VK-UK package deal and claims that it is preferable to its competitor because it accommodates a number of intuitions that VK-UK struggles to accommodate. I have reservations about both PK and PU, which I will not press here. At the same time, I will assume that the responses to Young Son and Landmark I have pointed to will indeed do the job for champions of VK-UK. This leaves Pritchard's argument that Fake Firefighters shows that UK fails left to right and Fake Barns shows that VK fails right to left. In what follows I will develop two ways of in which champions of VK-UK can handle these cases: the first one is to accept the counterintuitive consequence that agents in cases like Fake Barns and Fake Firefighters know, while the second one draws on my account of objectual understanding to offer an alternative account of understanding why that gets the cases right.

---

[1] Pritchard actually endorses a slightly different account of achievement so that his resulting account of understanding ends up being slightly different also. However, these differences are of no consequence for the purposes of this paper.

3 Response 1: Accepting the Counterintuitive Result

The first response on behalf of the champion of VK-UK I would like to consider consists in accepting the counterintuitive consequence that agents in cases like Fake Barns and Fake Firefighters have knowledge. To begin with, notice that this move will do the trick for champions of VK-UK. In particular, the problem for VK is solved, admittedly at the cost of accepting a counterintuitive consequence. At the same time, the problem posed for UK by Fake Firefighters disappears at no cost at all. After all, VK thus understood predicts that Ernie knows why his house burned down. As a result, UK predicts, correctly, that Ernie understands why his house burned down.

The remainder of this section will be devoted to arguing that the cost of accepting the counterintuitive result in these cases is itself manageable and that the resulting VK-UK bundle is at any preferable to Pritchard's alternative PK-PU package deal.

3.1 A Manageable Cost

In order to warm yourself up to the thought of accepting that agents in cases like Fake Barns and Fake Firefighters know, it may be worth noting that the intuition of ignorance is not universally shared. A number of people, perhaps most notably Ruth Millikan (1984), have claimed not to have it. What's more, as Tamar Gendler and John Hawthorne (2005) argue, the intuitions in cases that share the same structure with Fake Barns are highly unstable, which should also make accepting the counter-intuitive result more tolerable.

Notice also that the problem cases for VK constitute a fairly isolated class. In particular they differ from standard Gettier cases in that, as Pritchard himself rightly points out, the way luck enters the story is quite different in the two types of case. In standard Gettier cases—Havit/Nogot, Sheep etc.—luck "intervenes betwixt ability and success." (Pritchard 2009: 23) In other words, the problem here is, roughly, that something goes wrong in the process of belief acquisition and the agent, luckily, gets it right nonetheless. As opposed to that, in cases like Fake Barns and Fake Firefighters nothing goes wrong in the process of belief acquisition. Rather, the problem is rooted in the agent's environment. The agent is lucky because she gets it right despite being in an epistemically unfriendly environment in which she might so easily have got it wrong. Cases like Fake Barns and Fake Firefighters are thus importantly different from standard Gettier cases. At the same time, there is every reason to believe that VK will be able to handle standard Gettier cases. In fact, Pritchard himself claims that even his weak virtue condition on knowledge will handle these cases.

These initial considerations suggest that denying the intuition in these cases will constitute a surveyable cost. And yet denying the intuition of ignorance in cases like Fake Barns and Fake Firefighters will be viable only if we have a plausible explanation of why it should seem so intuitive that the agents in these cases lack knowledge. One explanation that seems particularly appealing to me exploits the following "safety heuristic":

> SH. In judging whether one knows, we assess how easily one might have been mistaken. If we judge that one might very easily have been mistaken, we judge (intuitively) that one does not know.

I would like to suggest that SH is a useful heuristic, one that makes judgements of knowledge and ignorance easy to make, while, at the same time, being highly reliable: most cases of ignorance will be cases in which one might easily have been mistaken and most cases of knowledge will be cases in which one might not easily have been mistaken.

At the same time, champions of VK may argue, SH is no more than a useful heuristic. After all, there is independent reason to believe that the safety principle according to which one knows that *p* only if one could not very easily have been mistaken about *p* does not constitute a genuine necessary condition on knowledge. To see this consider the following case:

> *Grandfather Clock*. Elmo's arch-nemesis, a powerful demon, has an interest that Elmo forms a belief that it's 8:22 by looking at the grandfather clock in the hallway when he comes down the stairs. Elmo's arch nemesis is prepared to do whatever it may take in order to ensure that Elmo acquires a belief that it's 8:22 by looking at the grandfather clock when he comes down the stairs. However, Elmo's arch-nemesis is also lazy. He will act only if Elmo does not come down the stairs at 8:22 of his own accord. Suppose, as it so happens, Elmo does come down the stairs at 8:22. Elmo's arch-nemesis remains inactive. Elmo forms a belief that it's 8:22. It is 8:22. The grandfather clock is working reliably as always.

Here, intuitively, Elmo knows that it's 8.22. At the same time, Elmo might very easily have been mistaken about the time. Had he come down a minute earlier or later, his arch-nemesis would have set the clock to 8.22 and Elmo would have been mistaken in his belief about the time.[2]

Given that SH constitutes a useful heuristic for making judgements of knowledge and ignorance, but no more than that, champions of VK have all it takes to explain the intuition of ignorance in cases like Fake Barns and Fake Firefighters. We realise that the agents in these cases might very easily have been mistaken and on the basis of SH judge, intuitively but erroneously, that they lack knowledge.

So, the thought then is that the cost of accepting the counterintuitive consequence that agents in cases like Fake Barns and Fake Firefighters know is an acceptable cost to the champion of VK-UK. Not only is the intuition not universally shared and has been argued to be unstable, but the range of problematic cases is also surveyable. Most importantly, there is a plausible explanation of why we should have a mistaken intuition in these cases in terms of SH.

---

[2] In Kelp (2009b) I argue that this case causes a problem even for the most refined versions of the safety principle on the epistemological market. For further counterexamples to safety see Neta and Rohrbaugh (2004) and Comesaña (2005).

3.2 VK-UK Versus PK-PU

So, which of the two package deals should we accept, VK-UK or PK-PU? One might think that PK-PU still has an edge over VK-UK because it does not accept any counterintuitive consequences and thus need do no explaining away. A closer look reveals that this argument would be too quick. As Pritchard himself notices, abandoning VK means losing the neat account of the value of knowledge that VK offered. In fact Pritchard finds himself forced to concede that knowledge is not distinctively valuable. Pritchard acknowledges that this is a counterintuitive consequence of his view but aims to take the sting out of it by arguing (i) that understanding rather than knowledge is distinctively valuable while, at the same time, (ii) understanding "tends to go hand-in-hand with knowledge" (Pritchard et al. 2010: 83) which explains why we would mistakenly think that knowledge is distinctively valuable. It transpires, then, that, by Pritchard's own lights, PK-PU also has counterintuitive consequences that need to be explained away. It's just that the counterintuitive consequences arise at another point in his theory. As far as counterintuitive consequences are concerned, then, the two bundles appear to be on equal footing.

Whether or not PK-PU itself has counterintuitive consequences that need to be explained away, there is reason to think that VK-UK is preferable to Pritchard's alternative on grounds of simplicity, elegance and uniformity in explanation. To begin with, VK, which countenances only a virtue condition on knowledge, is simpler and more elegant than PK, which countenances both a virtue and a safety condition. Moreover, the VK-UK bundle is also more uniform than the PK-PU bundle in that it gives a pure virtue theoretic account of both knowledge and understanding, while the PK-PU bundle combines a pure virtue theoretic account of understanding with a hybrid account of knowledge. (This may be of particular significance once one remembers that the basic commitment of virtue theory was that the primary focus of epistemic evaluation is on agents and communities rather than beliefs. By going hybrid, it seems that Pritchard has to give up this commitment.) VK-UK also offers a more unified account of the involvement of virtue in knowledge and understanding: for both the relevant cognitive success must be due to the exercise of competence. In contrast, Pritchard takes virtues to be involved in very different ways here. Moreover, by the lights of VK-UK, both knowledge and understanding enjoy the same kind of value, i.e. both are by their nature finally valuable. In contrast, Pritchard maintains that understanding is by its nature finally valuable, while knowledge isn't (although individual items of knowledge can be). Relatedly, Pritchard is committed to a version of epistemic value pluralism, while VK-UK is at least compatible with a version of monism according to which knowledge is the sole fundamental epistemic value. Unsurprisingly, I would also add that VK-UK value fits more nicely with the kind of knowledge based account of objectual understanding I favour.

Finally, it is hard to see how Pritchard can unify the thesis that knowledge is the aim of ordinary inquiry with the thesis that understanding is the aim of scientific inquiry. True, Pritchard (2009) offers an account of the aim of ordinary inquiry that would do the trick, *viz.* that understanding is the aim of ordinary inquiry. However, there is excellent reason to think that the thesis that understanding is the goal of ordinary inquiry is too strong to be plausible. In support of his thesis Pritchard considers a case in which someone finds his house on fire and is naturally led to inquire into the reason why it burned down. Pritchard points out that this inquiry will not be properly terminated until that person has come to understand why his house is on fire. I agree with Pritchard on this example. Crucially, the reason why inquiry here aims at understanding is grounded in the fact that the agent's curiosity is directed at the explanation of an event: the agent wants to find out why the house burned down. Notice, however, that very often our curiosity is directed at pure (i.e. non-explanatory) information. Suppose I am craving a certain type of chocolate. In this situation, I may want to know whether the store that's a ten-minute walk from where I am is still open and whether it has the type of chocolate I am craving in stock. It is of no interest whatsoever to me that the shop is still open because the owner has had an argument with his wife and is putting off going home or that they have the type of chocolate I crave in store because the delivery arrived a day early. Here the explanations are simply of no interest to me. Accordingly, it is very plausible that my inquiry can reach its goal and can be properly terminated even if I don't acquire understanding of why the relevant propositions are true. If that is correct, then it cannot be the case that understanding why constitutes the goal of ordinary inquiry.

It transpires that, by Pritchard's own lights, PK-PU does not have an advantage vis-à-vis VK-UK on the grounds that it gives a charitable account of all intuitions. While the present version of VK-UK explains away the intuition of ignorance in cases like Fake Barns, PK-PU explains away the intuition that knowledge is distinctively valuable. At the same time, VK-UK clearly outperforms Pritchard's alternative on theoretical virtues such as simplicity, uniformity and elegance. Indeed, I am inclined to think that the theoretical benefits VK-UK can claim against PK-PU are so great that even if PK-PU could give a charitable account of all intuitions, there would be excellent reason to favour VK-UK over PK-PU.[3]

4 Response 2: An Alternative Account of Understanding Why

One way of responding to Pritchard's argument against VK-UK is by accepting that agents in cases like Fake Barns know. While I think that this is a promising way of proceeding, I don't think that it is the only option available for the champion of VK-UK In what follows I will outline yet another way of resisting Pritchard's argument against UK. Here I will leave open how champions of VK-UK ought deal with Pritchard's argument against VK. In particular, the account offered here will be compatible with a version of VK according to which agents in cases like Fake Barns lack knowledge.[4]

[3] Of course, this is not to say that VK-UK has now been established. There might be theory that does better than VK-UK so understood.

[4] Some such accounts have been offered by Greco (2010), Sosa (2010) and myself (Kelp 2011).

4.1 Some More Data

Recall that according to Pritchard understanding why *p* is true belief that *p* because *q* that is due to the exercise of cognitive competence. Recall also that Pritchard distinguishes between two ways in which luck can affect one's true belief that *p:* 'intervening luck' where something goes wrong in the process of belief-formation and 'environmental luck' where the agent is in an unfriendly epistemic environment. According to Pritchard, understanding why *p* is incompatible with intervening luck but compatible with environmental luck, as cases like Fake Firefighters are supposed to establish.

As a first step I would like argue that understanding why *p* is not generally compatible with environmental luck. Consider the following pair of cases:

Shot in the Head. Zoe watches a man being shot in the head and die instantaneously. She comes to believe that he died because he was shot in the head.

Imminent Heart Attack. Zoe watches a man being shot in the head and die instantaneously. She comes to believe that he died because he was shot in the head. Unbeknownst to Zoe the man was also suffering from a heart attack that would have been the cause of his death had the shot been fired a second later.[5]

My intuitions here are that in Shot in the Head Zoe both knows and understands why the man died. In contrast, in Imminent Heart Attack, Zoe neither knows nor understands why the man died. The problem for Pritchard here is that it is hard to see how his account can accommodate these intuitions. True, Zoe is lucky to have got it right in Imminent Heart Attack. However, the type of luck that afflicts her belief is not Pritchard's intervening luck. After all, nothing goes wrong in the process of belief-formation. Rather, the problem here is that Zoe is in an epistemically unfriendly environment as the cause of the man's death is overdetermined. The relevant type of luck at issue in Imminent Heart Attack is thus environmental luck. Since, according to Pritchard, understanding is compatible with this type of luck we may expect PU to predict that Zoe understands why the man died.

We thus have two cases in which an agent's belief why *p* is afflicted by environmental luck that generate opposite intuitions concerning whether the agent understands why *p*. One might be inclined to think that this shows that intuitions about such cases are too unstable too provide solid data for theorising about understanding. If this is correct, the fact that VK-UK cannot accommodate the intuition of lack of understanding in Fake Firefighters might not carry any significant weight against the view. While I think this might eventually be the lesson to be learned from these cases, I am also convinced that, at this stage, it would be premature to draw this conclusion. The reason for this is that there is a structural difference between Fake Firefighters and Imminent Heart Attack, *viz.* that in Fake Firefighters Ernie's understanding is ultimately grounded in knowledge. After all, in Fake Firefighters, Ernie acquires his belief why the house is on fire from the firefighter, who in turn himself

[5] For a similar case see Grimm (2006).

knows why the house is on fire. In other words, Ernie acquires his understanding from a knowledgeable source. The same is not true in Imminent Heart Attack. Here Zoe acquires her belief why the man died first-hand, as it were. However, her belief does not qualify as knowledge and so is not grounded in knowledge in the way Ernie's belief is.

The crucial question now is how we can exploit this difference between the two cases in order to offer an alternative account of understanding why that accommodates all the relevant intuitions. While I believe that there is more than one way of achieving this, I would here like to focus on one particular way, which takes its lead from my account of objectual understanding.

4.2 *The Alternative Account*

I will begin by briefly rehearsing my proposed account of objectual understanding (call it 'KOU'). KOU places the following two principles linking knowledge and understanding centre stage:

> U-Max. If one knows everything there is to know about *X*, then one also understands everything there is to understand about *X*.
>
> U-Min. If one does not know anything about *X*, then one does not understand anything about *X* either.

While U-Max states that fully comprehensive knowledge is sufficient for maximal understanding, U-Min holds that at least some knowledge is necessary for minimal understanding. The further proposal is that no knowledge and fully comprehensive knowledge constitute the extremities of a spectrum. In between lie the various degrees of understanding. The quality of one's understanding of *X* can be measured in terms of approximation to fully comprehensive knowledge about *X*.

This account of degrees of understanding is coupled with a contextualist semantics for outright attributions of understanding. The crucial thesis here is that attributions of understanding are task-relative in the following sense:

> U-Out. An outright attribution of understanding of *X* is true just in case one knows enough about *X* to (likely) successfully perform a contextually determined task or range of tasks.

Task-relativity is the crucial aspect of KOU that I would like to use to provide an account of understanding why. I would like to begin with what I take to be an independently plausible suggestion, *viz.* that the relevant task for understanding why *p* consists in being able to give an explanation of why *p*.[6]

---

[6] I think that, ultimately, attributions of understanding why afford a contextualist semantics. Accordingly, a more precise version of this account would state that the task relevant to attributions of understanding why *p* consists in being able to give an explanation of why *p* that meets the explanatory demands at issue in the context of attribution. However, since for the purposes of this paper, there is no need to address the issue of contextualism about attributions of understanding, I will work with the simpler, non-contextualist version.

Next, I would first like to introduce the notion of a well-founded explanation:

WF. An explanation of *p* is well-founded if it is ultimately grounded in knowledge why *p*, that is to say, if it is grounded in a warrant why *p* that originates from a knowledgeable source, i.e. from a source that knows why *p*.

Again, there are various ways in which one might connect these two ideas. The one I want to suggest here connects very straightforwardly with U-Out:

U-Why. One understands why *p* just in case one knows enough to ensure (or make highly likely) that one would provide a well-founded explanation of why *p*.[7]

U-Why allows us to accommodate the intuitions in both Fake Firefighters and Imminent Heart Attack. To see this, notice first that, in Fake Firefighters, the explanation that the house burned down because of faulty wiring would be well-founded in the relevant sense if offered by Ernie. After all, Ernie has a warrant that the house burned down because of faulty wiring that originates from the fireman who knows why the house burned down. The question remains whether Ernie knows enough to ensure that he would provide this explanation. There is reason to think that the answer is 'yes'. True, Ernie doesn't know why the house is on fire (or so we are for now assuming). However, he does know a number of relevant facts, including that his house burned down, that he has been told that by a source he has no reason to distrust that it burned down because of faulty wiring, that this explanation is the most plausible one to him at this time and that he believes the explanation to be correct. Plausibly, Ernie's knowing these facts will ensure (or make highly likely) that Ernie would provide the relevant explanation of why his house burned down. Accordingly, U-Why can accommodate the intuition that Ernie understands why his house burned down.

At the same time, U-Why can also accommodate the intuition that, in Imminent Heart Attack, Zoe does not understand why the man died. Zoe does not herself know why the man died. At the same time, she herself is the original source of her warrant. As a result, Zoe's warrant why the man dies does not originate from a knowledgeable source. Hence she fails the well-foundedness requirement of U-Why.

It may be worth noting that U-Why also accommodates intuitions in a number of further cases. Consider the following two cases:

Ernie's Wife. Ernie phones his wife and tells her that their house burned down because of faulty wiring.

Fake Firefighters 2. Bert has also arrived at Ernie's house but hasn't talked to Ernie yet. He approaches a fake firefighter and asks him why the house is on fire. Making up an explanation on the spot the fake firefighter tells Bert that the house burned down because of faulty wiring.

---

[7] Notice that once one goes contextualist about attributions of understanding why *p* there are a number of ways in which one could accommodate WF in one's semantics. Most importantly, one could make WF part of the contextually determined explanatory demands. This would leave open the possibility of there being contexts in which the attributions of understanding are true even though the explanation the agent would provide is does not satisfy the well-foundedness requirement.

Intuitively, Ernie's wife comes to understand why the house burned down. U-Why can accommodate this intuition. After all, she will be in a similar epistemic position as Ernie (the main difference being that Ernie's wife knows that the house burned down on the basis of testimony rather than perception) and so knows enough to ensure (or make highly likely) that she would give the same explanation as Ernie. At the same time this explanation is well-founded as her warrant for why the house burned down originates from a knowledgeable source, i.e. the fireman.

As opposed to that, intuitively, Bert does not understand why the house burned down. Although he would give the same explanation as Ernie and his wife, in Bert's mouth this explanation is not well-founded. After all, the fake firefighter who offered it made it up on the spot and so Bert's warrant does not originate from a knowledgeable source.

5 Conclusion

We have seen that there are at least two of ways in which champions of VK-UK can resist Pritchard's argument. First, they can accept that agents in cases like Fake Barns have knowledge and offer an explanation of why we should mistakenly generate an intuition of ignorance in terms of the safety heuristic. The resulting view is preferable to Pritchard's alternative due to the extensive gains in simplicity, elegance and uniformity in explanation it offers. Second, even those champions of VK-UK who do not want to accept the counterintuitive consequence need not be moved by Pritchard's argument. An alternative account of understanding why—*viz.* U-Why—is available to them. This account is arguably preferable to Pritchard's because it accommodates the intuition not only in Fake Firefighters but also in Imminent Heart Attack, a case Pritchard is bound to struggle with. Pritchard's argument against VK-UK thus fails. Those philosophers of science who are attracted by UK need not be worried by Pritchard's attack against their preferred view. On the contrary they can plausibly extend their allegiances to VK. In this way, they will get the very appealing VK-UK package deal, which offers simple, elegant and unified accounts of both understanding and knowledge.

References

Achinstein, P. 1983. *The nature of explanation*. New York: Oxford University Press.
Comesaña, J. 2005. Unsafe knowledge. *Synthese* 146: 395–404.
DeRegt, H., and D. Dieks. 2005. A contextual approach to scientific understanding. *Synthese* 144: 137–170.
Elgin, C. 1996. *Considered judgement*. Princeton: Princeton University Press.
Elgin, C. 2006. From knowledge to understanding. In *Epistemology futures*, ed. S. Hetherington. Oxford: Oxford University Press.
Elgin, C. 2009. Is understanding factive? In *Epistemic value*, ed. A. Haddock, A. Millar, and D. Pritchard. Oxford: Oxford University Press.

Gendler-Szabo, T., and J. Hawthorne. 2005. The real guide to fake barns: A catalogue of gifts for your epistemic enemies. *Philosophical Studies* 124: 331–352.
Greco, J. 2010. *Achieving knowledge*. Cambridge: Cambridge University Press.
Greco, J., and J. Turri. 2011. Virtue epistemology. In *The Stanford encyclopedia of philosophy*, ed. E. Zalta. http://plato.stanford.edu/entries/epistemology-virtue/
Grimm, S. 2006. Is understanding a species of knowledge? *The British Journal for the Philosophy of Science* 57: 515–535. 17.
Grimm, S. 2014. Understanding as knowledge of causes. In *Virtue epistemology naturalized. Bridges between virtue epistemology and philosophy of science*, ed. A. Fairweather. Dordrecht: Springer.
Kelp, C. 2009a. Pritchard on virtue epistemology. *International Journal of Philosophical Studies* 17: 583–587.
Kelp, C. 2009b. Knowledge and safety. *Journal of Philosophical Research* 34: 21–31.
Kelp, C. 2011. In defence of virtue epistemology. *Synthese* 179: 409–433.
Kelp, C. 2013. *Towards a knowledge-based account of understanding*. Manuscript.
Kitcher, P. 2002. Scientific knowledge. In *The Oxford handbook of epistemology*, ed. P. Moser. Oxford: Oxford University Press.
Kvanvig, J. 2003. *The value of knowledge and the pursuit of understanding*. Cambridge: Cambridge University Press.
Kvanvig, J. 2009. Responses to critics. In *Epistemic value*, ed. A. Haddock, A. Millar, and D. Pritchard. Oxford: Oxford University Press.
Lackey, J. 2007. Why we don't deserve credit for everything we know. *Synthese* 158: 345–361.
Lackey, J. 2009. Knowledge and credit. *Philosophical Studies* 142: 27–42.
Lipton, P. 2004. *Inference to the best explanation*. London/New York: Routledge.
Millikan, R. 1984. Naturalist reflections on knowledge. *Pacific Philosophical Quarterly* 65: 315–334.
Neta, R., and G. Rohrbaugh. 2004. Luminosity and the safety of knowledge. *Pacific Philosophical Quarterly* 85: 396–406.
Pritchard, D. 2009. Knowledge, understanding and epistemic value. In *Epistemology*, Royal institute of philosophy lectures, vol. 64, ed. A. O'Hear, 19–43. Cambridge: Cambridge University Press.
Pritchard, D. 2012. Anti-luck virtue epistemology. *The Journal of Philosophy* 109: 248–279.
Pritchard, D., A. Millar, and A. Haddock. 2010. *The nature and value of knowledge*. Oxford: Oxford University Press.
Riggs, W. 2002. Reliability and the value of knowledge. *Philosophy and Phenomenological Research* 64: 79–96.
Riggs, W. 2009. Two problems of easy credit. *Synthese* 169: 201–216.
Salmon, W. 1989. Four decades of scientific explanation. In *Minnesota studies in the philosophy of science*, vol. 13, ed. P. Kitcher and W. Salmon. Minneapolis: University of Minnesota Press.
Salmon, W. 1998. The importance of scientific understanding. In *Causality and explanation*. New York: Oxford University Press.
Sosa, E. 2007. *A virtue epistemology: Apt belief and reflective knowledge*, vol. 1. Oxford: Oxford University Press.
Sosa, E. 2010. How competence matters in epistemology. *Philosophical Perspectives* 24: 465–475.
Strevens, M. 2006. Scientific explanation. In *The encyclopedia of philosophy*, 2nd ed, ed. D. Borchert. New York: Macmillan.
Woodward, J. 2003. *Making things happen: A theory of causal explanation*. New York: Oxford University Press.
Zagzebski, L. 2001. Recovering understanding. In *Knowledge, truth, and duty. Essays on epistemic justification, responsibility, and virtue*, ed. M. Steup. Cambridge: Cambridge University Press.

MIX
Papier aus verantwortungsvollen Quellen
Paper from responsible sources
FSC® C105338

If you have any concerns about our products,
you can contact us on
ProductSafety@springernature.com

In case Publisher is established outside the EU,
the EU authorized representative is:
Springer Nature Customer Service Center GmbH
Europaplatz 3, 69115 Heidelberg, Germany

Printed by Libri Plureos GmbH
in Hamburg, Germany